Environmental Humanities: Transformation, Governance, Ethics, Law

Volume 4

Sustainability, i.e. the demand for long-term and globally practicable lifestyles and economies, is increasingly being understood as the central challenge of our time. But just as today science is often simply equated with natural science, many people think only of the natural sciences when it comes to sustainability science. Undoubtedly, natural scientific and technical knowledge of problem relationships in dealing with nature, resources and climate is important. However, technical change does not happen on its own. In addition, the ecological challenges are simply too great not to aim for a behavioural change as well as technology. This is the starting point of this series of publications. Some questions are, for example, the conditions for individual and social change, the means or governance instruments and normative (ethical and legal) issues about the ultimate goals to be pursued. Transdisciplinary approaches should play a special role, i.e. approaches that do not operate from disciplinary boundaries but from questions of content without excessive subordination to established disciplinary dogmas. It is important to the editors that the present series stands for pluralism and expressly gives room to uncomfortable, unexpected and heterodox views and methods. In times in which sustainability research in particular (also) is increasingly influenced by the interests of clients, such openness seems necessary in the interest of truly finding knowledge.

Chijioke Francis Nwosu

Systemic Remediation

Approaching Sustainable Development Narratives from the Nigerian Afro-Igbo Mmadụ Perspective

Springer

Chijioke Francis Nwosu
Christian Social Ethics
Ludwig-Maximilians-University
Munich, Germany

ISSN 2524-5708 ISSN 2524-5716 (electronic)
Environmental Humanities: Transformation, Governance, Ethics, Law
ISBN 978-3-031-61908-3 ISBN 978-3-031-61909-0 (eBook)
https://doi.org/10.1007/978-3-031-61909-0

This Springer imprint is published by the registered company Springer Nature Switzerland AG
The registered company address is: Gewerbestrasse 11, 6330 Cham, Switzerland

This work is dedicated to Africa and all who believe in Africa
and to
The community of our common existence, where Mmadụ (the human person) remains the sole agent of sustainable development

Foreword

With a rapidly growing population of over 200 million people and a gross domestic product of 430 billion US dollars, Nigeria is one of the most dynamic countries in Africa. It is one of the world's leading oil and gas producers and is home to the second largest film industry in the world. Despite—or perhaps partly because of—its enormous wealth of natural resources, the country is torn apart by violence, corruption, poverty and inequality. Religious fundamentalism is also part of the country's signature, with the Islamist terrorist group Boko Haram in particular gaining international notoriety. After more than 60 years of independence, the colonial legacy is still omnipresent and is experiencing new exacerbations in the context of neo-colonial structures of resource exploitation. There is little sense of belonging together in a country made up of different tribes and religions.

All of this is an enormous challenge for Christian social ethics, which sets itself the task of analysing the causes of need and poverty and identifying leverage points for possible improvement. Chijioke Francis Nwosu, himself from the Igbo ethnic group in the south-eastern part of the country, approaches the topic in two ways: First, he uses selected examples to analyse structures of underdevelopment and injustice. He focuses, in particular, on the phenomenon of corruption, which he uses to identify central undesirable developments as "structures of sin" (John Paul II). As a normative counterconcept, he presents his very own interpretation of the concept of sustainable development. The focus here is on Amartya Sen's capacity building approach, which he combines with the indigenous guiding concept of Mmadụ. He describes this as the development of the person, which must be the goal of all development and all institutions.

Nwosu systematises this African-contextualised approach through the concept of human ecology, which can be found not least in the papal encyclical Laudato si': It is about a person-centred understanding of development that seeks to tap into the potential of human beings from the respective social and ecological contexts of life.

The present work was accepted as a dissertation at the Catholic Theological Faculty of the Ludwig Maximilian University of Munich in the summer of 2023. The methodological originality of its understanding of sustainability is the approach ex negativo, from the perspective of non-sustainability. Based on a penetrating analysis of undesirable developments and structural entanglements in dependencies, it identifies concrete obstacles on the path to sustainability. In doing so, Nwosu addresses both misguided attitudes at the individual level and structural interrelationships that the individual can only influence to a limited extent and that are sometimes deeply rooted in socio-economic conditions. Again and again, Nwosu distances himself from an idealistic approach that describes a morally ideal world without taking the many opposing forces into account with the necessary seriousness.

The case studies on the energy system in Nigeria, which are based on the author's own interviews and vividly illustrate the everyday problems of poverty and the economically desolate situation caused by the extremely inadequate energy supply system, give the work a problem-oriented, hands-on tone. One gets the impression that the author takes a knowledgeable look at people's everyday lives and the real problems of coping with life, which can form a good basis for social-ethical-pastoral work. It is basically absurd that in a country as rich in oil and gas as Nigeria, 80% of the rural population in many regions have no access to electricity. With the energy supply system, the author focuses on a very concrete field of action for sustainable development or non-sustainability in Nigeria.

The theory of the so-called resource curse is illustrated by Nwosu's analyses: the wealth of natural resources leads to a few who then seek to secure their power through violence and terror. Something similar can be observed in Angola, for example, which could actually be an extremely rich country thanks to its oil and diamond deposits, but which is also sinking into misery and violence. The criticism of economism is a recurring theme in the work, whereby Nwosu understands this above all as a decoupling of the pursuit of advantage from the awareness of social and cultural contexts and the associated duties. Ultimately, for him, economism is the expression of an image of man and a concept of society that does not correspond to the person-centred ideas of Mmadụ.

The socio-ethical guiding principle of work can be described in *Gaudium et spes*, a central document of Catholic social teaching from 1965, as "the root, support and goal of all institutions is and must be man as a person" (GS 25). However, personhood is not described here in the Kantian sense as autonomy but is characterised in more detail by the Igbo concept of Mmadụ (including Ùgwú-Mmadụ, which the author translates as "dignity") and Amartya Sen's concept of "functionings" as the development of human capabilities in their social, cultural and ecological networks of relationships.

The core idea of the entire work is derived from the Mmadụ approach: "Policies, therefore, should have as first principles the maximisation and harnessing of the people's capabilities into viable ends (functionings) that necessitate wellbeing and development in the subject, the Mmadụ, the human person who always stands at the receiving end—in Sen's terminology, the 'agent'" (p. 157f.). Everything depends on strengthening the human being as an agent and activating the specific abilities that define him as a human being. The precise and knowledgeable analyses of Amartya Sen's social philosophy show that the author has dealt with this intensively and thus gained a common thread for his entire work. It is about a person-centred concept of overcoming poverty. According to this concept, sustainability and poverty eradication are not primarily achieved through a better supply of goods but, above all, when people are empowered to actively contribute their potential to society, e.g. for food production, crafts, education or healthcare. Nwosu understands this development of personal qualities of active participation in shaping social life as "authentic freedom" (p. 168).

Overall, Chijioke Francis Nwosu aims for an ethics-centred understanding of development, which he outlines, following Christoph Stückelberger, with the terms "Empowerment, Transformation, Holistic, Integrity, Competence, Sustainability" (the initial letters form the word "ethics"). The ethics concept that Nwosu uses as a basis is Aristotle's eudaimonism, which he concretises through Nussbaum's model of "functionings" and the criteria of sustainability as conditions for a happy life. He describes the guiding goal of ethics as "authentic humanity" (p. 298) and integrates human rights aspects of Kant's ethics.

The comments on the "semantic nuances" of the Mmadụ concept (p. 331) are revealing. This is made up of the syllable Mma = "beauty, quintessence, core, nucleus, centre, pivot, hub, goodness, value, order" and the syllable Ndụ = "life, existence, reality, regeneration, existential mobility". The concept of personhood and human dignity is thus strongly linked to the idea of the beauty of life, which builds a bridge to ecological and aesthetic contextualisations. Overall, the explanations provide a good insight into African approaches to anthropology, which are vividly summarised, for example, by the sentence "I am because we are".

In all of this, Nwosu does not idealise indigenous traditions but also mentions previous problematic aspects, such as practices of infanticide in twins and other forms of human sacrifice. At the same time, he emphasises the destructive effect of colonial marginalisation of the cultural practice of the Mmadụ concept, which has destroyed the social cohesion of the indigenous communities as if with a knife.

The dissertation is a highly problem-conscious, independent, creative and methodologically reflective piece of research. It presents a highly complex topic coherently and cleverly chooses empirical indicators, normative guiding concepts and political program texts (mainly from the UN) to analyse socio-ethical challenges and propose solutions to problems. The interpretation of the concept of sustainability through the capability approach and the indigenous concept of Mmadụ as person-centred empowerment justice is an innovative and highly relevant discussion impulse for research in terms of development ethics. This is integrated into the discussion context of Catholic social teaching in a stimulating and coherent way. The analogies between different cultural concepts and thought patterns are reflected in a language-sensitive way. The intercultural linking of different anthropological and socio-ethical approaches is the greatest strength of the work, which represents a considerable, independent and perspective-rich research contribution.

The inductive approach of the work, which is essentially based on analyses of unsustainability and the "structures of sin", lends the dissertation a practical style that addresses socio-ethically controversial topics in a way that is relevant to everyday life and politics despite the fundamental question. The author remains aware, however, that all these ethical considerations require a high degree of patience and a willingness to take small steps towards improvement in view of the predominance of corruption, abuse of power and unresolved conflicts in Nigeria. I wish this work, which is an exemplary example of successful intercultural bridge-building, to many readers worldwide.

December 2023

Prof. Dr. Markus Vogt
Chair, Christian Social Ethics
Ludwig-Maximilians-University
Munich, Germany

Acknowledgements

I am grateful to a well-respected moderator, Prof. Dr. Markus Vogt, for giving me the opportunity to reflect and contribute to ideas and for guiding me in the path of independent research work. To Dr. Andreas Rentz and Ivo Frankenreiter, I say: Thank you for your recommendations, suggestions and corrections. I appreciate all varied forms of inspiration from my family, diocese and friends.

Objective, Hypotheses and Methodology

Objective of the Research Work

The decision to embark on this research is to contribute to human knowledge and scholarship not just theoretically, but instead in/with practical means of approach to the afro-Nigerian situation and with regard to the general development—human, economic, social, political, ecological, etc. Our objective is to apply perception while confronting the realities and the structure we know presently as Nigeria, evaluate this structure within the contemporary development strategies, standards and processes, observe signs of progress and hindrances peculiar to our research specimen, namely Nigeria, and then propose or/and make recommendations for improvement, discontinuation, remediation as well as restructuring where necessary. Ethics—Christian Social Ethics—would be the guiding research and scientific tool for the research work's scholarly contributions.

Research Hypotheses/Question

Our research work took as its hypothesis the fact that Nigeria's development index is counterintuitive to the known potential it possesses—human and natural. It reflected that there is a structural cause for it, and this "cause" has continued to keep the human agent (Mmadụ) within the system in chains of dysfunctional development process, hence, with little or no chances for well-being, capability development and authentic freedom.

Methodology

Our proposed methodology was (a) **Qualitative**: This research and work avoided being bottled up in the lone world and thoughts of the researcher. Our research work went out to observe, confront, dialogue and engage opinions—divergent as well as congruent—to the position of the work itself. Interviews and the assessment of documents with the views they hold were deeply relevant to our method of approach; (b) **Quantitative**: We gathered concrete facts and analysed collated data to support and dialogue with the proposed hypothesis of the research work. The data analyses were also controlled with the aid of comparisons with other independent scientific research works; (c) **Normative**: We focused on ethics in general and Christian Social Ethics in particular, in our use of a normative approach to evaluating our findings, interpreting the available data and proposing/recommending possible alternatives and innovations.

Contents

About the Author

Dr. Chijioke Francis Nwosu studied social ethics, applied ethics, theological ethics and education. His research experience has taken him across Nigeria, Germany, Austria and Switzerland. His areas of interest are social and human person's development ethics, sustainability and value ethics, social and political philosophy, as well as public and practical theology. He holds the Sustainability Award 2023 in the area of dissertation from the Selbach-Umweltstiftung, Munich, Germany.

Definition of Terms

Mmadụ	The afro-Igbo term for the human person.
Ùgwú-Mmadụ	The afro-Igbo terms for human dignity.
Structures of Sin	To be used according to John Paul II's ideology in applying the term, namely as an umbrella name for societal ills rooted in individual human failures.
Generalprävention, Integrationsprävention, Legitimatioinsthese, Vertrauensthese	Terms in the German study of the penal law (code) to be employed in our analysis of corruption and its fight in the Nigerian context.
A Moral Agent	An image of a human person living in freedom.
A Moral Patient	An image of a human person living in (man-made) bondage.
Fulanisation	Nigerian term to describe the violent political/religious activities of the ethnic group, the Fulanis.
Exnovation	Concept, the practice of innovating, modifying, abandoning, recreating an already existent (but obsolete, overtaken) innovation.
Almajiris	Adherents to a system of Islamic education practised in northern Nigeria.
Synergy	The combined power of a group of things when they are working together that is greater than the total power achieved by each working separately.
Systemic	A systemic problem or change is a basic one, experienced by the whole of an organisation or a country and not just particular parts of it.
Mmadụcology	an afro-Igbo linguistic coinage for human ecology.
Policy-Remediation	The research work's concept for a socio-ethical will to change policies and drive transformative changes.

Entelechy	Coined by Aristotle to describe the quality, purposeful end and completion of any genuine being as "being-at-work".

Abbreviations

ACF	Arewa Consultative Forum
ASUU	Academic Staff Union of Universities
BMPU	Budget Monitoring and Price Intelligencer Unit
BPE	Bureau of Public Enterprise
CA	Centesimus Annus
CSR	Corporate Social Responsibility
CV	Caritatis in Veritate
ETT	Exnovation, Transition and Transformation
EV	Evangelium Vitae
ICPC	Independent Corrupt Practices and other Related Crime Commission
IEA	International Energy Agency
INECSus	Institute for Ethics, Capacity Building and Sustainable Development
IPOB	Indigenous People of Biafra
LEEDS	Local Economic Empowerment and Development Strategy
LS	Laudato Si
MASSOB	Movement for the Actualisation of the Sovereign State of Biafra
MDGs	Millennium Development Goals
MEND	Movement for the Emancipation of Niger Delta
MM	Mater et Magistra
NBS	National Bureau of Statistics, Nigeria
NDA	Niger Delta Avengers
NDF	Niger Delta Forum
NDGJM	Niger Delta Greenland Justice Mandate
NEEDS	National Economic Empowerment and Development Strategy
NHA	National Health Act, Nigeria
OA	Octagesima Adveniens
OPC	Oodua People's Congress
PHCN	Power Holding Company of Nigeria
PP	Populorum Progressio
PPC	Public Procurement Commission

PPG	Planning, Planting and Greening
QA	Quadragesimo Anno
RJA	Reflection, Judgement and Action
SAP	Structural Adjustment Programme
SARS	Special Anti-Robbery Squad
SDGs	Sustainable Development Goals
SEEDS	State Economic Empowerment and Development Strategy
SS	Solicitudo rei Socialis
UN	United Nations
UNDP	United Nations Development Programme
UNFPA	United Nations Population Fund
UNICEF	United Nations Child's Fund (formally known as United Nations Children Emergency Fund
USAID	United States Agency for International Development
VG	Veritas Gaudium

Chapter 1
Nigeria and the People: Issues, Facts and Figures

Abstract The socio-political, economic, cultural and ecological Nigerian spaces are complex and dynamic. To truly understand Nigeria, it is crucial to delve into its precolonial, colonial and post-independent history. These historical influences, vital to understanding Nigeria's present development outlook, have united to shape the nation. Like many other African countries, Nigeria's development path has been marked by challenges, internal and external attacks, and man-made institutional dysfunctions. Predominant among these societal crimes is the issue of corruption. Assessing its nature, extent and effect on the Nigerian experiences provides a fascinating insight into the prolonged battle the Nigerian nation has faced for decades. While the fight against unbridled corruption within the system may not have seen total inertia, persistent questions surround the approach, method, leadership and institutions.

Keywords Nigeria · Politics · Colonialism · Institutions · Culture · Corruption · Development

1.1 Introduction

The Nigerian state is the most populous nation in Africa, with an estimated population capacity of about 200 million people. As a former colony of colonial Britain, Nigeria has its deep share of external influences, acculturations, political and economic exchanges and confrontations, as well as social and institutional intercommunications. These elements of social encounters left indelible marks of change in the culture and people of Nigeria. Thus, since after the "official" end of colonial rule in most African States, and in particular, Nigeria, people and opinions have always been deeply divergent in the appraisal of "*what it was*", "*what it later became*" and "*what the aftermath has been*". In other words, Nigeria still lives on with constant tensions and debates on whether the nation would have been better placed without the experience of "superimposed" external socio-political influences. However, whatever the debate may have become, the vantage point is that Nigeria has a divergent and

C. F. Nwosu, *Systemic Remediation*, Environmental Humanities: Transformation, Governance, Ethics, Law 4, https://doi.org/10.1007/978-3-031-61909-0_1

dynamic history, the concrete assessment of which must be referenced, should any valid statements about the general and present structure—political, economic, social, cultural and environmental—be accepted as verifiable facts.

This chapter presents, therefore, a concise assessment of a pre-colonial, colonial and post-colonial (independent) Nigeria. The journey to nationhood would be analysed with basic highlights of the core socio-political, socio-economic and socio-cultural encounters that have contributed to shaping the history of Nigeria to its present phase. Furthermore, the three basic eras of Nigeria's economic journey would be examined as corporate factors having much to contribute to determining the present status that Nigeria has acquired both on the national and international scenes. In the final analysis and to conclude the chapter, that single detracting factor ravaging most African countries, namely corruption, would be pre-empted and presented as an endemic factor within the polity and, therefore, a negative constituent element within the general social structure called Nigeria.

1.2 Nigeria: An Overview

Nigeria is located in the western part of Africa, sub-Saharan Africa, which at the same time is home to other sixteen (16) countries, namely, Benin, Burkina Faso, Cape Verde, the Gambia, Ghana, Guinea, Guinea Bissau, Ivory Coast, Liberia, Mali, Mauritania, Niger, Senegal Sierra Leone, as well as one British Colony, Saint Helena, Ascension and Tristan da Cunha, located in the South Atlantic (Fig. 1.1).

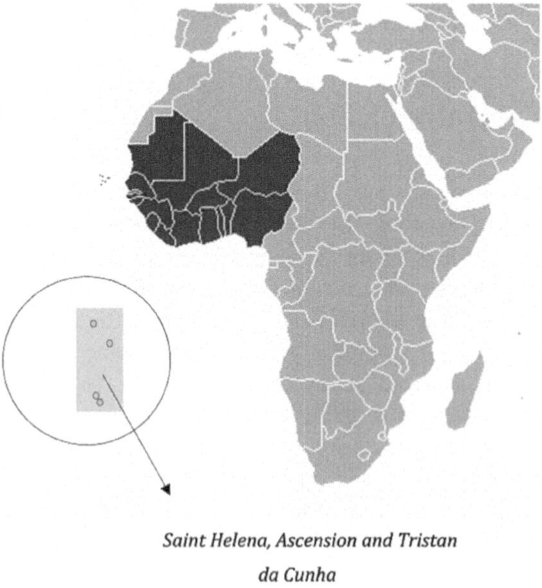

Fig. 1.1 Sub-Saharan West African region in the map of Africa

As a nation-state within the African continent, Nigeria was a former British colony that gained independence on 1 October 1960. It remains not just the most populous nation in Africa but takes the lead in nominal GDP of about $480.48 billion (2020) among all other African countries, with South Africa ($415.32 billion, 2020) and Egypt ($396.33 billion, 2020) sharing the second and third positions respectively. With an estimated population size of 180 to 200 million people, the current United Nations development indicator projects an estimation of over 260 million people by 2030. Geographical research and scientific monitoring of the Nigerian climate make the following classifications concerning its zones of vegetation, namely: Mangrove Swamp and Coastal Vegetation, Freshwater Swamp Forest, Low Land Rain Forest, Derived Savanna, Guinea Savanna, Sudan Savanna, and Sahel Savanna, while a few mountainous areas are found in the States of Plateau (Jos), Adamawa, Taraba and Northern part of Cross-River (The Federal Republic of Nigeria 2018).

Politically and as a democracy (even if only in theory), Nigeria has fought and still fights its way towards stable political, economic and social inclusions. This has left its history complex and intertwined with an intense timeline of both structural and administrative tensions. Such tensions are known to date, even among the extant works of literature, as the embedded absurdities—*structural sins*, which have given rise to Nigeria's structural artificiality and the consequent convolution of the various *parts* as federating *whole* (Ojo and Adebayo 2008). A subsequent analysis of Nigeria and its people, which follows in this chapter, would reveal evident and supportive data for the foregoing assumptions.

1.2.1 Pre-colonial and Colonial Nigeria

The name Nigeria never existed in pre-colonial Africa. The geographical location known as Nigeria today was home to human communities with identities, languages, cultures and conventions. There were recognisable socio-political administrative systems, even as evidenced by some extant documentation of the colonial administrations themselves. From the emirate-based, dynastic and aristocratic structure in the North to the monarchical and centralised structure of the south-west, from the hierarchical political structure of the Kingdom of Benin to the decentralised and egalitarian structure of the south-east—all these provided at least the very first platforms of an encounter between indigenous ideology and the colonial ideology. Thus, the then-existent groups, especially shortly before the arrival of colonial ideology, were not an ignorant group of men, women, and children locked off in their own world. There were observable pieces of evidence of "informal" contacts (predominantly economic contacts) with the Arab and European world. The arrival of the colonial ideology was to become an official as well as formalised advancement of this meeting of separate identities and cultures—the resultant *pros* and *cons* aside. Osaghae opined:

> What existed in the period before the establishment of colonial rule was a motley of groups whose histories and interactions interlaced as they were by external influences – principally trade with Europeans and with the Arab world – had nevertheless crystallised in three discernible regional formations by the end of the nineteenth century. (Osaghae 1998)

As mentioned above, the discernible regional formations would become (to date) the homes for the three largest ethnic groups, which served as socio-cultural agencies for the colonial experience. Again, the three regions inhabit the speakers and cultures of the three widely spoken languages in Nigeria, namely, Igbo in the South-East, Yoruba in the South-West and Hausa in the North. These concrete discernible formations would later play a combined role in working for the later independence of Nigeria from colonial administration through the provision of the architects of Nigeria's independence, namely Dr. Nnamdi Azikiwe (South-East—an Igbo), Obafemi Awolowo (South-West—a Yoruba) and Alhaji Sir Ahmadu Bello (North—an Hausa). Thus, pre-colonial Nigeria was a world inhabited by indigenous people who enjoyed their different identities, spoke their natural languages, lived their cultures and conventions, exchanged economic and social encounters, as well as safeguarded their boundaries and fought their tribal wars. Thus, during the periods dating from the 12th Century up until 1914, there was no nominal connotation associated with the name Nigeria. What existed were interspersed groups of indigenous people living in the kindreds, empires and kingdoms. They were not unconscious or ignorant about themselves and their identities, though they may have been unaware of other foreign cultures and conventions outside their circle, just like the colonial cultures that would later meet them were unaware until the encounter happened. To this effect, a completely new era of social, political, economic and cultural assimilations and dissimilitude was opened for Africa in general and in particular for Nigeria, with the historical Berlin Conference (1884–1885), which ushered in the official partitioning and scramble for same among the countries of Europe. The reason at first was to create a union within the European agenda and ideology in regulating activities, especially trade, by the individual participating countries in Africa. The detailed analyses of both the bilateral and the imperialistic consequences evident in the encounter are not the main focus of this academic work.

With the dawn of an administrative colonial presence in the then sub-Saharan West African Coast, today called Nigeria, the official colonial era began in Nigeria. Nevertheless, the British presence in the area was to be felt as early as 1861–1862, prior to the Berlin Conference, with the trade-driven establishment of the Crown Colony of Lagos (Stokke 1970). After the conference, attempts were made to integrate the ensuing resolutions into the administration by further establishing protectorates. Thus, the coastal areas were designated Oil Rivers Protectorate and later, in 1863, extended and re-named the Niger Coast Protectorate. In 1900, the Protectorate of Southern Nigeria was established but was later amalgamated with the Crown Colony of Lagos (1906) and re-named the Colony and Protectorate of Southern Nigeria. The Northern Protectorate already existed since 1900, given that it was the home of the colonial masters and their early administrative system built around local authorities. At first, the dual protectorates, which would later receive further internal provincial divisions, were considered two separate entities administratively. However, the impact of the European Missionaries, especially in the area of education, contributed to a change of course in the colonial administrative strategy. Consequently, because the missionaries were blatantly rejected in the Northern Protectorate, which was protective of its core Islamic culture, these missionaries found acceptance in the Southern Protectorate, where they embarked on programmes for education and the establishment of schools. The result was demonstrative, namely that the colonial administration was soon to be faced with a rapidly progressive and decentralised South in terms of education and development, as against the highly centralised emirates of the North and their mere tenacity to Koranic schools. Referring to a treatise with the northern emirs by the colonial government, Olav observed:

> [...] the British agreed to keep Christian missionaries out of the Muslim North. Later, it was an important feature of British policy to preserve the Muslim North in its Islamic purity by excluding Christian missionaries and thereby in practice to restrict education to that provided by the Koranic schools, excluding western education (Ibid., p. 18.)

The situation in the already more progressive southern part of Nigeria in reference to education and development did not take much time before it was considered challenging to the colonial administration. Consequently, the administration constituted a Selbourne Committee to make clarifications and recommendations on the question below, among others,

> [...] (i) Should the future administration of the three territories of Lagos, Niger Coast Protectorate (Southern Protectorate) and Niger Company's Territories (Northern Protectorate) be united under one head or if not, into how many administrations and with what seats of Government? On the basis of Customs Union, what Customs Union? What other, if any, taxation is possible? What estimate can be furnished of the cost of administration and of the probable revenue? (Uzoigwe 1898)

The committee, on its part, attempted this response:

> [...] (i) On the question of the future administration of the territories, Lagos, Niger Coast Protectorate and Niger Company's Territories, whether they should be united under one head, we are of the opinion that the object to be aimed at is the eventual establishment of a Governor General for the whole of the territories, resident in the territories. But we feel that the appointment of a Governor General is inadvisable at present. (Ibid.)

That the administration did not pay enough attention to the last phrase—But *we feel that the appointment of a Governor General is inadvisable at present*—and to other sensitive parts of the committee's recommendations, would come to be an indelible factor to the laying of the foundation of socio-political tension in the Nigerian structure. Thus, an ambitious direct response from the colonial masters gave rise to the amalgamation of the two protectorates in 1914 by the then Governor-General, Sir Frederick Lugard (1914–1919), under the single nomenclature, Nigeria, which was credited to a British journalist and writer, Flora Shaw (1885–1929), who later became the wife of Sir Fredrick Lugard. The predominantly proffered reason was for "administrative convenience", but it would come to lay a lasting foundation for malignant political and ethnic tensions within the Nigerian structure. Uzoigwe observed further,

> But the implementation of this recommendation has given rise to great difficulties for the future stability of Nigeria. It is the beginning of that preponderant position in size and population which northern Nigeria holds today, and which has made the Federation of Nigeria the most fragile of all federations. One of the accepted principles of a federal union is that no section of the country should be too big for the rest. This division not only surrendered a good deal of Yorubaland to the north, but also a small portion of Igboland. It would perhaps be unfair to expect the committee to have anticipated the consequences of their action more than half a century later. Moreover, their ultimate aim was not for a federated Nigeria but for a unified one. But divisions like this, when once made, always tend to have a permanent character. (Ibid., p. 475)

With the amalgamation in place, the administration faced a new problem of sticking to a unified method of translating their political ideology into functional administrative conventions within the affected protectorates and their respective cultural uniqueness. This gave rise to the system of *direct and indirect rule*. However, while the system worked predominantly well within the Northern Protectorate, its application in the South became grossly problematic, and the reason is obvious, namely that having attained the full status as a colony, the colonial administration found unhindered smooth-selling within the already highly centralised emirates of the north, where "[…] administrative and judicial control on territorial basis…, (Ibid., *Nigeria: An Introduction to the Politics, Economy and Social Setting of Modern Nigeria*, p. 17) has already been established by the governing Muslim elites. Added to this was the presence of a structured military and the practice of taxation. The administration introduced, therefore, a system of *indirect rule*, whereby,

> […] the colonial power ruled through the established traditional governments. In the case of the northern emirates this meant, inter alia, that the British deprived the emirs of their traditional de jure authority and left them with a de facto control that was ultimately based on the British monopoly of forces. (Ibid., p. 16)

This system became an ideological misnomer in the southern Protectorate, especially in the Middle Belt and (South)-East, due to the conventional de-centralised and egalitarian socio-political system already in place. To this effect, the colonial administration imposed its administrative set-up in a system of direct rule in the south, thereby sowing the first seeds for future cultural and political gaps (Ibid., p. 17). The direct-rule policy was exercised in a form of a *divide-and-rule* game, where malleable groups of hierarchies, incongruous with the culture and socio-political system of the South already in existence, were created by the colonial administration as leadership instruments.

With the direct-rule policy already in place, political and social tensions between the protectorates and among interest groups began to gain ground in the Nigerian system. This was also exacerbated by the conspicuous widening gap between the North and South with regard to "[…] social and economic progress, western education and political development" (Ibid., p. 18). Thus, the rapid development of the South due to the impact of Western education and the consequent rise in dissenting voices to the colonial methods and policies signalled an impending challenge to the powers that were. However, the colonial influence went on and brought some key changes, namely:

> Western education, the English Language, and Christianity spread during the period; new forms of money, transportation and communication were developed; and the Nigerian economy became based on the export of cash crops. Areas with lucrative crops such as cacao and peanuts (groundnuts) profited, while many people in different parts of the country have to migrate to work elsewhere as tenant farmers or use their newly acquired education and skills to work in cities as wage earners, traders and artisans. Two tiers of government emerged, central and local. The central government presided over by the governor-general and accountable to the secretary for the colonies in London, was more powerful but distant from the people. Local administration, where the colonial citizens typically experienced colonial authority, was based on the policy of indirect rule first developed in the north.[1]

The above status of the Nigerian society formed the socio-cultural, socio-political and socio-economic background upon which the nationalist movements that prepared the way for an independent Nigeria began. It would not be a big surprise to observe that the foundational nationalist ideology began in the south, where the activities of the missionaries and education played the key role in nurturing the prospective leaders of the nationalist ideas. In the final analysis, combined concerns about the increasing political and cultural tensions between the protectorates, added to the birth of a new nationalist front, prepared the ground for welcoming a politically "independent Nigeria". Thus, the administrative structure preceding had three visible regions, namely: The Northern Region and its Premier, Alhaji Ahmadu Bello (1954–1966); the Western Region and its Premier, Chief Jeremiah Obafemi Awolowo (1957–1960); and the Eastern Region with the Premier as Chief Michael Okpara (1959–1966).

1.2.2 Post-colonial and Independent Nigeria

The inception of a post-colonial Nigeria came on 1 October 1960, taking along with it all the colonial influences, changes and challenges. However, the Independence Day was greeted by all as a political freedom-at-last and the acceptance speech of the then Prime Minister of Nigeria, Sir Abubakar Tafawa Belewa (1912–1966), set the tone for an anticipated new Nigeria:

> [...] Today is Independence Day. The first of October 1960 is a date to which for two years every Nigerian has been eagerly looking forward. At last, our great day has arrived, and Nigeria is now indeed an independent sovereign nation [....] Words cannot adequately express my joy and pride at being the Nigerian citizen privileged to accept from Her Royal Highness these Constitutional Instruments which are the symbols of Nigeria's Independence [....] This is a wonderful day, and it is all the more wonderful because we have awaited it with increasing impatience, compelled to watch one country after another overtaking us on the road when we had so nearly reached our goal. But now we have acquired our rightful status, and I feel sure that history will show that the building of our nation proceeded at the wisest pace: it has been thorough, and Nigeria now stands well- built upon firm foundations [....] At the time when our constitutional development entered upon its final phase, the emphasis was

[1] Encyclopaedia Britannica. *Nigeria—Colonial History, Economy, People*. Available via https://www.britannica.com/place/Nigeria/Nigeria-as-a-colony. Accessed February 12, 2019.

> largely upon self-government [....] However, we were not to be allowed the selfish luxury of focusing our interest on our own homes [....] All too soon it has become evident that for us Independence implies a great deal more than self-government. This great country, which has now emerged without bitterness or bloodshed, finds that she must at once be ready to deal with grave international issues [....].[2]

The real witnesses to the transition from colonial administration to Nigeria's self-determination would be honest enough to delineate both internal and external concerns and tensions implicit in the speech. But the whole atmosphere lasted only for six years, from the Day of Independence, before the first outburst became a reality—this would be discussed under coup d'état and civil war. Nevertheless, the independent Nigeria kicked off with the duo—the Executive President, Dr. Nnamdi Azikiwe (Eastern Nigeria) and the Prime Minister, Sir Abubakar Tafawa Belewa (Northern Nigeria), with the premiers administering the two distinct regions (protectorates under colonial rule): Alhaji Ahmadu Bello (Northern Region) and Samuel Ladoke Akintola (Southern Region). A fourth republic was established in 1963—the Midwestern Region- with Dennis Osadebay as its administrator. The remaining personality among the founding fathers, Chief Jeremiah Obafemi Awolowo (1909–1987), served as the Finance Minister of the First Republic after succeeding Chief Festus Okotie-Ebo, who held the office between 1957 and 1966. Thus, with the three regions and the constitution of 1963, Nigeria became a Republic within the Commonwealth Nations, with a political system modelled to the Westminster parliamentary system, and this team mentioned above made up the administrative network of Nigeria's First Republic as shown below (Fig. 1.2).

1.3 The Journey to Nationhood: The Independent Nigeria and Its Republics

In the simplest understanding of the concept of nationhood, to become a nation means to be united by a common language, culture, conventions and economic life, and to be designated as a republic means that the socio-political powers embedded in the constitution and governance derive solely from the people. It is not an individual or personal hegemon or that of a country over and above the other. However, history and experiences have shown that nationhood remains a process that develops and being a republic is never a guarantee for stability, peace and harmony. In the following sub-sections, we try to explore Nigeria's journey so far towards being politically recognised as a nation and a republic.

---

[2] BlackPast, *1960—Sir Abubakar Tafawa Balewa, Independence Day*. Available via https://www.blackpast.org/global-african-history/1960-sir-abubakar-tafawa-balewa-independence-day/. Accessed February 12, 2019.

Fig. 1.2 The foundational clue of Independent Nigeria—The first republic. *Illustration mine*

1.3.1 The First Republic (1963–1966)

The first republic, at its inception, faced a number of challenges. The first was the need to decolonise the constitution available to them at independence. The structural pattern of the independence constitution was, in essence, an agglomeration of the preceding ones within the colonial political structure, for example, the Richard's Constitution—1946/7; MacPherson Constitution—1951; Lyttleton Constitution—1954; and Robertson Constitution—1958. Thus, there are two essential features of the Constitution of the First Republic: (a) The constitution borrowed much from the Westminster Model, which no doubt is understood about a young independent nation with a group of elite leaders groomed and educated within the colonial administrative system. Of course, the then Prime Minister at independence made it clear that association with the people of the United Kingdom has made tremendous goodwill on both sides possible. Consequently, their system of government would automatically become our heritage (Harriman 2006); (b) Secondly, the constitutional structure left monstrous centrifugal federalism, which to date has become the cause of ethnic-based regional tensions (Nwosu 2016). Instead of promoting autonomy and self-development of the sectors and individual ethnic groups, the implementation of the constitution became problematic since the imported British model, its consensual political culture and its advanced degree of industrialisation—the latter achieved partly at the expense of the colonies at the periphery—ensured the "[...] proper operation of parliamentary government there. (But) none of these conditions was present in the post-independence Nigeria" (Graf 1989).

The *centrifugal federalism* structure, therefore, enthroned a hegemonic system of some dominant-majority groups against the minority ethnic groups. It did not take long for the minority groups to begin their agitations and protests. Hence, it became clear that,

> [...] the very political boundaries set by the colonial administration were arbitrary and scarcely related to the subject of peoples' historical experiences. The system of indirect rule absolutized regional and ethnic differences, while colonial capitalism, for reasons of its own internal logic, sectorised and regionalised the national economy. The results were inter-ethnic competition, the formation of a dependent successor elite, continuing dependent development, and hence the creation of the major preconditions for neo-colonialism following formal independence. (Ibid.)

With the above state of affairs, the administrative clue of the first republic found itself in several political misdeeds ranging from charges of election rigging, thuggery and political intimidation, and above all, Nigerian politics was raised to the status of "winner-takes-it-all". These brought discontent and disappointment among the citizenry. The minority groups became scandalously alienated from the claimed democratic political system. They were forced to watch in broad daylight how resources taken from their soil and opportunities that should be equitable for all were mismanaged and "*nepotised*" through ethno-regional media into other areas and benefiting groups, over which they have zero control. An increasing number began to nurse negative feelings about governance since it has become, in the Nigerian case, a mere instrument for class rule. As the political class throve and grew immensely wealthier amid abject poverty and destitution of the majority of the masses, pronounced resentment towards government and the governing class began to evolve. Consequently, a concrete response from the masses became the general strike of 1964, which weakened further the already fractured administrative system of the first republic, sending signals for impending interment. Furthermore, the whole scenario became worsened by the dishonesty, malpractices, bribery and corruption that marred the election of that same year, 1964. There were attempts on probes, charges and countercharges against culprits and suspects. However, such political moves gave rise to political riots, crimes and murder and brought the society of the first republic to a total breakdown of law and order (Ibid., *Ethical Approach to Abuse of Common Good and Human Rights in Nigeria*, pp. 9–10).

By the wake of 1966, Nigeria was in a deluge of lawlessness, and within the administration itself, distrust, power conflict and total ethnic bias. This was made clear when the executive President himself, Dr. Nnamdi Azikiwe, felt assaulted after issuing an outstanding order to bring in the army to stem the tide of a near-drifting society but was reminded that his position as an executive President is titular. Therefore, it does not confer on him the powers to issue such orders as trigger military operations, which was then, as a matter of fact, an institution heavily dominated by his northern counterparts. This development degenerated into ethnic tensions, and by January 1966, a group of young military officers—the majority Easterners—led by

Major Gen. Kaduna Nzeogwu, conducted a bloody but aborted coup (14–15 January 1966), which claimed the lives of the Prime Minister, Sir Abubakar Tafawa Balewa, the Premiers of Northern and Western Regions, Alhaji Ahmadu Bello and Chief Samuel Ladoke Akintola, respectively, the Federal Minister of Finance, Chief Okotie-Ebo and several prominent army officers. However, after the coup was aborted, the administration of the first republic was replaced by a military government led by Major Gen. Aguiyi Ironsi—an Easterner (Ibid., p. 10).

1.3.2 The Second Republic (1979–1983)

The ousting of the first republic took another thirteen years of military rule before another republic—the second republic—was reconstituted in Nigerian history. Within this 13-year episode, both bloody and administrative events spanned the fragile Nigerian society, namely, the persistent grip of northern military officers on power, the plot of the immediate military administration after the 1966 coup to overstay in power, the bloody and terrible three-year civil war that claimed millions of lives and properties; the second military coup that ousted the administration of Yakubu Gowon; the assassination of the leader and mastermind of the second military coup, Gen. Murtala Mohammed (in power 1975–1976); and the final road-map to transition to civilian rule, which ushered in the second republic through the military administration of the then military Head of State, Gen. Olusegun Obasanjo (1976–1979). The second republic, therefore, was built on the 1979 Constitution, which transited from the Westminster model of political structure, where the power centres on the Prime Minister, to the American political structure, where the President becomes the leader of the republic. Thus, by the 1979 election and the clear ethnic and regional inclinations implicit in the results, the Nigerian socio-political atmosphere, already overburdened by strains of political, social and economic instabilities, became desperate, at least, to rescue itself from military escapades and further civil clashes. Consequently, there were good turnouts during the elections and the strongest region with the highest number of votes took the leadership mantel of the second republic. Below is an illustration (Table 1.1).

The above result shows the stench of ethnic tensions that were present within the Nigerian structure. Hence, each of the candidates could be seen winning only within the core regional states of their ethnic provinces, with Alhaji Shehu Shagari (33.80%), a Northerner emerging as the winner and Chief Obafemi Awolowo (29.20%), a Southwestern politician and Dr. Nnamdi Azikiwe (16.70%), a Southeastern politician taking the second and third positions respectively. The geopolitical structure, which has its origin in colonial machination, played a vital role in influencing the nature of that result.

With Shehu Shagari leading the administrative clue of the second republic and despite the belligerent and haggard polity that was in place, the administration seemed focused at the beginning with its approach to "[…] federalism, its extensive mandate to lawmakers, law executors and law adjudicators, its codes of conduct and its federal

Table 1.1 Voting percentages according to states in the 1979 elections

1979 Presidential Elections in Nigeria –Major Candidates from Three Different Ethnic Groups

State	ShehuShagari (NPN) %	NnamdiAzikiwe (NPP) %	ObafemiAwolowo (UPN) %
Anambra	13.50	**82.58**	0.75
Bauchi	**62.48**	4.72	3.00
Bendel	36.19	8.60	53.23
Benue	**76.39**	11.71	2.57
Borno	34.71	1.35	3.35
Cross River	64.40	7.66	11.76
Gongola	**35.52**	4.35	21.67
Imo	8.80	**86.67**	0.64
Kaduna	**43.12**	4.72	6.68
Kano	**19.94**	0.91	1.23
Kwara	**53.62**	0.52	39.48
Lagos	7.18	9.57	**82.30**
Niger	**74.88**	1.11	3.69
Ogun	6.23	0.32	**92.11**
Ondo	4.19	0.86	**94.51**
Oyo	12.75	0.55	**85.78**
Plateau	**34.73**	49.17	5.29
Rivers	72.65	14.35	10.33
Sokoto	**66.58**	0.92	2.52
Total	33.80	16.70	29.20

Source Emefiena and Agudiegwu (2015)

character [...]" (Ibid., *The Nigerian State: Political System in the Post-Colonial Era*, p. 77), this administration almost appeared flawless. Thus, there were a number of recorded achievements, such that nothing could have stopped the second republic from assuaging the high hopes and expectations bestowed on it by the Nigerian populace, given the palpable revenue accruing for the nation due to the rise in oil prices in the international markets between 1979 and mid-1981. However, a political syndrome of *Elite Class Struggle*, later to be known as *godfatherism*, began to set in—a form of class struggle, which "[...] cannot be totally divorced from the long years of military rule that created a class of Big-Men or Big-Man rule in Nigeria [...] informed by the usual flagrant disdain for rules by groups and individuals in the process of seeking political offices."[3] To this effect, the process of party formation under the Second Republic became heavily biased by elitist influences, which made party politics an avenue for social domination and enrichment. Nevertheless, the regime was cut up midway by the oil glut of the early 1980s, which added to a new wave of social crises that saw mismanagement and corruption being celebrated as an everyday mode of life. There were strike actions, religious unrest and resurgent ethnic bickering. At a point, one of the Nigerian dailies described the Nigerian political scene as government by the rich, for the rich and in the name of the people, since under the watch of the administration and amidst excruciating poverty and destitution

[3] Animasawin, Gbemisola, *Godfatherism and Nigeria's Fourth Republic: The Pyramid of Violence and Political Insecurity in Ibadan, Oyo State, Nigeria*. Research Report, Working Papers Series 27, IFRA-Nigeria, 2013. 3, pp. 1–24. Available via http://hal.archives-ouvertes.fr. Accessed December 6, 2021.

among the citizenry, politicians who "[…] continued to consume flagrantly, raise their emoluments and live in an unprecedentedly lavish style" (Ibid., *The Nigerian State: Political System in the Post-Colonial Era*, p. 95).

By 1983, the regime and its administration lost out on favour with the masses. Although the President himself was returned as the leader, his party and the general administration were weakened by the loss of so many seats to the opposition at the state level. Thus, during its second tenure, the administration was forced to lead with minority status. As would be expected, deep clashes of ideologies and policies took centre stage and led to a collapse of a functional polity. By the end of that same year of re-election, precisely on 31 December 1983, the regime was brought to an end in a bloodless coup led by Gen. Mohammadu Buhari (in power, 1983–1985), who cited the reason for the ousting as the response to the attitude of corruption, inept and insensitive leadership of his predecessors as well as the drifting atmosphere of immorality and impropriety (Ibid., *Ethical Approach to Abuse of Common Good and Human Rights in Nigeria*, p. 17). With the toppling of the second republic and the sole reason for intervention, Nigeria's constitution as a republic was again suspended and this was followed by another period of political tango with the military.

1.3.3 The Third Republic (1992–1993)

Nigeria witnessed another fourteen years of military rule before another reconstitution of the republic. Within the 14-year episode, the political stage administered by the military began again with Gen. Mohammadu Buhari, whose regime lasted between 1983 and 1985 before he was removed by yet another military coup masterminded by Gen. Ibrahim Babangida. Babangida's administration extended through 1993 before it was halted by his recklessness in cancelling the 12 June 1993, elections, recognised both nationally and internationally as free and fair, and which was to bring in Moshood Abiola as an elected President. This election was to bring the Third Republic into power. However, the attempt to constitute the third republic was a drama at the stage of military capriciousness—very short, heated, bloody, deadly, reckless, and, above all, disappointing and dehumanising. The ensuing events after the cancellation of the election showed a planned and malicious "manhunt" targeting individuals and persons who were labelled threats and political enemies. There were negative social scares that infested the general polity and Osaghae described them this way:

> The strikes, riots and demonstrations that greeted the annulment and continued long afterwards, brought about a virtual standstill. The already fragile economy was seriously affected, and the resurgence of strong ethno-regional forces rocked the country's very foundations, with the strong possibility of a civil war or disintegration. (Ibid., *Nigeria Since Independence, Crippled Giant*, p. 252)

Amidst frustrations and confusion, as well as the influence of pressure groups, the newly elected but militarily proscribed President himself, Moshood Abiola, returned from self-exile and decided to put up a fight to reclaim his mandate. With his vice-presidential candidate, Babagana Kingibe, he filed a suit in a Lagos high court, challenging the annulment of the June 12 elections that were supposed to bring them to power, thereby seeking the nullification of the Interim National Government (ING). In a monumental judgement delivered on 10 November 1993, Justice Mrs. Dolapo Akinsanya ruled and declared Decree No. 61 null and void since it was signed into law when Babangida had no more legislative powers to do so. Consequently, ING was declared null and void, and a court order was issued for the restoration of the annulled June 12. The stand-offs following all these rumbles continued and culminated eventually in the labelling of the president-elect as an impostor, his eventual arrest and imprisonment. He later died in prison and his wife, Mrs. Kudirat Abiola, who, in the subsequent years, became very outspoken and vehement against injustice, would later be assassinated by unknown persons.

However, the chaos and uproar that Babangida's action caused forced him to step aside for an interim government under Mr. Ernest Shonekan (a short period of administration, which lasted barely three months (26 August 1993–17 November 1993) and which pro-democracy legal luminaries argued existed in a legal vacuum before the military seized power again for the fifth time. This fifth military coup brought an administrative junta led by the leader of the coup, Gen. Sani Abacha. Within the following five terrible years, Nigerian society became a jungle infested with all manner of dictatorship, abuse of human rights, political killings, imprisonments and maiming, as well as corruption in orchestrated forms. Perhaps the climax came when the regime committed a dual administrative recklessness in 1995, namely: (a) the trial of about forty-four people and the subsequent death sentences meted to them for an alleged 1995 coup, which many regarded as the President's conscious strategy to eliminate his opponents; (b) and the killing of the activist and freedom fighter Ken Saro Wiwa alongside other eight Ogoni sons from the Niger Delta. Furthermore, the regime did not stop at that. Rather, the dictator himself felt threatened by the presence and opinions of the former military Head of State, Olusegun Obasanjo and his former Deputy, Yar'adua and decided to tag them suspects in connection with the alleged coup. To this effect, he arrested and imprisoned them. Many journalists, activists, pro-democracy groups, Labour Union leaders and politicians of opposition groups were taken as hostages and prisoners. Outrage and outcry were heard and felt in both national and international communities, and this attracted worldwide condemnation and sanctions, with Nigeria losing its membership in the Commonwealth. This development of events also made the restiveness in the Niger Delta area exponential. Thus, by 1998, the whole nation degenerated into a situation where people lived inauthentic lives and were under the daily grip of fear of the unknown since daily lives among the citizenry remained hanging in the balance at any particular moment. However, the junta, Gen. Sani Abacha, died in an undisclosed circumstance (or was allegedly killed) on 8 June 1998, bringing the regime and the confused Nigerian situation to an end. A transition government, still a military regime led by Gen. Abdulsalami Abubakar, was put in place that same year. This eventually paved the way for Nigeria's Fourth Republic.

1.3.4 The Fourth Republic (1999 till Date)

Abacha sought to perpetrate himself in power by his offer and proposal to the Nigerian masses to unmask himself of his military status and stand in for a new election as a civilian. However, his evil plans became more dreaded among the masses, who were becoming aware of his intimate association with the former dictator of the defunct Zaire Republic (1971–1997) (later reverted again to the Democratic Republic of Congo), Mobutu Sese Seko. Abacha's ambition died with him in June 1998, paving the way for the constitution of 1999 and the leader of the first regime of the fourth republic, Olusegun Obasanjo, who was lucky to have survived the prison term handed upon him by the dictator, Gen. Sani Abacha.

There were strides to reconstruct a decimated society at the very beginning of the fourth republic. The administration's competence in handling the military and clearly defining their powers and limits was remarkable—credit to the new leader, Olusegun Obasanjo, who was himself an ex-military man. There were key reforms that were taken up by the fourth republic, which to an extent, tried to manage the socio-economic and socio-political situation of Nigeria, keeping it at least domestically and policy-wise malleable, though internationally unstable and unsustainable. Worthy of mention was the administration's appreciative effort in digitalising communication systems and institutions and making them one of the most stable in Africa. Secondly, the administrative clue of the fourth republic chose, at least within their very first tenure, resource persons with innovative ingenuity. Consequently, there were sporadic waves of key institutional reforms and policy implementation. However, the socio-cultural, socio-political and socio-economic turmoil persistent in the general Nigerian geopolitical atmosphere lingered its ugly head intermittently. Chief among them, and one of the causes of administrative imputations against the regime, was a mismanagement of the critical rising tensions in the Niger Delta area, which was already exacerbated by the preceding regime. Nwosu described it this way:

> This happened on November 20, 1999, barely six months into the regime's first tenure in office, after some unidentified hoodlums or some militant groups or youths (as was claimed) abducted six policemen on duty in the Niger Delta area and later killed them. In response, the President gave a 14-day ultimatum to the government of Bayelsa State in the Niger Delta where the event took place, demanding that the perpetrators be tracked down and brought to book. But within these fourteen days, the innocent citizens of Bayelsa State, men and women, young and old, women and children were reduced to a near genocide, with the deployment of some hundreds of (armed) soldiers to a village called Odi in Bayelsa State. Several dozens of unarmed civilians were (therefore) sacrificed in this operation (carried out under the military slang, "Operation Hakuri II"). Anger and bitterness were again rekindled. (Ibid., *Ethical Approach to Abuse of Common Good and Human Rights in Nigeria*, pp. 24–25.)

The above-described *Odi massacre* brought a strong dent in the image of this administration and the regime had to grapple with that reality for the rest of its stay in office. However, both national and international opinions were, to an extent, sceptical about giving credit to the general fight for freedom and emancipation by the indigenous people of the Niger Delta, especially after the gruesome murder of policemen on duty by alleged Ogoni youths, which was condemned as murder and disregard of the victims' right to life. Two wrongs may hardly make a right.

Furthermore, with the situation boiling and tumbling in the Niger Delta area, the administration settled more for confrontation and intimidation as a strategy. Added to this was another strong appearance of the Nigerian factor, political and economic godfatherism at the administrative scenes of the regime. Thus, by the election of 2003, which would usher the administration into its second tenure, the *"might of the godfathers"* and their *"rigging machinations"* made the 2003 elections one of the most corrupt in the history of Nigeria as a nation. Reporting the state of affairs, Odey has this to say:

> President Obasanjo and the PDP that he leads saw the handwriting on the wall: that people were fed up with his inept, parasitic and corrupt regime and were determined to employ the advantages of democracy to remove that government. But they were not prepared for such a change. Hence, they resorted to the use of the army, the police and political thugs who were given definite instruction to intimidate and compel Nigerians to bow to their demolishing antics. The president and his party brought in the army and the police, not to "supervise the elections for us" but to terrorise us into submission [....] We have to be honest enough to acknowledge the fact that rigging of elections has been raised to a culture in Nigeria and that some of the other parties that contested for leadership rigged. However, the problem is that President Obasanjo's PDP introduced a new and very dangerous dimension into rigging. In this new dimension, those who did not belong to the ruling PDP were disfranchised long before the elections. When the elections finally came, they were told that they had no right to vote and that there was no point voting because no matter what the actual results of their voting turned out to be, whatever the PDP superpowers (godfathers) said must be. (Odey 2003)

By 2007, President Obasanjo himself made a formal appeal for an amendment in the constitution to the Senate and the lower chamber, the House of Representatives, which would allow him to extend the president's tenure of office to a third time. To this end, there were clashes of opinions and camps of opposition and supporters. However, in the end, this first administrative clue of the fourth republic had to relinquish power on the 29 May 2007, after an election that was still considered fraudulent and hijacked by a strategic clue of godfathers, gloomed by the fourth republic. Thus, another group of administrators led by Umaru Musa Yar'adua, a Northerner, took over and continued the Fourth Republic.

The Yar'adua administration first chose a Vice President, Dr. Goodluck Jonathan, from the Niger Delta geopolitical area as an initial political response to the restiveness, agitations and militancy ravaging the area and which has been negatively impacting the Nigerian economy. This political strategy gradually led to an official Amnesty Programme, which succeeded to an extent in removing lethal weapons from militant groups and paved the way for the establishment of a recovery, restitution and development programme to be administered by the Niger Delta Ministry of the federal government through the Niger Delta Development Commission, NDDC (already founded in 2000). The programme had four technical words as its focus and approach, namely: Disarmament, Demobilisation, Rehabilitation and Reintegration—DDRR (Ibid., *Ethical Approach to Abuse of Common Good and Human Rights in Nigeria*, p. 130). The acceptance of this programme, within which a road map of development approaches by the federal government was designed, brought relative succour to the Niger Delta. The dual focus of the programme was the government's determination to invest development funds in this area, first, for the provision of basic amenities to improve the quality of life and, secondly, for the implementation of necessary policies to combat environmental degradation due to oil exploration. Thus, with the initial seriousness in approach to this programme, Yar'adua's administration was able to diversify its attention to some other ailing structural and infrastructural matters within the general polity (Ibid., p. 25). Yet, it did not take a much longer time before the Nigerian factor took a grip on the programme. Nwosu summarised this thus:

> […] the administration of the programme has been heavily politicised and made a selective position, where the federal manipulators would often like to (field) the person of their choice so as to continue unabated the circle of corruption. In this way, transparency and accountability gave way only to structured legislative and political interference from the federal machinery. This situation manifests itself mainly (through) the practice of inflated contract awards and preference for substandard materials for proposed projects. Executed projects have always low lifespan which can never justify the huge amount quoted and ratified as the gulped amount for the project. With these ills living in the mallow of this body trusted with these responsibilities, the whole programme for sustainable development, well spelt out as a blueprint (of the commission), has suffered and is still suffering man-made destabilisation. (Ibid., p. 133)

Amidst varied administrative challenges, however, the president's health situation, which was made a "*secret under oath*" within his cabinet, seemingly made him and his administration a weakling. To this effect, political opportunists exploited the situation and began their corrupt and nefarious exploits. By May 2010, the president's health deteriorated further, and he died on the 5 May 2010. Until his death, his long battle with a terminal sickness was never communicated to the citizens and this provided, for a while, a strategised opportunity for the political miscreants, hijackers and looters within his administration. These took advantage of his health incapacitation and frequent administrative absence for their own selfish galore. With the death of Yar'adua while in office, *luck fell on Goodluck*, the vice president, who continued the administration's first tenure until the following election period in 2011.

By 2011, there was scepticism about the nature and composition of the administrative clue that would continue the Fourth Republic. The North and its elites, already disappointed by the tragic fate of Yar'adua and his administration, were bent on reclaiming the presidency. Thus, when Goodluck Ebele Jonathan, the former vice president of Yar'adua, declared his candidacy for the 2011 elections, there were serious and frontal dissenting voices from the northern forum against his moves. However, the ruling party's (PDP) lot, amidst divergent and stringent oppositional voices from the north, still fell on Goodluck, who was empowered to stand in for the election and subsequently build a new administrative clue for a third regime within the fourth republic. In reaction to the feelings of neglect and disempowerment, the religious- and politics-embedded Islamic extremists in the North—later to be called by the name Boko Haram, began an official disdain of the political situation through overtly organised terrorist activities and intermittent violent strikes in the forms of suicide bombing, looting, raping and kidnapping. This situation in the North would contribute positively to the majority support for the candidacy of the Southerner, Goodluck Ebele Jonathan, and a repudiation of the candidacy of the Northerner, Muhammadu Buhari, during the 2011 elections. A second contributory factor was the alleged inflammatory utterance of the northern candidate, to the fact that if North should lose the presidency in 2011, the government of his opponent, Goodluck, would be made ungovernable. Nevertheless, the electorate exercised their franchise and elected Goodluck Ebele Jonathan, who was to lead the administrative clue of the fourth regime in the fourth republic.

With the electoral victory of the PDP under Goodluck Jonathan (a Southerner) and Vice President Namadi Sambo (a Northerner) in 2011, the new administrative team braced for action amidst social, economic and geopolitical turmoil. The violent situation in the north was escalating with increasing losses of lives and properties. The onslaught reached its climax with the "Chibok Story" when the extremists abducted 276 schoolgirls on the night of 14th through 15th April 2014. This became a nightmare for the whole of Nigerian polity and sparked global outcry and solidarity with Nigeria, with every corner of the globe condemning the action and lending their voices in demanding the release of the innocent girls under the #tag "Bring Back Our Girls". However, before this abduction, the administration of Jonathan had begun a massive development approach in the North through the provision of educational infrastructure to absolve the teeming number of children and young adults who roam the streets and crannies of the Northern states under the name *Ahmajiris*—in other words, a concept taking from the Arabic root, المُهَاجِر "al-Muhajirun", which describes a person who searches for Islamic knowledge by being in constant migration away from his or her home. To this effect, this administrative policy became an outright enemy of the extremist groups and its overt rejection formed one of the key agendas of the group Boko Haram.

With the charged atmosphere surrounding the entire governance, Jonathan's administration suffered a near loss of control over the military, unlike the regime of Obasanjo, which began the fourth republic. There were clear and outward indications that his constitutional duty as the Commander-in-Chief of the Armed Forces was not being accorded to him and, therefore, not being respected. To reinstate his administration's commitment and concern for the security situation in the country, he decided to invest in combat equipment (Arms) to reinforce and re-empower the armed forces. However, this effort was sabotaged through an alleged corrupt Arms Deal involving his fellow Southerner, Olisa Metuh, who was accused of illegality, misappropriation and embezzlement. Again, the administration's attempted effort for economic recovery through a proposed removal of a corrupt practice known as the *Fuel Subsidy* ignited a nationwide protest, tactfully sponsored by his political opponents and so many beneficiaries and exploiters of the so-called national cake, the crude oil. All the efforts to put the prospects of the proposed policy for the benefit of the entire nation across the masses fell on deaf ears. Thus, the policy could not see the light of day. Consequently, within the first quarter of 2014 preceding the already planned election for the second tenure of the regime, added to the assaults being felt by the masses due to incoherence in speech and eloquence by the wife of the president, the frustrations were very conspicuous and the political enemies grabbed the opportunity and labelled the administration "clueless".

Amidst this confusing state of affairs, Jonathan's administration convened a National Conference that was inaugurated on the 17th of March 2014, creating (to the administration's good intention) a platform for dialogue and reconstruction. An excerpt from the president's speech at the opening of the conference reads as follows:

> [...] The strongest nations in the world today also went through their own formative stages; some for decades and others for centuries. We must learn from them that nationhood will not happen overnight, especially given the circumstances of our birth as a nation [....] The National Conference is therefore being convened to engage in intense introspection about the political and socio-economic challenges confronting our nation and to chart the best and most acceptable way for the resolution of such challenges in the collective interest of all the constituent parts of our fatherland [...,] We cannot continue to fold our arms and assume that things will straighten themselves out in due course, instead of taking practical steps to overcome impediments on our path to true nationhood, rapid development and national prosperity [....] This conference is open for us to table our thoughts and positions on issues and make recommendations that will advance our togetherness. The issues range from forms of government, structures of government, devolution of powers, revenue sharing, resource control, state and local government creation, boundary adjustment, state police and fiscal federalism, to local government elections, indigeneship, gender equality and children's rights, amongst others [....].[4]

The speech, however, contains strong and meaningful clauses, which could aid one to believe that its eventual manifesto could also be implementable. Nevertheless, there was again a plurality of views and opinions for and against the conference: for some and the majority from the opposition camp, it is the president's last-minute delay tactics and crafty approach to corner the minds of the masses and thereby influence the approaching election. Others saw it as a belated decision and another moment to waste the Nigerian financial resources. Handful opinions among the masses still took it as a step in the right direction to at least create a common forum for dialoguing with diversity. Furthermore, because of the imminence of the election and coupled with the fact that Jonathan's administration was bent on developing a manifesto with the aid of the conference as a blueprint in their campaign trail for the second tenure, the agenda of the conference was rushed—just barely four months (March 17, 2014–Mid-July, 2014) of interaction among the delegates on critical issues of deep national concerns in a complex structure like Nigeria. Deliberations on mainstream issues of Nigeria's geopolitical and socio-economic problems were heated among participants but not sufficiently and pragmatically handled, so many questions remained unanswered. The period that dated from the close of the conference to the date of the election, 28 March 2015, knew no effort by the administration to dissect the key deliberations and recommendations of the conference and bring them to the electorate at the grassroots. Consequently, there were conclusions that the national conference was again bygone

---

[4] Premium Times Nigeria, *President Jonathan's Remarks at the Inauguration of The National Conference*. March 17, 2014. Available via https://www.premiumtimesng.com/national-conference/national-conference-president-jonathans-remarks-inauguration-national-conference/. Accessed October 15, 2019.

and should again be abandoned the way its preceding ones have been treated within the Nigerian system.

In the last days and periods preceding the day of the election, it was very clear that the president himself had lost many previous supporters from his southern home, especially in the south-south, whose electoral mandate was tactfully appropriated by the opposition party, APC, in the 2015 elections, with the choice of their son, Prof. Yemi Osinbajo, for the seat of the vice president. Again, the last days preceding the election saw a massive coalition of all the northern states, and in unity, they were already speaking with one voice about their commitment to retrieving the presidency seat. To this end, Jonathan's administration retained the choice of a Northerner as vice president, bereft of the anticipated effect. Amidst this critical situation, the election took place and for the first time in the history of Nigeria since independence, a sitting president lost to the opposition. In his decency and sense of purpose, he conceded and handed over power to the fifth civilian regime within the fourth republic, to be led by Muhammadu Buhari as the president and Prof. Yemi Osinbajo as vice President.

The administrative agenda of Muhammadu Buhari, especially from his campaign manifesto, would focus, in the main, on the fight against corruption, the fight against insurgency and policy implementation for socio-economic stability. It would be remembered that one of his vehement accusations against the administration of Goodluck Jonathan was their inability to terminate the terrorist group Boko Haram and to restore peace and safety of lives and properties in the Northeast. Nevertheless, since the inception of the administration in 2015 to date, the situation in the Northeast has not vanished and Buhari's administration has even witnessed more cases of kidnappings, violent onslaughts and insurgencies than the administration of his predecessor. However, it does not mean that Buhari's administration has not implemented frontal policies against Boko Haram and the general security situation in the North. To the credit of the administration and its leader, who was himself an ex-military personality, the regime seems to be in control of the armed forces and enjoys more of their respect, unlike his predecessor, Jonathan.

Sharp, uncomfortable and distracting feelings were conspicuous among the Nigerian citizens during the regime's first tenure regarding the administrative clue and the personnel that built up the cabinet. Between 2015 and the regime's end of its first tenure, 2019, there were clear celebrations of incompetence by some of the administration's ministries and their leaders, which aided the accusation against the administration of Muhammadu Buhari as being partisan and ethnic through his lopsided choices of personnel, yet, within a federal system, he passionately promised to defend. There seems to be a response to addressing the issue of incompetence. However, the administration remains adamant about inclusive democracy, especially with a political structure like Nigeria under the heavy pang of social and cultural diversity.

Again, there existed deep negative feelings of repugnance, especially from the south, the middle-belt region and some other southern minority groups, concerning the administration's suspicious silence and blatant refusal to proscribe the "banditry-group" known as Fulani Herdsmen. This has presently taken centre stage as the new wave of violent conflict with a socio-cultural and socio-political agenda, which many describe as the *fulanisation of Nigeria.* The most current Nigerian situation depicts a situation of anarchy, where herdsmen, under the watch of Buhari's administration, access lethal weapons and are using them daily on innocent citizens, most of them helpless farmers, who say "no" to the destruction of their farmlands and means of livelihood, through illegal besiege of herdsmen and their cows in the name of "open grazing". This has elicited confrontational stand-off situations between the North and South and tension-filled debates in both the upper and lower chambers of the National Assembly, bordering on a constitutional demand for the adoption of open grazing for every part of the federation. The situation is precarious and the International Crisis Group reports:

> These conflicts have exacted a heavy humanitarian toll with thousands killed and tens of thousands displaced. Some estimates suggest about 2,500 were killed country-wide in 2016—a toll higher than that caused by the Boko Haram insurgency over the same period. In Benue, one of the hardest-hit states, Governor Samuel Ortom reports more than 1,878 people were killed between 2014 and 2016. Tens of thousands also have been displaced. From January 2015 to February 2017, at least 62,000 people were displaced in Kaduna, Benue and Plateau States; in the absence of Internally Displaced Persons (IDP) camps, most seek shelter in other, poor, rural communities, straining their already scarce resources. The fear of conflict alone can drive residents to relatively more secure urban and semi-urban areas. Since both authorities and donors often ignore these conflicts, affected localities receive far less support from the National Emergency Management Agency (NEMA) and international agencies than those impacted by the Boko Haram insurgency [...] For women and girls, the impact is frequently magnified. The relatives of men killed in the violence often evict widows from their farmland. Moreover, post-conflict economic and social disenfranchisement renders women and girls even more vulnerable to sexual and economic predation [....] The economic toll has also been huge. According to a 2015 study, the federal government was losing $13.7 billion in revenue annually because of herder-farmer conflict in Benue, Kaduna, Nasarawa and Plateau States. The study found that on average these four states lost 47 per cent of their internally-generated revenues. In March 2017, Benue State Governor [...] asserted that attacks by herders coming from more northern States, and possibly also from Cameroon and Niger, had cost his State ₦95 billion (about $634 million at that time) between 2012 and 2014.[5]

The administration's body language has not yet shown interest in any pragmatic approach, perhaps, in exploring more advanced, scientific, innovative and sustainable methods in animal husbandry as an alternative response to quelling the conflict. Consequently, many accuse the president himself of a double standard since he shares the same tribe with the "Herders" and has been a herder economically. Moreover, the administration condemns the activities of groups agitating non-violently for self-determination, labelling such groups terrorist groups and their leaders, insurgents to the Nigeria State. This constitutes one of the most current standoffs being organised against this administration.

In the main, the issue of Biafra, with the undying spirit of emancipation accompanying the movement and which once led to a civil war between 1967 and 1970

---

[5] International Crisis Group. *Herders against Farmers: Nigeria's Expanding Deadly Conflict.* Crisis Group Africa Report, No. 252, 19. Sept. 2017, p. 7. Abuja/Brussels. Available via https://www.crisisgroup.org/africa/west-africa/nigeria/252-herders-against-farmers-nigerias expanding-deadly-conflict. Accessed October 15, 2019.

(the Nigeria-Biafra War), still rages both in ideology and in the spirit of activists and groups, has not died and would not die. Recent socio-political activities have seen confrontations between the federal government and these groups, especially in the southeastern states. Lives and properties have been lost and the administration of Muhammadu Buhari shows no readiness for any dialogue. The last straw has already broken the camel's back with the most recent development, where the federal government made a re-arrest of the exiled leader of the group Indigenous People of Biafra (IPOB), Nnamdi Kanu, and repatriated him into the country from Kenya. This action of the federal government, added to the continued detention of the said leader, has awoken a strong sense of solidarity within the southeastern states, where many sympathisers demand his release and the possible initiation of a process for no further delay regarding the independence of Biafra. It may sound utopic, but it is as serious as it is passionate in the minds and lives of the Southeasterners.

Amidst the above-mentioned layers of socio-political and socio-economic concerns plaguing the present administration and the challenges they pose to the general development and the well-being of the polity, the global crisis with the Covid-19 pandemic also has deep observable fiscal challenges for the Nigerian economy. The general and grassroots effects of the implemented phases of lockdowns within the country, which of course, were curative responses to the unannounced outbreak of the pandemic, would not be anything near its end. The current indigenous scientific assessment of the International Food Policy Research Institute gave an analysis of impacting socio-economic changes within the macro and micro sectors thus:

> We first consider the impact during Nigeria's five-week lockdown period in Abuja FCT and in Lagos and Ogun States, as well as the Kano State lockdown. (Available data) shows the impact on aggregate GDP and its components. The largest losses, in absolute and relative terms, are recorded in the industrial (-38.8 percent) and services (-39.2 percent) sectors, with GDP losses in the services sector (USD 10.8 billion) accounting for over two-thirds of the recorded 34.1 percent loss in national GDP (USD 16.4 billion). Losses in the agricultural sector are small in absolute terms (USD 1.2 billion), but relatively large in relative terms (-13.1 percent) considering that the sector is exempt from any direct restrictive measures [….]. (Andam et al. 2020)

In addition, a broader Agri-Food observable and impacting change is further detailed below by the institute (Table 1.2).

The analyses of and the fact presentation above are referenced here to prove that the administration of Muhammadu Buhari must be *playing the ostrich* if possible remedial administrative blueprints are not designed and implemented to contain the already wobbling economy.

Furthermore, and perhaps considered as Nigeria's impending nightmare, is the spree with which this administration rushes into external borrowing, yet Nigeria consumes more import-products than it produces and exports. Experts warn about an impending *debt-tsunami* that may besiege the country in the near future if there is no rethink, no retreat, or at least no debt-servicing economic strategies and policies in concrete terms. At present, the external and domestic debt rates in Nigeria stand at $28.57 billion and ₦16.02 trillion, respectively, as against $7.3 billion in external debt and ₦8.84 trillion in domestic debt of the previous administration at the beginning of

Table 1.2 Estimated change in GDP of Nigeria during the Covid-19 lockdown period

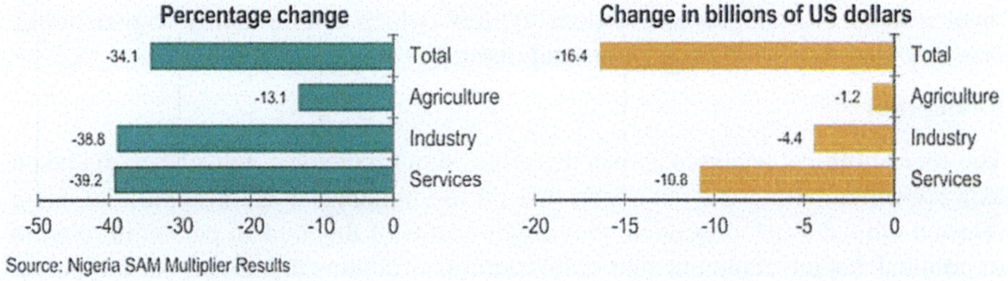

Source Reproduced with permission from the International Food Policy Research Institute (IFPRI). www.ifpri.org NIGERIA IFPRI

the administration of Muhammadu Buhari. Nevertheless, both local and international experts may not have placed Nigeria on the red line of financial borrowings, but the domestic socio-economic and socio-political realities elicit enormous concerns about the predictable negative impacts of the present practice. To this end, there should be no better proof for the concrete manifestations of Nigeria's current economic conundrum as to the realities of rising conditions of hunger, increasing youth unemployment rate, brain drain, geo-ethnic and geopolitical imbroglio, increasing rates of group agitation and the demand for self-actualisation, unhealthy competitive tendencies leading to the survival of the fittest syndrome, youth crime and other forms of forced criminality, forced emigration of helpless youths in search of acceptance and integration elsewhere, susceptibility of young girls and women to sexual abuse and trafficking, etc.

As an academic work, the present book would be unscientific if it is predominantly bent on spelling only dooms for our evaluative macro-specimen, Nigeria. Thus, despite the painful difficulties and challenges being experienced within the general polity under the present administration and exacerbated by allied global economic and pandemic crises, there are also administrative efforts and policy-based approaches to some sensitive areas, namely: the continuous countercombat to contain the insurgency in the North-East; a *yet-to-be-confirmed* assessment of the administration's claim to the fight against corruption and application of punitive justice; noticeable efforts to revitalise and diversify the transport sector in the recent past through a gradual comeback of the railway system, however, this development is alluded to yet with strong reservations due to the approach and manner which have deep geo-ethnic and geopolitical undertones; and a few other areas. Nevertheless, some of the efforts being made by the administration are being blurred and marred by four critical and sensitive administrative recklessness, namely: (a) the administration's playing of the *bedfellow* with the killer-herdsmen and the refusal to declare them bandits and terrorists; (b) the obtrusive nature of the administration's selective personnel appointments into political offices, grossly infested with ethnic hegemonic lines; (c) the regime's flare for external borrowings on the one hand, and on the other hand, the observable nonchalance towards re-positioning and re-basing the domestic economy; and (d) the application of weaponised intimidation, arrest and imprisonment instead of dialogue on freedom fighters, which intermittently trigger police brutality and abuse of human rights and dignity.

Evaluation

The socio-political analyses above, therefore, sum up the overview of Nigeria taken as a specimen, alongside our focus first on the human person, *Mmadụ* and using relevant ethical tools to evaluate the present status of this human person in relation to political, social, economic and socio-cultural structures that go into development processes. One of our critical guiding hypotheses is: How and to what extent has the current Nigerian structure and system helped or contributed to thwarting the moral process of supporting the *human person* towards the realisation of his or her *full potential*, in other words, the *human capabilities* within the Nigerian setting?

1.4 Four Cardinal Socio-Political and Socio-Economic Factors in Understanding Nigeria's Fragile Polity

The journey of all known human societies towards nationhood has always been a process, and within this process, events and experiences impact the general history of such a journey integrally. To this effect, this book reflects upon four cardinal concrete events that have shaped the Nigerian system and structure to their present status as proof of the pieces of evidence it has. They are not just accidents. They rather team up to build concrete situations that emerge and persist in their emergence within the general process of growth and development.

1.4.1 Protracted Military Presence and Control of Power

It is believed that the experience of military rule in Nigeria has been too incessant and, therefore, contributed to the fragility and instability of the present structure. The chain that trailed a sequential suspension of the constitution and democracy whenever the military strikes contributed so impactfully to bequeathing to the present structure a still crawling democracy and deformed constitution. Known democratic systems that have achieved relative stability followed predominantly (not absolutely) sequential and/or systematic ideology application, re-application and integration. Thus, the known system of imposition of decree associated with all forms of military rule contradicts the core tenets of democracy in all ramifications and impacts the general polity. In other words, "*to decree*" serves the purpose of imposition, dominion, exclusion, oppression, etc. The constitutional basic rights and norms are in this regard contravened. The pessimistic picture of the truth about Nigeria's resentful experience with the military and thus, the weakening effect of its incessantness to democracy could be captured in Nwabueze's five phases of Nigeria's democratisation bid, namely, "[…] era of colonial autocracy and absolutism, the emergence of constitutional democracy and the return of autocracy and absolutism under military government, restoration of constitutional democracy and second coming of military autocracy and absolutism" (Nwabueze 1989).

Behind varied opinions and independent research studies, predicated upon the backdrop on whether the periods of military rule have helped to increase the country's External Reserves (Nwosu, Ohakwe 2013) for whichever plausible or accidental reasons; or the position that some of the military interventions helped in some periods of transition to civilian rule, this academic work opines that the Nigerian experience has constituted, in the main, the bane to democratic consolidation in Nigeria. A mere aid to the transiting of power does not explicitly guarantee a *consolidated* transfer of power. Consequently, the Nigerian system experiences to date the political *merry-go-round*, where the entire political and leadership settings witness the successive return of its former military rulers but in coded civilian clothing. Thus, the question remains: what has been the advantage? Simply put, have the protracted, or, as the case may be, the intermittent political appearances of the military within the political economy of Nigeria been able to bequeath consolidated democracy to her system? This position—the case of non-consolidated democracy—taken in relation to the military and politics in Nigeria, and which would constitute one of its basic arguments to proving that related structural dysfunctions contribute to the defrauding of Mmadụ, the human person, and invariably to the stagnation of sustainable development in Nigeria, becomes further comprehensible with an understanding of what consolidated democracy should mean for Nigeria. Hence, proving that the military's protracted presence in Nigerian politics has denied the country a consolidated democracy means that there exist varied punctures in real behavioural and institutional changes, which are drivers of democratic citizenship and culture (cf. Ibid., *The Military and the Challenge of Democratic Consolidation in Nigeria: Positive Scepticism and Negative Optimism*, p. 12). In other words, consolidated democracy reflects both the attitude and character of a particular political economy and must "… include such divergent items as popular legitimacy, the diffusion of democratic values, civilian supremacy over the military, the elimination of authoritarian enclaves, judicial reform, and the alleviation of poverty and economic stabilisation among others…" (Ibid., p. 11).

When brought to the Nigerian scene, these salient factors become, policy-wise, deficient. To this effect, we maintain that a relatively consistent form of democratic governance would be requisite for more stable political institutionalisation. However, this *over-presence* of the military in the Nigerian case becomes evident and illustrative, as seen in the table of Statistics of Civilian and Military Government Regimes in Nigeria attached to the appendix. Again, given that the political world is not oblivious to the fact that sustenance in the democratic process involves a varied interplay of both negative and positive factors, our point of departure in the Nigerian case remains the consequence-based frequency at which the democratic process of the country since independence has been interrupted, hijacked, mal-administrated and militarily influenced. In the final analysis, this has been attributed to the inability of the Nigerian

State to develop a viable and functional policy to address her peripheral capitalist system, seen as a crisis of social justice within the structure. Kieh and Agbese opined:

> [...] the military engagement-disengagement nexus is deeply rooted in the crisis of social justice in Nigeria. The military has intervened and disengaged from Nigerian politics in the past because of the inability of Nigeria's peripheral capitalist system to meet the needs and aspirations of the Nigerian masses [....] (yet) The problem of political instability (unconsolidated democracy) in Nigeria remains unsolved (Kieh and Agbese 1993).

Our take, therefore, is that the wound of inconsistency regarding nation-building and the instability it has brought into the Nigerian system remains a challenge that stirs Nigeria to the face and needs to be addressed. No amount of military intervention can restore democracy if a nation does not see it as urgent to live out the tenets of democracy as part of its identity and character. Thus, it is anticipated that within this lingering socio-political dilemma, the recommendations of this academic work would be developed and applied.

1.4.2 Oil, Eco-Disaster, Corruption and Planned Inhumanity

Until 1956, when crude oil was officially discovered in commercial quantities in Nigeria, the nation was predominantly dependent on agriculture and the export of cash crops. However, this discovery and the subsequent paradigm shift to the "petro-economy" have deeply impacted the general Nigerian system and character. Nigeria's dependence on oil exploitation and exportation accounts for almost 80 per cent of its revenue. Nigeria currently occupies the first position among the oil-producing countries in Africa, with the production capacity pegged at approximately two million barrels of oil per day. This places Nigeria at the eleventh position in the global space (2021 data). Added to this, Nigeria's proven oil reserve is estimated at 37.2 billion (2011). Nevertheless, experts warn that the network and dynamics of Nigeria's economic policy remain shaky and unbalanced due to the near suffocation of and overdependence on the oil sector. In other words, the overdependence of the mainstream economy on oil exportation and the revenue accruing from it cause imbalance and neglect of other sectors that help to drive and integrate the general economy. Simply put, a nation's strong degeneration into a mono-economy invariably stifles the possibility of diversifying the economy and the known non-diversified economies are often targeted prey to periods of economic recessions—both domestically and internationally. Concrete pieces of evidence are an integral part of the Nigerian story, given that the country has experienced three intermittent recession periods within the past eighteen years. Using "**Q**" to represent the quarters, the first two quarters, which represent the First Recession Period, occurred before 2004, showing **Q1** = −7.59% and **Q2** = −0.18%. The country was able to move out of recession after these two consecutive quarters of negative Growth in 2004. The Second Recession Period occurred between 2016 and 2017 with five different quarters: **Q1** = −0.67%, **Q2** = −1.49%, **Q3** = −2.34%, **Q4** = −1.73% and in the year of exit from recession,

Q5 = −0.91%. The Third Recession Period, also impacted by the global Covid-19 pandemic, produced three sequential quarters of recessionary statistics, from which the economy was expected to recover by the end of 2020, namely **Q1** = 1.87%, **Q2** = −6.1% and **Q3** = −3.62%.[6]

In the Nigerian case, the reason has always been the fluctuating global oil market, which impacts her revenue during major falls in oil prices. To this end, we reference this situation as a current Nigerian trend, which constitutes a major challenge to the Nigerian economic structure. Researching independently and reporting under the sub-section "Rohstoffreichtum—Fluch oder Segen?", Tetzlaff remarked that the Nigerian case remains disappointing since during the *boom-days* of the oil markets, no effort was made to build and stabilise the country's reserve fund. Consequently, as the oil price began to fall and eventually collapsed, the nation was forced to cut its budget, leading to an excruciating increase in the prices of the commonest daily commodities, thereby putting unbearable pressure on the poor masses. Added to the situation was the recklessness with which the country embraced external borrowings, which further encumbered the country's debt crises. Thus, whereas the debt level stood at $24 billion in 2000, it came up to $69.6 billion in 2015. This has happened despite a number of debt reliefs gained from the International Monetary Fund (IMF) and World Bank. Currently (pegging the date as March 2021), both the domestic and external debt amount to about $87 billion and many opinions think that the end is not yet in view (Tetzlaff 2018).

Given that the chances of the 1970s and 1980s during the major oil boom eras were mismanaged and lost, the general system and management have since been a sorry spectacle. To date, Nigeria manages a crude oil-based economy and infrastructure that is at the same time derelict. Considered to be the biggest in the continent, namely the Nigerian economy with a GDP of about $432 billion (stand, December 2020), the fat figures, unfortunately, have little or no impact on the everyday lives of the common men and women on the streets of her slumps and suburbs. Meredith gave an informative description thus:

> Vast sums (of the 'petro-dollar', oil money) had been spent on prestige projects, to no advantage. A total of $8 billion had gone on constructing a steel industry complex based at Ajaokuta that had yet to produce a single bar of steel. Billions more had been sunk into an ultra-modern capital (city) at Abuja, complete with glittering hotels and office towers, that the ruling elite enjoyed using but that brought little benefit to ordinary Nigerians [….] Even worse were the vast sums siphoned off through corruption. (Meredith 2011)

In addition to the foregoing internal socio-political and socio-economic quagmire with regard to oil and its exploration, Nigeria, just like other developing economies that supply raw materials, remains constantly exposed and defenceless to fluctuations in demand and price on the raw material markets. This has a strong weakening effect

---

[6] Proshare Nigeria Limited, History of Nigeria's GDP and Recession. CBN Communique No. 133 of MPC Meeting, Nov. 23–24, 2020. Available via https://www.proshare.co/articles/outlook-2021-eyeballing-major-economic-indicators-in-2020-exchangerate?menu=Economy&classificati on =Read&category =Reviews%20%26%20Outlooks. Accessed November 22, 2021. * Proshare Nigeria Limited, founded on December 20, 2006, is a niche financial information service hub with a key focus on delivering market intelligence, data, etc.

on the general polity, namely: (a) unstable budget and development plan with overdependence of import capacity, weak internal revenue, fluctuating policy actions as well as poor debt servicing capacity; (b) the destabilisation of manufacturers' revenue, the weakening of investors efforts and general job opportunities, as well as the systematic instigation of political instability; and (c) deprivation of job opportunities, especially when the processing of such raw materials is done in foreign lands, thereby weakening further the work market, and most often, the general exploration and exploitation of such raw materials cause huge ecological challenges (Stockmann et al. 2010). Furthermore, when the developing economy in question has no capable policy to diversify the general economy, the whole system creates chains of interrelated problems like debased development status, cheated export relationships, low investors' interest and exploited job market, non-innovative production, weak growth impulse, further consolidation of structural dysfunction and, of course, the most conspicuous of the general overt effects, poverty (cf. Ibid.).

To this effect, it is to be critically noted that the exploitation of oil in Nigeria and the techno-explorative activities of varied multinational companies in the oil-producing states, especially the Niger Delta, have for decades recorded scores of adverse ecological deluges. This has long kept Nigeria on national and international watchlists as a nation with predominantly man-made ecological disasters. An official independent study and assessment of the United Nations done within the last decade of Nigeria's history summarised the situation thus:

> The assessment has been unprecedented. Over a 14-month period, the UN Environment team examined more than 200 locations, surveyed 122 kilometres of pipeline rights of way, reviewed more than 5,000 medical records and engaged over 23,000 people at local community meetings. The environmental restoration of Ogoniland (in the Niger Delta) could prove to be the world's most wide-ranging and long-term oil clean-up exercise ever undertaken if contaminated drinking water, land, creeks and important ecosystems such as mangroves are to be brought back to full, productive health. The report's key findings are alarming both in terms of human health protection and environmental protection.[7]

Within the period of our research, these ecological challenges, amidst conflicting interplay of efforts of the one administrative regime to contain the challenges on the one hand, as against the detractive attitude of another regime, on the other hand, are still very remote from being over. Hence, the Niger Delta area remains an ailing part of the Nigerian socio-political and socio-economic structures inhabiting helpless citizens—the indigenous people of the Niger Delta—who, for a while now, have been saying "no" to the destruction of their local economy by the activities and neglect of the Nigerian State with the allied multinational companies. This remains a major source of conflict and tension within Nigeria's socio-economic sectors.

As could be seen in many situations with all African nations that inhabit varied forms of natural resources, the cancer of corruption seems to be the detractive negative reality alongside the existence of such natural capitals. The Nigerian case has

---

[7] United Nations, *Environmental Assessment of Ogoniland. Site Factsheets, Executive Summary and Full Report*. The United Nations Environmental Programme, UNEP, Presentation of the Research Study, August 1, 2011. Available via www.unep.org/resources/assessment/environmental-assessment-ogoniland. Accessed November 23, 2021.

never been different. Thus, we delineate the practice of corruption in the Nigerian system as a major negative catalyst in keeping her structure and system fragile and malfunctioning. A more detailed examination of corruption would be elaborated on in another sub-section to better explore the insight it has to offer our general overview. In the meantime, we view the Nigerian oil sector and the crime being committed by the Nigerian approach as a moral question. Ethically speaking, it is likened to the parable of the goose that lays the golden egg and in context with the Nigerian case, it is the moral imbalance associated with rentier attitudes and politics, in other words, the mentality of maximising rents without concrete efforts to ensure responsible, innovative and sustainable productivity. To this end, the neglect of the goose has led to one of the worst ever witnessed and planned inhumanity in the Niger Delta area, where the combined machinations of both the federal government of Nigeria and their allied multinational companies sacrifice human dignity and human rights at the altar of meditated greed. The inhumanity speaks out in concrete terms with the level of ecological and environmental degradation currently evident in this area. Nwosu examined the situation thus and posited two fundamental questions:

> [...] the Nigerian Niger Delta presents a horrible picture as a representation and confirmation of (what degradation and dehumanising) mean. Assessing the degree of impoverishment in this area brings to the fore a people being battered by poverty and underdevelopment, a people whose governmental plans and policies for sustainable infrastructural development would always hit the hard rock of non-implementation due to corruption and mismanagement; a people confronted always by devilish manipulations, politicisation, inadequate funding as well as embezzlement... from those who are charged with the responsibility of executing those projects [...] Within the framework of this problem, two questions (motivated by the) principles behind the agitation of this ethnic minority group stand clearly, namely: Why is it that abundant human and natural resources have had so little an impact on poverty alleviation in the Niger Delta region? (Secondly), What reason can we advance to substantiate the fact that practically all past development plans and efforts to address the needs of this region have failed? (Petersen and Wehrmann 2015)

The two questions, as posited above, remain persistent in the Nigerian case, and thus, they offer the ground to investigate both the ideological and administrative structures that contribute to the questions remaining unanswered or not sufficiently attended to in the Nigerian system. Socio-political and economic structural dysfunctions in any known polity elicit a form of *chain reaction* which continues to detract such a system, keeping it malfunctioning, retarded, stagnated, unsustainable, etc. The situation worsens when the "social cancer" called corruption becomes almost conventional.

1.4.3 Geopolitical and Geo-Structural Imbalances: Towards a New Developing Anthem Called Restructuring

In presenting the geopolitical structure of Nigeria, which gained independence in 1960, it is argued that the geopolitical divisions bequeathed to the country at independence have remained striking and foundational to the fragility that threatens her existence. The early history of the concept of geopolitics, basically with connections to colonial and imperial thinking and ideology, and which remains central to its meaning, describes the concept as "[…] the role of geography in international political relations […] (focusing) on strategic aims of states" (cf. Ibid.). Geopolitics contend with hegemonic tensions in spaces and places, especially in international political relations, but not limited to that (Ibid.). Today, the conceptual development of the word geopolitics has evolved to include Critical Geopolitics, which aims at understanding spaces and places constructed by varied political actors (practical geopolitics), strategic institutions (formal geopolitics) and mass media (popular geopolitics). It would "[…] critically reflect the binary scheme of 'Us' versus "Them" (the representation of "self" and "other") as well as geopolitical images such as "East", "West" […] "Third World" (today replaced with the concept, *the global south*), which are perceived as instruments to construct a simple model of the world, which can then be used to advise and inform foreign and security policymaking" (Ibid., *Geopolitics: Terms—Concepts—Critical Perspectives*). However, an appeal is made to one of the criticisms against critical geopolitics to prove an inbuilt fragility originally caused by colonial master's geopolitical mistakes, namely that "Critical (geopolitical) thinkers […] need to acknowledge that they are never detached observers, that is to say, they can know the world only through that conceptual schema provided by our culture and languages" (Nwodo 2021).

The above-mentioned possible ideological mistake has remained fundamental to the detrimental geopolitical imbalance evident within the Nigerian system since it was made and which, under the guiding principles for a functional federal character, remains an unfortunate part of Nigeria's history. Therefore, in the present Nigerian system, the arbitrary, forced and convenience-based fusion of regions for purely administrative purposes made the colonial policy a detached observer in this context. This has been detrimental to the general political and administrative system in which Nigeria operates to date. Indigenously based cultural and lingual identities, as political and social histories show, play fundamental roles in the development of organic societies and institutions, which respond better to general growth and development. To this end, therefore, a timeline of events and happenings within the Nigerian state has held a persistent premonition to the general political and administrative structure that Nigerian has, pointing to the fact that the geopolitical map that remains operative after over sixty years as an independent state has for long rubbed the nation of tension-free socio-political, socio-economic, socio-cultural as well as socio-ecological development.

At the time of our research, there began a wave of revolution-based tensions cutting across almost all known cultural and language groups in the country, with individual groups calling for a total breakaway from the present structure under the revolutionary mantra now known as *Restructuring*. Many among the ruling elites, especially the powerfully rich clue, who control politics and the economy, view this agitation as a threat to their comfort zones. Thus, they calumniate anyone joining the dissenting voices as a plotter, with the motive of destabilising the Nigerian state and unity. However, the "hoi polloi" who experience the daily impact of this structural imbalance persist in every lean way to remind the Nigerian system that a systemic change is duly inevitable. To this effect, all proponents of restructuring remain united in a structural agenda that regards the present Nigerian constitution—the 1999 Constitution of the Federal Republic of Nigeria—as fraudulent and, therefore, has failed to address the nation's social, economic, political and multicultural complexities. The demands are clear: the restructuring, redefinition, reinvigoration and empowerment of the regional identities as the proper response to the power devolution originally believed to work better for the Nigerian peculiarities. This will help reinstate the regional sense of autonomy with regard to natural resources and the development of domestic security networks that could function better administratively. The unitive goal would be "[…] to emphasise more on selfless service rather than individual enrichment, promote genuine unity instead of ethnic bigotry and challenge our capacity to exploit our abundant potentialities to make life more abundant for our people" (Falana 2018). As one of the cardinal factors that would be present in the prospective field of application for our research, our opinion aligns itself with this demand, which, in the main, involves a policy-based decentralised and democratised economic powers. Unfortunately, the Nigerian situation has been predominantly hijacked and appropriated by the minority ruling class through the liquidation of public assets and the consequent robbery of same in the name of privatisation (Siep 2019). This debate remains a wound to be healed.

1.4.4 Fanatic De-secularisation, Religions Fundamentalism and Sectional/Regional Insurgencies

Nigeria and its complex structures present a syncretic form of atmosphere where religious beliefs, religious denominations, fanaticism, fundamentalism, sectionalism, regionalism, politics and political affiliations, all get mixed up in the one social atmosphere of a struggling and fragile state. In most Nigerian cases, individual crises do not have purely isolated causes. Most political feuds and bigotries have predominantly ethnic and religious insinuations. Religious affiliations, on the other hand, are getting unnecessarily politicised. Fundamentalists are threatening to impose their religious conventions and convictions as mandatory laws in secular states, while sectional and regional insurgencies are being weaponised as both political and religious strategies. The proliferation of churches, beliefs, religious sects and denominations presents an atmosphere of anarchy. This exists as a reality alongside a consequent breed of opportunists who monetise these strategies, using them as effective weapons to exploit and defraud their victims. All these, in the main, have combined to enthrone a strong and negative de-secularisation of the Nigerian state and the institutionalisation of quasi-theocracy, whereby the masses are cowed into believing that the minutest social, national and general life issues and challenges must always be tied to the invisible hands of some spiritual and esoteric powers and, therefore, must be subjected to long prayers, speaking in tongues, deliverance sessions, etc. This development has rubbed the Nigerian society of authentic development while enthroning triviality, loss of self-confidence, banality, brain-drain, wonder-seeking attitudes, low self-esteem, etc.

The above-described state of affairs within the Nigerian social atmosphere is identified as one of the challenges of both human and societal development that requires ethical and principle-based confrontation. The disdain for one another, especially in matters of national identity seen between the predominantly Islamic North of Nigeria and the predominantly Christian South, remains a cardinal issue pointing to both structural and systemic frictions and failures. In a broader sense, these frictions play themselves out as well within and among groups of the same religion, thereby giving rise to sects, conservatives, liberalists, hard-liners and soft-liners, as well as fundamentalists propagating religious radicalism. All these groups make up the one religious and faith-based atmosphere of Nigeria and the impact has been endemic. Of special interest, predominantly due to its influence on the political, economic and social-cultural spheres of Nigeria as a geopolitical entity, is the impactful experience with the Islamic fundamentalist group, Boko Haram. A concise reference to the group sheds an insightful light on the topic of discussion.

Aware of the fact that religious violence has been part of Nigerian history, the emergence of the group Boko Haram shook the very foundation of Nigeria's socio-cultural spaces. Indeed, the ideology of this group has been proscribed and their activities have been named terrorism. This is because their strategies and approaches elicit the characteristics found in the definition of modern terrorism, as seen in Ludwig Siep. This definition points to the fact that modern non-state terrorism aims specifically at the largest possible number of defenceless people, and this is, according to Walzer, in the form of indiscriminate killing because of belonging to a regime, a people or an alliance, a religion or a race. The purpose is to exert psychological pressure on the state or the public. Therefore, the gigantic reinforcement by the media and the feeling of the omnipresent threat become an indispensable means of propaganda (Nwosu 2020a). Nwosu, in collaboration with this definition, articulated five factors as normative grounds for viewing the Boko Haram groups as terrorists:

> This definition reveals at least five (5) characteristics of terrorist acts through which they can be differentiated from any other (conventional) use of force, namely: 1) The group is **non-State** (without Sovereignty), therefore illegitimate; 2) Their **Target** is indiscriminate and therefore wrong; 3) The **Perpetrator(s)** always have fundamentalist agenda; 4) The **Aim** remains to disrupt and distract the opponent "indirectly"; and 5) The **Propaganda** is predominately the media. Within the aforementioned characteristics, therefore, lie the tools with which to look at the group, Boko Haram in the Nigerian experience. The open ethical question to answer remains: Is the group a threat to peace or has the group any justification for its violent actions? (Ibid.)

Important here would be to describe the group and their ideology. The entire world has come to know the group with the coinage *Boko Haram*, a concept which aligns with one of their core agenda. The designated full name of the group, which simultaneously sums up their ideology, is *Jama'atu Ahlus Sunnah Lid Da'awati Wal Jihad*—this is translated as *People Committed to the Prophet's Teachings for Propagation and Jihad*. In other words, the members make up a committed group of adherers to the teachings of the Islamic prophet, Mohammed, and who are bent on propagating the teachings at all costs, even to the point of declaring Jihad, which is the official war against the infidels. Their most popular name Boko Haram, therefore, is an umbrella name taken from their frontal declaration and total rejection of the so-called *Western education, Western influence* which they regard as an evil and enemy of Islamic principles. This informs their mainstream demand for an Islamic State and society where only the Koran, the Hadith and the Arabic language would be the means and mode of the social mores within their caliphate (Smith 2016).

The antecedents to the upsurge of this group, based on varied and sometimes conflicting opinions, are both politically and religiously influenced. Politically it is opined that the group is bent on reforming the North back to the model of Islamic legacies established through the armed jihad led by the then Fulani Islamic cleric, Usman Dan Fodio, which the group considers being neglected by the corrupt elites of Northern Nigeria. Thus,

> […] many in northern Nigeria have come to see democracy (or western influence) as a system that keeps them poor and enriches undeserving, corrupt leaders. In Maiduguri (the Nigerian home base of the group), the wealthy take up residence in heavily secured mansions while the poor fetch water from wells, and (at the same time) sign-posts at roundabouts are written in Arabic proclaiming 'Allah is the Provider'. (Ibid.)

Smith emphasised, "It was amidst this atmosphere that Mohammed Yusuf (the first leader of the group) began to lead his followers" (cf. Ibid., *Boko Haram: A Threat to Peace or A Legal Agitation? The Political Ethical Implications*, pp. 8–9). Religiously, the group rages war against all forms of infiltration by other religious beliefs, especially Christianity, in the core Muslim States. Thus, as has been seen in their activities, many victims of their abduction who profess other faiths and beliefs are often forced, under the threat of decapitation, to convert to Islam and recite the Koran (Ibid., pp. 9–10).

Since our research is in Christian Social Ethics, we regard the complex moral dilemmas surrounding the ideology and activities of this group as a cardinal challenge to the existence of Nigeria and to any effort to ensure sustainability in the Nigerian State. The specifics of their approaches, as articulated by Nwosu, show the concrete experiences with the group which elude every appeal for moral justification, namely: (a) *The group is non-state or non-governmental*: The insurgency of the group with the use of violence and other lethal means has no conventional legitimacy. In other words, since the body itself has no legitimate sovereign authority, its application of violent force remains illegal and treasonable; (b) *Their targeted group(s)*: Ninety-Eight per cent (98%) of the targeted victims of Boko Haram insurgency in Nigeria since their inception in 2009 have been the innocent populace through broad daylight massacre, suicide bombings and the kidnapping and taking hostages of men and women, especially young girls for ransom, rape and marriage—the most outrageous was the kidnapping of 276 helpless young school girls on the night of 14–15th April 2014, in a Secondary School in the town of Chibok, Borno State, Nigeria. To date, some of those girls are still missing. Therefore, there is no moral ground for justification since their targeted groups can never fit into the specifications of the "political code" and the "non-combatant" principle; (c) *The Perpetrator(s)*: The leadership of Boko Haram and the group's agenda have a fundamentalist undertone. As fundamentalists, they preach and hold on to their subjective agendas and are bent on imposing them on their listeners by all means. Thus, their general approach is often synonymous with violence and intimidation; (d) *Their Aim*: Boko Haram aims at an enforced influence on the Nigerian sovereign state towards a possible secession of some parts of its federal territories for the establishment of an Islamic Caliphate. Since this aim has not yet been realised due to counterpolitical and military violent forces, they resort to guerrilla and faceless attacks that claim hundreds and thousands of the lives of helpless innocent citizens—all in the continued distraction of their direct assailant, the Nigerian State; (e) *Their Propaganda*: The most effective propaganda of the Boko Haram, just as the present world witnesses among other terrorist groups, is the use of the media. They broadcast their agenda, challenge the sovereign state and create constant tension in the society through threats of impending attacks and counterattacks. The actual locations of their video and voice recordings remain always shrouded in mystery. The specific case of Boko Haram shows sporadic video shots of how they maim and kill their victims, especially through decapitation (Ibid., p. 12).

All the above discussed five characteristics and their evaluation show that the general existence of Boko Haram raises legitimacy and moral questions. Their actions correlate with the core characteristics of all terror-based groups of our modern times, and to describe the group as a violent religious sect, perhaps with a subsidiary political agenda, would be more suitable for the discussion. However, the dilemma remains, and the imposed challenges to the general polity have been persistent and devastating to the development process in Nigeria. Nwosu opined in the final analysis:

> To set the records right, it is important to remark that the terrorist activities of Boko Haram are chiefly directed at the so-called infidels, but not limited to that. The Nigerian experience records the killing of members of the same Islamic faith by the radical group. Thus, the core moral dilemma is: How does one strike the balance between their quest for the restoration of core Islamic culture and mores on the one hand, and on the other hand, their ruthless slaughter of their own members in the name of revolution? Furthermore, the activities of the group violate all possible three conditions for potential justification of the actions of such "illegal" militants, namely: a) if the use of violence were to be the last option; b) if the group has the voluntary support of the populace; and c) if the group has at least a sensible orientation towards the principle of *jus in Bello*—but on the contrary, there are ample other options, the masses do not support them, and their direct object of attack remains indiscriminate and obeys no conventional rules (Ach, J. S.; Bayertz, K.; Siep L. (eds.) Grundkurs Ethik 2: Anwendungen, 2011, p. 148). That is why the group evokes so many open political and moral dilemmas with their manner of approach. (Ibid., *Neo-Colonialism: The Last Stage of Imperialism*, p. 234)

The Nigerian populace still awaits the concrete defeat of this fundamentalist group. However, in the meantime, there have been strategised military confrontations between the group and the Nigerian army, which to an extent might have helped to displace the group from holding on to their base within the geopolitical area of Bornu State (North-East). Nevertheless, the group seems to maintain a faceless identity with ubiquitous locational manipulations, which has become the greatest bane in the fight against the group. Current pieces of evidence show their persistence in constant readjustments, regrouping and recruitment of members even from the bordered neighbouring countries, as well as their sporadic guerrilla warfare from outside the country. The confrontation goes on.

The case of regional and sectional insurgencies tells the same story of a sick geopolitical structure. The mainstream antagonists represent the primary regions of the country, the administrative fusion of which constitutes to date a colonial mistake, namely: (a) the Arewa Consultative Forum, ACF (2000) in the North, which represents the proponents of "one Nigeria" and maintenance of the status quo; (b) the Oodua People's Congress, OPC (1999) in the Southwest, which represents proponents of not strictly divided Nigeria, but an equitable regional political and cultural identities; and (c) the Movement for the Sovereign State of Biafra, MASSOB (1999) and additionally now, the Indigenous People of Biafra, IPOB (2012) in the Southeast, which represents the proponents for the frontal abandoning of the concept "Nigeria" and declaration of independence for the regions and sections based on their true identities and cultures. This persistent strife within the geopolitical spaces, added to the evident endogenous character of the individual groups defending their unique territories, ideologies and identities, speaks volumes about the structural challenges that Nigeria faces. Our research is not ignorant of these challenges. Thus, its contribution would be to open an independent strategy in response to the current Nigerian structure as it is. The focus would be on grassroots dialogue with *Mmadụ*, the human person, who, amidst these political, social, economic and cultural dysfunctions, remains the victim of inbuilt *structural sins*, **which impede the human ability to attain his or her full potentialities within the general development process**.

1.5 Nigeria and Socio-Economic Development: How Developed? How Sustainable?

A fundamental character of a society is the integral relationship involving the social, political, economic, ecological and cultural institutions, which unite to conceive the structural pattern upon which such a society survives. Referring to the word "integral" means evoking an idea of symbiosis among these core elements, which should mutually diffuse into one another in developing and sustaining the general life of a society, particularly the well-being of the human person, Mmadụ. To be regarded as a *societal constant* among all other elements is the economic aspect and constitution of any society's existence, which remains the concrete platform upon which all other elements constellate. The Nigerian socio-economic development has been waveringly dynamic. However, the quality and content of its dynamics may not have been sustainable enough to aid Nigeria in translating its rich human and natural resources into a people-oriented and integrative economy and development. Consequently, many other nations who shared the same economic and development status as Nigeria in the last two decades proved to have left Nigeria disappointingly behind within the general indices of development. We examine the challenges by looking at three theses, which better articulate the challenges associated with the Nigerian socio-economic structures.

1.5.1 Thesis 1: Nigeria's Economy Lacks Diffusion and Suffers a "Missing Middle"

Both the trans-sectional and trans-national debates of the pre-independent and early independent African nations sought to initiate policies that would be able to respond to the economic and development standards viewed at the same time as intimidating, which almost all the colonial African states had to face when compared to their western colonial powers. The majority of opinions in these debates, especially the group of proponents of Pan-Africanism led by the then President of Ghana, Kwame Nkrumah, thought that the most practical approach to bridging the gap would be to advance a massive campaign for purely industrialised Africa. Thus, some of the African states and their governments began varied attempts to implement this Pan-African call to redeem the African status in the world of economy and industry. However, the first attempts were massively hijacked by various imperialist policy chains, which further prepared the stage for external detractions. Nkrumah observed unequivocally,

> The African countries are faced with the need to turn subsistence economies into organisms that will generate viable and improved conditions of living for their populations. However, many African governments, instead of getting together in united action which would stimulate maximum capital accumulation and the construction of a solid overall African economy, are granting concessions for the working of mineral, agricultural and forestry resources whose purpose is the drawing off of output to sustain and enlarge the industries and economies of the imperialist countries. Not one of the investing syndicates has any intention of founding in any one of these countries an integrated industrial complex that would give impetus to genuine economic growth. Nor are the returns on the export of primary products from mining, agriculture and forestry likely to provide to any important extent the looked-for capital for investing in industrial foundations. (Gelb et al. 2015)

This quest to industrialise found meaning as well in the Nigerian system, boosted more by the discovery of crude oil reserves in commercial quantity. Thus, Nigeria's era of industrialisation recorded swift transitions from its pre-independence status to post-independence macroeconomic structures, probably intended to respond positively to employment demands, maintain the federal character of the Nigerian state, and of course, give the country competitive character among other economies of the world. However, the development of events to date records and shows pieces of evidence of a common characteristic of most African economies, which still battle colonial aftermaths and influences, namely the lack of *socio-economic diffusion* leading to the so-called *missing middle*. This is a result of the embrace of industrialisation and, with it, the type of productivity enclaves that followed, which were heavily borrowed and imitated concepts. In other words, the foundation of Nigeria's industrial institutions has remained extremely exogenous, thereby stifling the mutual diffusion of economic sectors to help create the needed "middle". Gelb, Meyer and Ramachandran, with reference to the economic development of sub-Saharan Africa, observed:

> Why have "productivity enclaves", islands of high productivity in a sea of smaller low-productivity firms, not diffused more rapidly to create the "middle" so frequently observed to be missing in Africa's economies? Why have productive firms not led a process of equalising growth, absorbing land and labour to employ a larger share of the workforce? […] (The awaited convergence, therefore, is seen as the movement) towards the upper end of the productivity spectrum (high productivity) […] seen as a process of diffusion, wherein the economic agents driving the high-productivity parts of the economy actively extend the reach and scope of their businesses until they encompass the bulk of the economy – and ultimately increase economic growth and output per capita. (Ibid., p. 120)

With the above observation, the exogenous status of the agents dealing with Nigeria's productivity enclaves and the resultant detraction of the domestic convergence among the mainstream economic sectors remains an initial characterisation of the Nigerian economy. As a consequence, the first two decades of post-independent Nigeria witnessed painfully either the insolvency or total extinction of some of those productivity firms, especially due to the exit of the exogenous know-how and the drastic shortage of endogenous manpower. However, despite the fact that the country's economic policies, with regard to industrialisation and mechanisation, have remained operational, the production capacity and output have remained rather diminutive. Some debilitating factors combine as the root causes, namely, unreliable power supply, insecurity and unbridled corruption in both private and public spaces. These were prevalent and predominant in the first two decades of post-independent Nigeria. The eventual constellation of the effects of these negative factors worked *detractingly* against the true convergence and diffusion needed for both international and domestic competitiveness in the economy and markets. This *below-performance* characteristic of the economy is yet to be conquered in the Nigerian structure and system, despite the attempts made through the Structural Adjustment Programme of the early 1980s and the recent so-called globalisation narrative, which claims an opening of the global market as the platform for equitable trade and fairness in the distribution of goods and services. The observation made in this context by Gelb, Meyer and Ramachandran with reference to African economies is explicable:

> Despite the reforms that African countries have undergone since the structural adjustment phase of the late 1980s, McMillan and Rodrik (2011) found that between 1990 and 2005, Africa's structural change ran counter to the expected pattern of structural convergence. In contrast to rapidly growing Asian countries, labour moved from high- to low-productivity activities. They also show that economies with a revealed comparative advantage in extractives are at a disadvantage. The larger the share of natural resources in exports, the smaller the scope of productivity-enhancing structural change. Even though "enclave" extractive sectors may have high labour productivity relative to the rest, they cannot absorb the surplus labour from agriculture; much of this has thus moved into low-productivity services. More recent data suggest that there may be a turnaround, with a positive growth contribution from structural change over 2005-2010 (McMillan 2013). Even so, over the long term, globalisation appears not to have fostered a desirable pattern of structural change in Africa. (Ibid., p. 123)

This failure to converge, as observed above, remains an ongoing battle even in the Nigerian system. Thus, despite the country's prime place among other African countries with a GDP of about $480.48 billion (2020), the wide economic gaps, both domestic and international, which are consequences of exploitative economic patterns, remain persistent and resistant to redeeming the country from its deficits and excessive borrowing syndrome. In other words, instead of the economy aiding the growth, increase, and stability of production firms and productivity itself, the current economic structure remains too weak to create employment opportunities and absorb the labour force. Specific factors are responsible for this with reference to the general status of the African economy, thus:

> Firm-level survey responses on the business environment, including access to finance and land, and the availability of power supply and skilled labour, have some explanatory power in explaining this difference, as do foreign ownership, export status of firms, startup size and size of the market. However, even after controlling for these variables, about 60 percent of the size gap between African firms and those in other countries remains unexplained. (Ibid., p. 125)

The above-named inhibiting factors remain a replica of the Nigerian economy, though there are varied unilateral efforts by individual States of the Federation to introduce forms of rescue-and-recovery economic programmes so as to respond temporarily to domestic problems. However, the federal structure of the Nigerian economic development stifles such individual efforts, making mutual convergence and diffusion a hard nut to crack. One concrete example with reference to most African economies and cancerous to Nigeria is the debilitating effect of the so-called indirect cost, which ravages the federal institutions that provide public services. Regarding transport and cargo dwell costs, our referral experts bemoaned the observable huge loss of revenue typical of most African business environments, Nigerian inclusive, thus:

> Transport costs not only reflect problems in the supply of physical infrastructure, but also more complex political economy issues related to business environment. In many African ports, cargo dwell times are about two weeks, compared to under a week in Asia, Europe and Latin America (Beuran et al. 2012). Long dwell times are not just the result of port inefficiency but may be linked to businesses' inventory management practices: cargo is sometimes purposely left at the port when the cost of clearing it immediately is high. There are storage-cost savings, especially when firms cannot immediately sell their imported goods. Terminal operators may also benefit from longer dwell times by receiving informal payments [...] African firms (economies) appear substantially less productive, relative to firms (and economies) in comparator countries when these 'indirect costs' are included. (Joseph 2015)

With regard to *Thesis One*, we develop the hypothesis that the status of Nigeria's developing and struggling economy needs a proven structural change, which would be able to integrate large-scale domestic entrepreneurship, invest in and emphasise competence in skilled labour, as well as reduce indirect revenue losses due to corrupt practices which cause the economy not to converge and diffuse. We keep this in focus as an area for the application of *ethical remediation*.

1.5.2 *Thesis 2: Nigeria's Economy Wrestles Imperial Exogenous Tracks*

One of Nigeria's varied aspirations at its independence was to provide economic leadership within Africa and beyond, given her rich human and natural resources. However, it became a prospective goal that twisted the cause of events, namely, that almost all the main economic infrastructures and public service institutions, which it inherited from the colonial administration, either eroded or went extinct. Typical examples were Railway Services, Health Institutions, Water Corporation Services, Road Construction and Maintenance Services, etc. *Thesis Two* argues that an indelible gap was initially created within the general economic development process due to the near total dependence of institutional and public service on *exogenous tracks*. Thus, for Joseph, "A revolution in institution-building and maintaining behaviours must accompany the upsurge in infrastructure investments" (Ibid., p. 271). Paths towards recovery have been an ongoing policy in Nigeria, yet the scar remains and weighs sporadic influences on the general economic process. By recovery, we mean the concrete resolve towards an endogenous, multifaceted, consistent and purposeful process, whereby a people's aspiration and goals constantly undergo revaluation within their indigenous social actors and social fabric, "[…] the evolving stock of historical, cultural and institutional knowledge, as well as scientific and technological infrastructure" (Fofack 2015).

The persistent mark of these exogenous tracks has been the poor and weak management of and harnessing of both local and global knowledge such that, despite the laudable efforts of local innovators to explore and promote indigenous socio-economic and technological prowess, the ideological preference to and craze for foreign goods and services continue to keep endogenous economic development underperforming. Fofack observed that many of the countries in sub-Saharan Africa have been victims of economic processes that are deeply extroverted. He opined:

> Sadly enough, the majority of countries in sub-Saharan Africa have fallen under […] models (of) development process (which) have been extroverted – not endogenously developed, with the development thinking emanating not indigenously from within Africa but paradoxically, from the global industry of development experts largely based in donor countries [….] The extroverted development model has perversely suppressed national aspirations, and in the process, reduced the local development actors (policymakers, private sector operators, intellectuals, and civil society organisations) to bystanders of a process that they should have been leading. In addition to ignoring and crowding out local expertise and indigenous knowledge, the global industry of development experts has identified a monolithic and unified development goal for Africa as a whole, notwithstanding the continental diversity and heterogeneity of societies and economies. That goal has been seen as mainly increasing the national income of countries, measured in terms of changes in gross domestic product (GDP). (Ibid., p. 295)

This inbuilt flaw has become the main criticism against the GDP-Model of economic measuring and monitoring, which extant economists and philosophers like Amartya Sen and Jean-Paul Fitoussi consider insufficient to explain the general progress of a society. The simple reason is that the mere increase in the records of GDP could only be arbitrary in integrating other essential factors that define the life of a society, namely, the relationship between income and consumption, the quality of happiness, freedom and well-being which must impact the life of both the human and ecological agents of the society as the determinant factors in measuring real progress of the society. In the Nigerian case, the GDP statistics at the international level still have a persistent battle to fight, namely, to ignite a diffusible "multiplier effect" so as to like other basic factors that help define real progress and well-being for her teeming population. This explains, in concrete terms, why the increase in the GDP of the country has yet to impact the unemployment rate within the country remedially. Thus, in the foregoing context, economists would prefer to allude to the negative correlation between economic growth and development. This negative correlation led to what Fofack again described as the two pitfalls of Africa's economic development, namely "[…] the acquiescence to the globally dominant ideology and development mimicry" (Ibid.). Consequently, any remedial steps must be strategised in a manner to ensure that "[…] the development aspirations are collectively shared and all the national treasures (human and financial resources) are strategically deployed to serve the overarching development objective—self-reliance" (Ibid., p. 272).

The externalities that characterise the African economies in general and the Nigerian economy, in particular, remain an indelible feature of the post-colonial assumption that "[…] African countries should aspire to become like their former colonial powers, which have since mutated into the status of donors in the development industry" (Ibid., p. 273). Thus, endogenous innovative ideas settle to imitate and copy and never to challenge and compete. This donor ideology still reflects and informs the modern idea of globalisation, which in principle preaches liberalisation yet, in practice, remains predominantly hegemonic and capitalistic. The donor forces, which have become even more sophisticated in our time, enslave to date the Nigerian economy, with China and its modern strategy of conquest (a modern form of imperialism) in Africa topping the list. The result has been conspicuous, namely, that an overwhelming majority of so-called Africa's big economies are only bystanders and conditional co-operators at the receiving end (Ibid., p. 274). As one of the big African economies, the Nigerian state drifted away for nearly four decades to a monolithic economic development almost totally based on oil and gas exploration and exportation. To this effect, the Nigerian state in the post-independence era played its role in piloting the economic process dominated by the structuralist's emphasis on a strong state's influence and participation in the economic thinking process, whereby "[…] there was a strong belief in the power of state-led industrialisation. In the short term, addressing rural poverty and achieving food security were top among the development priorities of young nations emerging from colonialism" (Ibid., p. 276). However, the failure to legitimise strong diversification policies with reference to economic factors of production in the Nigerian case keeps the economy hanging in the balance even within the neoliberal emphasis and reliance on market liberalisation ideology—more akin to the Washington Consensus. Fofack illustrates these two economic development paths, which are not unfamiliar to the Nigerian experience, with the facts below (Table 1.3).

The above-mentioned Nigerian economic shortcomings are not just arbitrary. They are rather as scientifically based as they are historical and geopolitical. They

Table 1.3 The evolving path of (economic) development theory

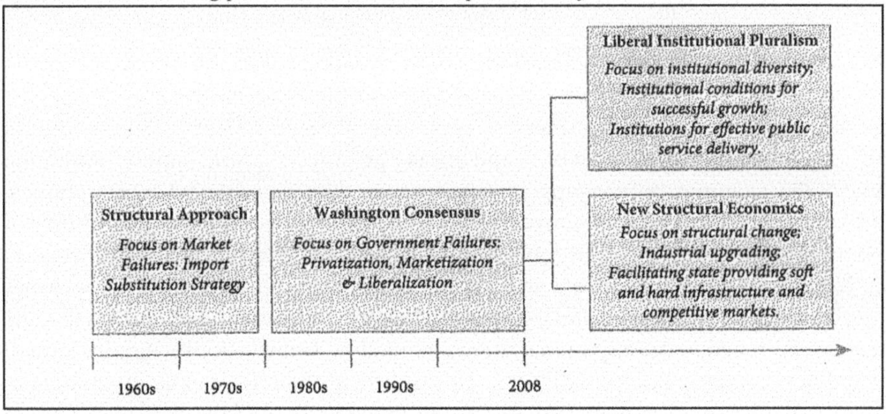

Source Ibid., p. 274

are proven facts characteristic of the discrepancies between the global North and South economies. A network of asymmetrical factors is responsible, namely, first, the initial and conspicuous deficit of human capital, basic physical infrastructure, as well as the relevant institutions of learning and training in most African countries at independence. To this end, capital became the "… binding constraint to closing technological and infrastructure gaps" (cf. Ibid., p. 275). Secondly, and rooted in the Nigerian situation, the inherited colonial legacy of resource extraction and rents was, from its very beginnings, never designed to benefit local and regional economic integration. Consequently, the persistent practice of *interior-to-cost* or from the *periphery of the colonies to the colonial empires* associated with resource extraction deepened the exogenous character of the general economy and widened the dependency tracks toward imperial capitalism (Ibid., p. 275). However, a concrete policy to bridge this gap in Nigeria, with regard to oil and gas exploration, gave rise to the mega refineries located within the country. Nevertheless, the general system of operation still reserves predominantly the rentier character of most multinational business relationships. The Nigerian structure is a first-hand experience and the liberation struggle persists. Thirdly, the Nigerian economic system is yet to decolonise its embrace of the trending neoliberal economic ideology, which emphasises the market and the liberalisation of the capital account. In Nigeria, this has long instituted an unguarded spree of external borrowing and the widening of exploitative donor-recipient relationships. This is the direct residual of implementing foreign-based policies and ideologies with a wide range of peculiarities that tend to find them incongruous. Fofack, agreeing with Williamson 1990, Adedeji 1999 and Rodrik 2006, analysed the situation thus:

> In Africa, the package put forward to advance the neoliberal agenda and also respond to macroeconomic imbalances and adverse TOT shocks which afflicted the continent included a range of policies: public sector downsizing, public expenditure cuts and switching, privatization of SOEs, deregulation, trade and financial liberalisation, dismantling of state marketing

> boards, eliminating agricultural subsidies, tightening monetary policy, as well as unification and increased competitiveness of exchange rates. (Ibid., p. 277)

There is no doubt that these policies may have recorded applicable economic praxis and successes. However, what stands peculiar to the Nigerian case is the persistent struggle to interpret and translate economic plans to reflect its socio-cultural and socio-economic realities within and among proven economic options. Although some of the economic policies appear superimposed by the privileged big economies, for example, Nixon's administrative policy of cancelling the direct convertibility of US$ to gold in 1971, which eventually opened the era of freely floating currencies and balance of payments crises, and which, as a consequence, pushed many struggling economies into default and international debt crisis, our research insists that the greatest challenge to the Nigerian system and method is the lack of the political will for indigenous economic possibilities. To this effect, the Nigerian neoliberal experience is yet to help the country achieve the prospect behind the first goal both in the Millennium Development Goals agenda through the year 2015 and the Sustainable Development Goals agenda through the year 2030 respectively, namely the fight against extreme poverty and hunger—a situation which extant scholars have described as the "[...] economic tragedy of the twentieth century" (Ibid., p. 280). Given this lingering failure, therefore, our research sees the need for a defined "New Structural Economics", which applies neoclassical economics to understanding the real drivers of economic structures within its development evolution. In other words, "[...] the need to take into account structural features in the analysis of economic infrastructure for growth and integration of local and regional markets" (Ibid., *Inclusive Growth and Developmental Governance: The Next African Frontiers*, p. 213). This approach would work to synthesise both state failure and market failure. In the final analysis, the persistent exogenous tracks of the Nigerian economy elicit the negative economic trend known as Dutch disease, whereby an almost exclusive dependence on natural resource export (in the Nigerian case, Crude Oil) and, therefore, nearly absolute dependence on externalities inflicts decline in domestic and local economic stability like manufacturing and agriculture. The turning point ensued when opening access to external influences became a mandatory precondition for most struggling economies facing recurrent balances of payments and debts in the hands of creditor nations. The Nigerian economy keeps struggling under this economic quagmire. However, there are both scholarly and quasi-politically induced efforts to reignite concerns and debates on remedial paths towards recovery.

1.5.3 *Thesis 3: Nigeria Operates* Non-inclusive *Economic and Development Governance*

That many African primary-resource-based economies and development indices are still described as "non-inclusive" remains a fact sustained by concrete pieces of evidence taken from both global and domestic experiences. The fact that the multidiverse continent with rich natural resources and inhabiting the second largest percentage (about 17.51%) of the world's population after Asia remains, at the same time, the home for the greater number of people living in or below poverty levels, is viewed as a global discordant. With reference to Nigeria, the imbalances in the economic processes speak very loudly in confirming this evidence. Thus, the developing Nigerian economy is yet to integrate economic policies with concrete and generative institutional transformation fully, diffuse economic benefits to reduce poverty levels, de-monopolise access to economic factors hijacked by the elites, institute justice and transparency in the formal sector and, above all, embark on a massive campaign and policy tracks as to initiate a rediscovery of its agricultural potentialities so as to drive the necessary diversification practices needed within the economy. The third thesis sees in the above-mentioned economic discrepancies elements that are yet to be addressed and harnessed properly to make economic development more inclusive than non-inclusive. Scholarly opinion observes two crucial frontiers to be breached, namely, "[…] patterns of growth that do not disproportionately benefit elites while the material condition of the greater majority stagnates; and acquiring modes of governance that are developmental in the sense of steadily optimising performance and output" (Ibid., p. 214).

The above two patterns, thesis three would argue, have a proportional relationship with economic and development governance. The Nigerian case records varied palliative economic and political experiences, beginning with the so-called wasted years of oil boom misappropriations between 1960 and 1973 in her history. Added to this were the incessant interruptions of political governance through social unrest, military coup d'état, as well as religious bigotry. To date, this syndrome of epileptic economic and development governance persists. Joseph, with reference to the continent of Africa, and her gradual degeneration and dependency on development aid, opined: "The great central question is what can anyone do to try to improve the quality of governance (in Africa). How can development aid have any hope of actually doing good for poor people in poor nations if we don't understand what kinds of assistance policies (that) can encourage positive political development?" (Ibid.). This central concern about governance remains the bane of Nigeria's economic and political experiences. The elite minority influence and hijack of the economic process leave the general developmental trajectory deeply oppressive and corrupt. Their presence within the corridor of power and governance has long taken the shape of *handing-over-the-same-baton* for further perpetration and consolidation of the status quo. To this effect, the gaps between the rich and poor widen geometrically, alongside plaguing social ills that keep daily lives almost hanging in the balance. Relationally, this eroding form of governance system has affected institutional provision and administration in the Nigerian system and, with reference to Africa, Joseph opined that no African country, arguing from concrete and recorded experiences on the ground, can "[…] achieve sustained and transformative growth, involving the shift from largely subsistence agricultural economies to globally competitive ones, and the generation of well-paying jobs for their urbanising populations, without significantly improving ways of managing their collective affairs through appropriate institutions" (Ibid., p. 216).

Therefore, our third thesis would argue that growth-enhancing governance would be directly proportional to more inclusive economic and development policies, whereby the growth-enhancing factors must, as a matter of principle, develop and apply the right and functional institutional governance capable of positive and transformative outputs. When the reverse becomes the case, society would be set on a pedestal where both the factors of production and the general provision of goods and services become disproportionately the benefit of the *selected few*. Nigeria's geopolitical structure, in relation to her experience of the colonial administrative imbalances, has for long made practicable growth-enhancing governance elusive. The rich diversity that Nigeria has, added to its domestic economic possibilities suffered, along the line, huge policy-blindness. The result of this has become the negligence to explore and develop the heterogeneous line of her economic powers unique and multidimensional to the various ethnic and cultural groups that make up her society. The consequence is a near-mono-economic culture reduced only to the language of oil and gas exploration. Nevertheless, even with the constellation of the economy around the *petro-dollar*, the economic and development governance in practice keeps eliciting a non-inclusive character of governance that marginalises and detracts. However, recent upturns in global economic events tend to impact the Nigerian system and experiences and have awakened a general feeling of distaste against governance apathy within the whole system and structure unconditionally. To this end, two prospect-based remedial steps would be considered as being open and applicable in the Nigerian case, namely (a) an indigenised and applied form of *Developmental Patrimonialism*, which would systematically dislodge the elite minority's monopoly of economic factors, goods and services; secondly, (b) a constitutionally based governance revolution that would reflect a "seven-point" agenda for inclusive economic and development governance.

With reference to the first proposal, the current practice of revenue allocation, whereby the individual federating states in Nigeria depend heavily on the monthly allocations coming from the centre—a revenue distribution form that has its source from the oil and gas exploration and exportation—may qualify as a form of developmental patrimonialism. However, it is heavily intractable and misappropriated by rent-seeking public officials in the Nigerian system. This is due to the deficiency of both central and regional monitoring and control systems that are supposed to ensure the purposeful and corresponding delivery of allocated grants. Thus, if developmental patrimonialism means that generated rents are systematically and administratively harnessed and managed through techno-bureaucratic means for national interest and not just the enrichment of the few (the Rwandan experience taken as an example) (Ibid., p. 220), then the present Nigerian system needs an overhaul. It is all a matter of sound policy application which can drive the democratic potential for inclusivity. Joseph, with reference to Bratton and Logan, described the general situation in Africa thus,

> [...] the full potential of democracy—including the promise of accountable governance—has yet to be fulfilled. Economic growth is still elusive, corruption remains widespread, and aid dependency continues to frustrate recipients and donors alike. As global leaders contemplate the massive increases in international assistance to the continent (of Africa), questions have,

> therefore, been asked about indigenous capacity to absorb an influx of new funds without exacerbating old problems. A smart aid approach would seem to require that African political leaders are held accountable—not only to donor agencies but, more importantly, to their own people—for sound policy choices and the effective use of resources. (Ibid.)

The above observation is true about the persistent challenge that Nigeria faces, which calls to question the country's inability to experience relatively stable governance since after her independence. However, we must emphasise that amorphous groups of economic innovators and developers are beginning to emerge. The likes who would be able to lead the expected revolution for recovery. Our research considers it pertinent to partner with such prospective opportunities. With reference to the second proposal, a 7-Point Agenda would be requisite for Nigeria's developing economy to serve the inclusive national interest. The agenda must be predominantly endogenous since it must depend on the Nigerian citizens themselves to grasp "[…] their political rights as citizens, notably to regularly claim accountability from leaders" (Ibid., p. 220). Inspired by the scientific study by Bratton and Logan, the agenda could be thematised thus (Fig. 1.3).

The above-presented schema represents what our *third thesis* tries to illustrate as being deficient in Nigerian economic and development governance. Thus, taken from the illustration, the four cardinal focuses remain, first, the clamour to utilise and invest in indigenous knowledge and expertise, which at the same time remains open to engage the open world of opportunities—ENDOGENEITY; secondly, the resolve and the political will to revitalise actions towards practice-oriented democratic governance, which in turn supports and encourages representation and participation—DEMOCRACY; thirdly, the development model would have socio-political and socio-economic drivers so as to be realistic, integrative and people-oriented—DEVELOPMENT; and finally, an endogenous, democratic and people-oriented development agenda would contribute to the general process that would work to

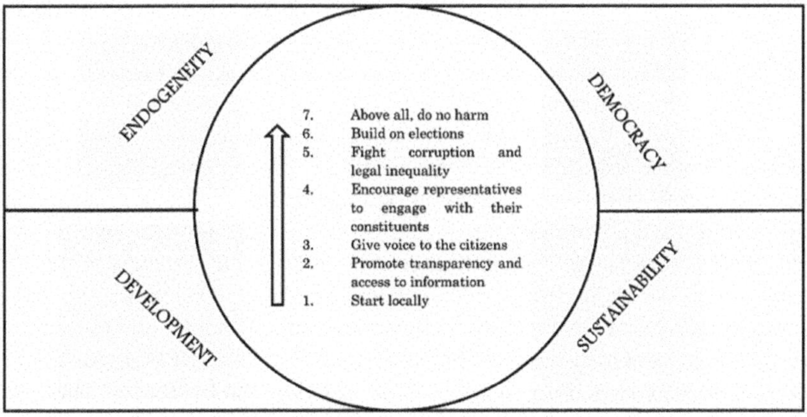

Fig. 1.3 A seven-point agenda to monitor inclusive and development governance. *Illustrations mine*

create well-being and make sustainable development the real basis for freedom—SUSTAINABILITY. Therefore, to purify the Nigerian economic and development governance of the lingering non-inclusive character, the governance system needs to study and apply an integrated model of the schema given above and, perhaps, in an indigenised form representative of Bratton and Logan's (2009) observation and study of the dominant African situation, thus:

> *Start locally*: [...] Transfer authority over local affairs not only to locally elected officials but to citizens themselves; *Promote Transparency and Access to Information*: [...] Support processes of public budgeting, which can engage the public while reducing the diversion of public funds; *Give Voice to Citizens*: [...] insist on public popular input into public policy processes; *Encourage Representatives to Engage with their Constituents*: [...] Consider providing incentives for greater engagement...publicise the constituency service of every elected representative; *Fight Corruption and Legal Inequality*: [...] Encourage citizen engagement in the fight against corruption; *Build on Elections*: [...] Use elections as vehicles for publicising the broader rights and responsibilities of citizenship; *Above all, Do No Harm*: [...] Purge aid programmes of requirements that cause governments to concentrate more on accountability to foreign agencies than to their own citizens. (Korff 2009)

The above-evaluated core areas within the Nigerian system must be revisited and improved upon. Every prospective effort must brace itself up to be confronted by the premonitions expressed in the postulated three theses to the character and mode of economic and development governance given our central specimen of research, namely, Nigeria. The assumption is far from an outright resignation to fate due to policy failures within the Nigerian state. It is rather an acknowledgement of a system that is still developing and still needs the political prerequisites to be more inclusive and people-oriented in character.

Evaluation

Economy and politics as two essential drivers of development, in other words, sustainable development. The evaluation done on the Nigerian economy in this chapter keeps in focus the concrete situation of this development factor, which directly affects the governance processes and the development conditions of the central agent of development, *Mmadụ*, the human person. In the case of Nigeria, the question would be asked: Is there any dominant detracting factor viewed as the bane of sustainable economic and development governance? Though interrelated with varied combined factors, the answer would be found in that single socio-economic and socio-political illness, *Corruption*. Thus, our evaluation delineates this syndrome, corruption, as endemic within the known Nigerian social, economic and political institutions. To this end, the degree of corrupt practices as seen in the Nigerian system would continue to plague every concerted effort to channel economic and development governance to its purposeful end, namely, that the human agent—Mmadụ—could, within the combined socio-economic, socio-political, socio-cultural as well as socio-ecological factors, realise his or her full potentials for well-being and happiness. Therefore, a concise analysis of the theme "corruption" would be insightful and informative for our research focus.

1.6 The Special Case of Endemic Corruption in Nigeria

While corruption is not exclusive to Nigeria or Africa in general, its degree and mode of practice in Nigeria have been a matter of both domestic and international concerns. As the largest oil producer in Africa and the eighth world's largest exporter of crude oil, Nigeria's wealth of human and natural resources, combined with over-average congenial weather conditions, has the capacity to put her at heights within sustainable development dynamics. However, the country and the general system battle endemic socio-political and socio-economic cancer, namely corruption. The special case with Africa, in other words, with the various countries of the global South, which are natural havens to varied quantities of natural resources, added to the manner and form of administering these resources tied to the colonial influences and experiences within the continent, make the theme corruption an endemic one. Our research and opinion evaluate the case of Nigeria not just as a mere social problem but, as Korff opined, an ethical challenge for both human society and the church (Ibid.).

1.6.1 Corruption: An Overview of Schools of Thought and Definitions

Views, opinions and definitions on the theme of corruption are predominantly as multiple as there are human experiences. However, there are basic characteristics of the concept that remain central in all varied forms of it, namely: it is predominantly a social phenomenon; it involves both individual or groups of human persons; it is found within systems, institutions and relationships; it involves action and opposite action or reaction; it is always partially or totally detracting of the targeted object or victim, and it predominantly violates moral and ethical principles and conventions. Thus, defined from the point of view of its etymological Latin root, *corrumpere,* from the verb *rumpere*, corruption as a concept becomes associated with such negative (German) terms like "*verderben*", meaning degeneration, moral decay; "*vernichten*", meaning, to destroy, to wreck an ethical structure; or "*bestechen*", meaning to bribe something, to buy or to be bought over illegally (Ibid., p. 311).

Institutions of thought have varied opinions and approaches to the complex theme of corruption. One of the earliest opinions on the subject is attributed to Aristotle and, subsequently, to the Aristotelians, who associated corruption with the word "*Entartung*", meaning degeneration. Thus, for the Aristotelians, corruption leads to social degeneration and moral deterioration, which eventually detracts from the commonwealth of the society, leading to a decline of the entire State. For the *Moralists*, corrupt actions are generally rooted in individual persons' inclination to deviate from the norms, which has consequences, not just for the individual persons but also for the community and the common good. The *Culturalists* approach the term corruption with reference to established societal conventions, which may not be intrinsically wrong but may suffer abuse and misapplication. Directly oppositional to the positions of both moralists and culturalists is the opinion of the *Functionalists*. For them, corruption as a social phenomenon may not be reduced outright to a mere moral judgement of the moralist or the trivial argument of the culturalists. The existence of corruption, they argue, is temporal and can even be a channel leading to a more developed society.[8] In other words, its existence elicits positive and negative consequences that impact a given society in both directions. This would remain intensely debatable and controversial. Furthermore, the Post-Functionalists attempted to reconstruct the functionalists' position by rejecting an outright ascription of any positive effects of corruption. Thus, they argue that the persistence of corrupt practices even within the so-called developed and developing countries and economies—perhaps in more controlled and monitored forms—makes it incongruous to argue for a direct positive character of corruption. To this end, it would be trivial to argue that the good emerged from the same evil that detracts from the good.

Giving attention to corruption as a terminology, we examine a handful of extant definitions that discuss the same subject matter, though with varied but related areas of interest and emphasis. Considered in recent times, an integrative definition is that of Transparency International, which defines corruption as "[...] the misuse of resources or power for private gains" (Kurer 2015). Although some scholars have considered the use of the word "misuse" with regard to the definition as being merely legalistic and could appear parlous and "[...] inoperationable and thus beyond measurement [....]" (cf. Ibid., p. 34), the definition captures the most common and predominant form of corruption as it is known and experienced across societies—a practice of substituting rules for the sole interest of a particular individual or groups of individuals (Ibid.). Oskar Kurer discussed the term "public interest" in the definition of corruption and opined:

> In recent years the public-interest definition has been resurrected, albeit in the slightly different guise of corruption in a democracy. Corruption, it is argued, 'is best understood in terms of transactions that subvert the impersonal processes of democracy' [...]; or, in a somewhat different version, 'one of the most sinister forms of political corruption in a democracy is when the "democratic transcript" is betrayed: that is, when members of the political class act in such a way as to prevent or circumvent the exercise of accountability [....] The public interest is identified with an ideal form of democracy where corruption damages this 'democratic transcript'.[9]

The above reference to *democratic transcript* brings to the fore a central character associated with corrupt practices in general, namely, the appropriation for selfish interests, abuse or violation of the principle guiding the use and administration of public goods and services. The act itself could be collaborated or solely individualised.

---

[8] United Kingdom, *Why Corruption Matters: Understanding Causes, Effects and How to Address Them*. Evidence Paper on Corruption, Department for International Development, January 2015. Available via https://assets.publishing.service.gov.uk. Accessed March 5, 2021.

[9] Switzerland, *Korruption Vermeiden: Hinweise für im Ausland tätige Schweizer Unternehmen*. Staatssekretariat für Wirtschaft. 2017, Bern: SECO Resort Pub., pp. 7–8.

With a network of experiences in the study and monitoring of economic crime, money laundering and corruption, Duyne views corruption as "[...] improbity or decay in the decision-making process in which a decision-maker consents to deviate or demand a deviation from the criterion which should rule his or her decision-making, (all) in exchange for a reward or for the promise or expectation of a reward, while these motives influencing his or her decision-making cannot be part of the justification of the decision" (Nwosu 2020b). This attitude of deviance strikes at the very nature of practices, relative as they may be found among societies and cultures, but which are predominantly considered offensive to some set of rules or conventions. Nwosu explained further:

> The act of deviating amounts to abuse in both public and private spheres [...] Such deviant attitude usually has an initiator, in our context, an official or public office holder entrusted with executive, legislative, judicial or non-deliberative powers, for example, in dictatorial and autocratic regimes and states. The issue of failed and abusive decision-making as the identified image of corruption gives an integrative understanding of corruption as not just deviations in the official State-based political, economic and bureaucratic transactions but rather reveals the expected role that social morality should play in matters of morals and responsibility. (Johnston 2005)

Acknowledging the difficulties posed by the terminological terms used in varied definitions of corruption, Michael Johnston, while defining corruption as "the abuse of public roles and resources for private benefit" (Ibid.), emphasised that such terms as abuse, public, private, or benefit, could be deeply ambiguous since they are known matters of contention across societies and cultures. Thus, he preferred instead an understanding of corruption with a focus on *systemic corruption problems*, which usually have damaging effects on a given societal system. The general characteristics associated with such systemic problems were articulated by Johnston thus:

> ... uses of and connections between wealth and power that significantly weaken open, competitive participation and/or economic and political institutions, or delay or prevent their development. Systemic corruption problems disrupt the developmental idea...in a variety of ways symptomatic of underlying participation and institutional problems. Looked at this way, corruption is (therefore) both a symptom of development difficulties—and thus a useful diagnostic tool for understanding what those problems are—and a reinforcing cause. (Warren et al. 2015)

The reference to a *systemic-based inhibition of development ideas* is a key expression, which, as would be proved later, better describes the concrete situations that are found in Africa generally and Nigeria in particular.

Warren, on his part, attempted a definition of corruption from the perspective of analysing its existence within a democracy. The contention is that corruption can invariably equal a *failed* democracy. To this end, corruption becomes the practice and multiplication of exclusions instead of inclusion which legitimises democracies. In other words, "[...] corrupt actions, practices and institutions undermine the means through which inclusions are enabled" (cf. Ibid., p. 49). Furthermore, because modern democracies involve differentiated domains of institutions, corruption, in its political affiliations, tends to be duplicitous. Table 1.4, according to Warren's concept, gives a concise picture of corruption domains and consequences.

Table 1.4 Differentiating domains and kinds of corruption

Corruption in democracies

Table 3.1 Differentiating domains and kinds of corruption

Domains of inclusion	*Executive*	*Judicial*	*Legislative*	*Public sphere*	*Civil society*	*Markets*
Norms of inclusion by domain	Public trust	Equal standing and protection; impartiality	Responsive representation	Mutual persuasion	Generalised trust and reciprocity	Fair and voluntary exchange under conditions of full information
Kinds of corruption	Abuse of public office for private gain	Rights, truth and fairness-seeking processes undermined by power or money	Responsiveness to money or power in ways that could not be publicly justified	Duplicity, deception, dissimulation	Particularised trust leveraged for group gains	Rent-seeking enabled by market and information constraints
Damage to democratic inclusions	Loss of collective agency	Eroded rights	Failed representation	Failed public deliberation	Loss of social capital	Unfair distribution; distorted or dampened investment

Source Ibid., pp. 48–49

With reference to the table above, Johnston opinionated:

> Democratic inclusions are effected through complex institutional and social ecologies that include constitutional states (differentiated into legislative, administrative and judicial functions); robust public spheres of opinion formation from which collective decisions and actions draw their legitimacy; civil societies that cultivate citizen capacities and form alternative venues of collective action; and market economies, which are correlated with but ambivalently related to democratic institutions. Each domain potentially contributes to democracy, and each has modes of corruption associated with its democratic potentials. (cf. Ibid., *The Socio-Economic and Political Implications of Unbridled Corruption in the Nigerian System: An Ethical Approach*, p. 5).

Within international politics and relations, corruption is approached multilaterally, focusing on the activities of donor organisations, international NGOs, multinational companies and corporations, foreign banks, as well as tax (secret) havens, etc. Within these groups, there are networks of illicit and illegal activities ranging from bribery, forgery, money laundering, and manipulation of official data, coupled with thuggery usually employed as intimidation and cover-up (Ibid., The Socio-Economic and Political Implications of Unbridled Corruption in the Nigerian System: An Ethical Approach, p. 4). The aforementioned illicit and illegal activities happen in the public (political) and private spheres, representing the two main forms of corruption. Nevertheless, corruption in all its manner and forms bears three main characteristics, either wholly or partly, namely, (a) it is public-office-related, (b) it is mainly for illicit and selfish purposes, and (c) it is always a moral violation.

1.6.2 The Two Spheres of Corruption: Public (Political) and Private

The above general thought patterns describe eclectically the central theme of corruption studied under two spheres, namely the public or political sphere and the private sphere. In other words, the contemporary approach to corruption recognises two forms of it, namely, public (political) corruption and private corruption. Nwosu attempted a distinction as follows:

> Globally, corruption is viewed as having both public and private nature. The common denominator would be that corruption is considered in all ramifications as an act that is offensive, either to the direct interested parties or to an indirect victim or a group of victims. It is to be noted that with increasing and rapidly changing (situations of) economic liberalisation and the politics of privatisation, the two spheres of corrupt practices narrow down to one another, making sharp distinctions more difficult. Corruption in public spheres is more pronounced and prevalently discussed and debated. This is because its main focus remains politics and public socio-economic goods and services linked to public and political offices. Thus, many opinions view both attributes "public and political" as being synonymous in relation to corruption. On the other hand, corruption in private spheres exists within other parts of the socio-economic (spheres) that are not directly political; examples are corporate organisations, business ventures, companies and enterprises, etc. (Smith 2015)

In relation to our research focus, we would concentrate on political (public) corruption, which would better aid our work in examining the Nigerian socio-political and socio-economic structural system, and how the deep presence of corruption has for long remained the bane of both human and social sustenance in the development dynamics.

1.6.3 Basic Allied Forms of Public/Private Corruption

As applicable to many post-colonial African countries, Nigeria for one, the most common public (and perhaps implicitly private) forms of corrupt practices identified through research and empirical studies are initiating commission for illicit services, unwarranted payment for public services, (illicit) gratuities, string-pulling, illegal levies and tolls, side-lining, misappropriation, as well as embezzlement of public funds (Ibid., p. 61). Smith analysed the different forms thus:

> [...] commission for illicit services refers to a user's payment to officials who then grant access to unwarranted advantages to (the) user, (especially on an area involving) a process supposedly (to be) based on competitive bids [...]; Unwarranted payment for public services involves an official forcing a user to pay for a service that is ostensibly provided for free or inflating the cost of a routine service [...]; an (illicit) gratuity is also a kind of payment for services but usually after the fact and commonly couched in the idiom of 'thank you'; [...] String-pulling refers to using social and political influence to promote favouritism, offering preferential access to employment, education and a range of other opportunities, particularly those allocated by the state; [...] (Illegal) levies and tolls are relatively stark forms of tribute that persons in power can extract from ordinary citizens (a form of extortion); Side-lining

> refers to the use of public or company resources for private purposes, (most common being) using official vehicles for personal travel, running a private clinic in a public health facility or using university resources to conduct a private consulting job; misappropriation (and/or embezzlement) is the use of public materials [...] not simply [...] for private purposes but (such materials are) expropriated entirely, usually in a manner more concealed than side-lined [...]. (Ibid., *Handbuch der Wirtschaftsethik*, p. 314)

In the Nigerian context, Smith's analysis proves to be deeply expository. Bribery intersperses the whole of the Nigerian system in the form of "*do ut des*", meaning, I give so that you give as well. In this way, the voluntary exchange of benefits, services or rewards is initiated between two actors or groups of actors, however, illicitly and undeservingly (Ibid., *The Socio-Economic and Political Implications of Unbridled Corruption in the Nigerian System: An Ethical Approach*, p. 16). The origin of the expression "419" is seen in the Nigerian penal code section that prohibits fraud. Nigeria, with time, has become endemically notorious, both domestically and internationally, in matters of digitalised, fraudulent, unfair and illegal forms of deception. The concrete forms are money laundering, internet scams, etc. The case of *racketeering*, in other words, extortion, is a daily experience practised within public positions of authority related to economic, political and social institutions. While the relationship between the actors in bribery is voluntary, that of extortion is parasitic, compulsive and detracting. In other words, only the perpetrating actor gains, and in the Nigerian system, the minority elites and politicians. A massive form of misappropriation and embezzlement of public funds happens in Nigeria to an alarming degree. These public funds, goods and services would end up in private pockets or being stashed away in foreign bank accounts. A domestic strategy within the Nigerian experience is the so-called *budget-padding*, or, simply put, the inflation of budget proposals that would never see the light of day. However, they would all be recorded as funded while, at the same time, the financial benefits end up in private pockets. This has become the commonest system of self-enrichment for politicians. Nigeria is trapped as well in the web of a *quid pro quo* practice. This is associated with the expression "*give and take*". The widely practised form of it is clientelism, where a patron who is politically, economically, socially, religiously or culturally strong bestows an undue favour to the "client" with whom he or she has affiliations. The regular form of it is nepotism, and in the Nigerian political system, an added form known as *godfatherism* emerged.

In all these varied forms of corruption, one finds them endemic in the Nigerian system such that the whole societal structure has remained intertwined with such negative practices that have almost become conventional and normal. To this end, we stand on the premise that the endemic nature of corruption in the Nigerian system qualifies as an infestation and a bane to the integrative development of the nation itself. A relevant elaboration, therefore, is a look at a "Seven-Thesis" experience-based proposition on corruption and how it exists and affects society.

1.6.4 Corruption in Nigeria: A Positivistic Seven-Thesis Approach

One of the contentions of our research is that corruption remains one of the banes of growth, development and sustainability. To approach the corruption discourse now in a positivistic way is to present facts based on experiences. Of course, the varied definitions of corruption are independent and collective opinions formulated from observations and experiences. As mentioned before, the appeal to a *seven-thesis* format here is to identify the central characteristics of corruption-infested societies and polities and their (almost) universal consequences, which apply to Nigeria. The aim is to "[…] prove the reality of the case at hand, access the effects and examine the difficulties in the fight against it" (Toyo 2006). Our theses are informed by the Centre for Constitutionalism and Demilitarisation (CENCOD), Lagos, Nigeria, established on 5 December 1998, and was focused on the need to initiate a recovery process in the entire civil society after protracted military interferences and incessant suspension of the constitution. Professor Eskor Toyo, a retired Professor of Economics at the University of Calabar, Nigeria, bared his mind in a "Thirty-Five Theses" (cf. Ibid., *The Socio-Economic and Political Implications of Unbridled Corruption in the Nigerian System: An Ethical Approach*, p. 17) article on the known experiences with corruption. We examine the theses and attempt forms of application with some originality.

Thesis One: *In all known socio-political and socio-economic histories, it is the population of the working people, the common men and women on the streets who are the victims of corruption*

Because public or political corruption operated predominantly at the elite levels, the consequences bear more on the common people in the society. The infested Nigerian society works only to benefit the perpetrators since, looking at the structure on the ground, it does not matter for those who benefit from the system if the public school system is functional enough or not. After all, they have their own kids in the best and expensive private schools at home and abroad; they do not care if the healthcare system and institutions are functional since they have access to medical trips to developed countries still with the same tax-payers money left under their control; they care less about the security of lives and properties of the masses, after all, they all have their private and personal security networks heavily maintained from common coffer. The population of the common people—the working class and the economically disadvantaged—remains always at the receiving end.

Thesis Two: *Corruption attacks and appropriates the social surplus unjustly*

In economics and policymaking, social surplus refers to the values, utilities and welfare that a given polity or a system stands to gain through investments in and consumption of goods and services produced or acquired. However, in a corrupt-infested system, Nigeria's investment policies suffer major attacks from elite managers leading to detractions for selfish gains. Consequently, the common socio-economic welfare, which the society should offer with the distribution of goods and services, and in which the citizens should have a fair share, becomes "*corban*"—in the context under discussion—it means devoted to and appropriated by the politically and economically powerful.

Thesis Three: *Corruption thrives in predatory societies*

A predatory society experiences violations of respect for the rule of law and disregard for equality before the law. Thus, the discretionary powers and authority of the rulers and the so-called powerful elites within such societies become monopolised by the few and can often degenerate into forms of dictatorship. Such a system harbours networks of safe havens where the practice of corruption happens under a myriad of classified information, in the main, information designated to defraud, steal or misappropriate, etc. However, access to such information is strictly reserved only for the initiators and perpetrators. The political killings of 19 October 1986, Dele Giwa; November 1995, The Ogoni Nine; 9 June 1996, Kudirat Abiola, etc., all in Nigeria were abuses that convict Nigeria as a corrupt and predatory society (Rawls 2001).

Thesis Four: *Economic and Social Nepotism have "multiplier effects" for corruption*

The use of the macroeconomic word "multiplier effect" would be reduced to the word "inducement" in the social and political spheres. To gauge multiplier effect in simple macroeconomic principle, the formula ME = CI/CS, meaning, Multiplier Effect = Change in Income/Change in Spending, indicates that the change experiences in both income and spending sectors of the economy would almost be directly proportional to the type and quality of change expected in the general polity. Applied to our fifth thesis, a society's economic and social "income" could be represented as the combination of covert and overt societal effects directly proportional to what the same society chooses to embrace and practice or, perhaps, compelled to practice—for example, nepotism. In other words, a system's peculiar attitudinal and behavioural approaches in policy and decision-making have obvious multiplier effects on the general system and structure of that system—our research uses the word attitudinal spending. Nepotism, blamed often on the colonially bequeathed geopolitical imbalances in the Nigerian state, remains one of the most effective corruption-inducing factors, which predominantly detracts from the practice of recognising and empowering merits, competence and innovation.

Thesis Five: *Corruption has eluded all military interventions*

Concrete historical and political experiences have proved that even in societies, the history of which has become deeply intertwined with military interventions and coup d'états and where almost all, except for a few, claim to have struck for the sole purpose of rescuing a particular society from corrupt-infested regimes and systems, corruption has never ceased to exist in such societies. Instead, many of the intervening forces became even more corrupt than the situation they claimed to rescue. The Nigerian experience of this has two phases. The first phase spanned from 1966 through 1979, and the second phase from 1983 through 1999. While the military regimes left ruptures of corrupt-infested polity behind, the democratic civilian regimes and society since 1999 have yet to write a trusted and formidable history about transparency in the fight against corruption. It shows, therefore, that even with the application of military forces and decree policies, the fight against corruption goes beyond the threat of power and authority. There must be systems and strategies that should be able to influence attitudes and mindsets.

Thesis Six: *Corruption detracts social justice*

Social justice, especially in the opinion of John Rawls, has its roots in the *difference principle*. This is proposed in Rawls' second principle of justice, namely: "Social and economic inequalities are to satisfy two conditions: first, they are to be attached to offices and positions open to all under conditions of fair equality of opportunity; and second, they are to be to the greatest benefit of the least-advantaged members of society".[10] Reference and appeal to attach the idea of fairness to positions and opportunities open to all sound already threatening to societies where corrupt elites dominate the economic and political scene. To this end and from concrete experiences accruing from both previous and contemporary epochs, this mode of "feudal dominance" has remained dominant in detracting from social justice. The Nigerian cases, therefore, are violations of its constitutional responsibility as enshrined thus:

> The state shall, within the context of the ideals and objectives for which provisions are made in this constitution [...] control the national economy in such a manner as to secure the maximum welfare, freedom and happiness of every citizen on the basis of social justice and equality of status and opportunity; [...] (d) without prejudice to the right of any person to participate in areas of the economy within the major sector of the economy, protect the right of every citizen to engage in any economic activities outside the major sectors of economy. (Ibid., *Thirty-Five Theses on Corruption*, p. 8)

In general, our sixth thesis would prove that instead of economic factors being utilised to bridge the gaps of inequality, corruption translates them into weapons for *instrumentalising* the human agents themselves.

Thesis Seven: *Wherever capitalism is not moderated and controlled, it becomes an inducement to the corruption of the human heart*

Capitalism is arguably an exploitative and oppressive economic system, especially where there are no democratic and liberal structures to guard against the excesses of monopoly. Such monopoly, or rather exploitation, according to Toyo, happens in five major ways, namely,

> [...] the exploitation of the wage or salaried workers in the capitalist market; the exploitation of consumers; the exploitation of society by the use of its resources to promote the selfish interests of a few, the general predatory drive which the capitalist process imparts to the whole society, compelling all decision-making actors in the capitalist market to seek so-called 'success', or even survival, through one form or the other of exploitation, and exploitation by

---

[10] The Federal Republic of Nigeria, *On Fundamental Objectives and Directive Principles of State Policy*. Constitution of the Federal Republic of Nigeria, Article 16, Section 1.1, Sub-Section b and d, Chapter II, 1999.

> the use of state power and governmental positions in favour of the bourgeois accumulators and others parasitic groups. (Stahl et al. 2017)

To this end, capitalism, just like nepotism, can induce social and economic class struggles, characterised by the general struggle to survive, to control or resist to be controlled. The Nigerian case, just like in almost all the colonial African territories, which are reservoirs for primary resource products, is made more complex by neo-capitalistic imperialism involving multinational agencies and allies.

In the end, therefore, what is left to be emphasised is the fact that corruption remains an endogenic part of almost all social, political, economic and cultural systems. However, experiences have shown that there is yet to be a unanimous approach to the best form of studying, appraising and defining it. Recent research works and opinion groups refer to the need to approach corruption studies in contexts and with reference to cultural differences and conventions. Our research also gives this opinion concise attention.

1.6.5 Corruption: Issues of Interculturality in Approach

It may be surprising to note that most of the countries, basically the developing countries and economies, which seem to have begun the application of legal and organisational anti-corruption frameworks, still experience very high rates and levels of corruption. Some experts identify the real problem as that of uninformed reform policies and interventions, which are deeply weak in understanding the local contexts in which they operate (Ibid.). Consequently, there have been the most recent calls and demands to change strategy and focus on contextualising the factors that drive corruption. Such a way of thinking is deeply informed by the fact that some socio-cultural practices taken (or probably accepted) to be social conventions by a group of people who share common cultural affiliations begin to experience conflicts of interest and understanding when they are confronted with most generalised forms of anti-corruption reform modalities. Such conventions may not be intrinsically moral deviations, but they can be effective drivers of implicit excesses that may lead to corrupt practices. The document of the Basel Institute of Governance observed:

> [...] there is a need to re-think the practice of anti-corruption on the basis of evidence about the factors that drive corruption in different contexts, including the incentives, expectations and agendas across all groups in society [....] In particular, it is necessary to understand the motivations shaping the behaviour of ordinary citizens who, although most sharply suffering the costs from the misuse of public funds, may nonetheless tolerate or even participate in corruption [....] Thus, while much has been written in the literature about the potential of bottom-up collective action to stand up against corrupt government and state officials [...], this potential may only be unleashed if the factors shaping attitudes towards corruption among citizens are properly addressed. (Ike and Onyia 2018)

Keeping the issue of cultural relativity in focus, therefore, would not make an intrinsically morally wrong act to be right. In this context, the purpose would be to help develop more integrative and effective policies and institutions, which are not just theories of ideas, but responsive strategies that are not foreign to individuals' and communities' daily life experiences. In this way, and on the one hand, the danger of double standards in the fight against corruption would, to a greater extent, be minimal. On the other hand, recognising and accepting the cultural relativity in the approach to the fight against corruption would eventually open reform policies that would prove more effective in the fight against persisting drivers of unbridled corruption, especially in developing countries and economies. Nigeria provides a good example.

1.6.6 Corruption: The Need for Conceptual Confluence Amidst Intercultural Differences

Approaches to the issue of corruption may be relativised. However, relativity does not necessarily destroy the objective character and meaning of a thing. As a fundament, the approach to the issue of corruption must focus on those elements that define the very nature of corruption as a deviation, amoral, unethical and detractive. Such elements may be relatively identified in the general worldviews of individual cultural institutions, but their normative statuses remain intrinsically objective. The conceptual confluence about corruption in all its forms and amidst cultural relativity, therefore, is the focus on its nature as a detraction, whereby the undue advantage intentionally taken directly, indirectly or cooperatively by the one, detracts, exploits or infringes the rights and privileges of another individual, a group of individuals or the common good. This makes it pertinent to view corruption as a moral question.

1.6.7 Fight Against Corruption: The Lingering Ethical Dilemma in the Nigerian Case

The reality of the existence of multiple socio-political institutions in the fight against corruption alongside the incessant plunge of the Nigerian polity into corrupt practices establishes an ethical dilemma, which casts doubt on both the general policymaking process and the manner of policy implementation. Nigeria specialises in investing time in the glorification of *political clauses* (policies), the implementation of which provides renewed opportunities for further corrupt practices in the hands of state and public officials. Currently, the enforcement of the Anti-Graft (anti-corruption) War in Nigeria has given rise to such law enforcement agencies like the Budget Monitoring and Price Intelligencer Unit (BMPU); the Independent Corrupt Practices and Other Related Crime Commission (ICPC); the Public Procurement Commission (PPC); the Bureau of Public Enterprise (BPE), as well as other attempts to salvage the image of Nigeria within the international community through the ratification of regional and international anti-corruption conventions and conferences. Despite these multiple strategies, "[…] the fight against corruption is regarded as unintegrated, uncoordinated, dishonest and a pursuit of personal vendetta" (Bauermann 2005). The Nigerian case is considered appalling amidst a teeming number of its youths who have lost perspective within the system and only invest time in learning and practising corruption in its most recent and escalating form, known as internet scamming or "*Yahoo Boys*" affairs.

The above-examined ethical situation leads to the question of the best normative approach to the issue of corruption in Nigeria: Why is the fight not working, or why have efforts remained less effective? What factors are responsible for the persistence of corrupt practices in the Nigerian society—a society privileged with both rich human and natural resources? The opinion of Baurmann would prepare the platform upon which we would attempt an answer with regard to the normative approach. Thus, attempting to explain the reason why varied approaches to the fight against corruption seem not to be working, Baurmann opined,

> [...] Diese Erkenntnis trifft für Korruption in besonderen Maße zu, weil es sich bei Korruption in vielen Fällen um ein Verhalten handelt, das in einem hohen Grad sozial eingebettet ist: Subkulturelle Normen können korruptes Verhalten rechtfertigen und von bestimmten Personen geradezu fordern und seine Praktizierung wird typischerweise durch ein gruppenspezifisches soziales Kapital erleichtert, das gleichzeitig die Aufdeckung, Verfolgung und Sanktionierung von Korruption durch staatliche Sozialmoral, durch die korruptes Verhalten geächtet wird und die als Handlungsorientierung sowohl unter den Bürgern als auch den Vertreten der staatlichen Organe ausreichend wirksam ist, sind deshalb die Chancen gering, Korruption dauerhaft auf ein erträgliches Maß zu beschränken. (cf. Ibid., p. 166)

The awareness (Erkenntnis), in the opinion of Baurmann, which needs to be created in any fight against corruption is that the norms and policies anchored on the social life of human society must be made to motivate people intrinsically towards voluntary readiness to comply with such norms and policies. Hence, following Baurmann, Nigeria with its policies and approaches must first answer three relevant questions, namely (1) How integral or justifiable is the approach? (2) How legitimate is the approach? (3) How trustworthy or reliable is the approach? Without a balanced integration of these three relevant issues, Nigeria remains a failure in the fundamental objectives of state policy.

1.6.8 Ethics and Corruption: The Socio-Ethical Normative Approach

For a law or norm to be relevant, it must have the power of coercion. Ethics itself, therefore, must be validly normative to be able to qualify as an evaluative means and approach to social moral questions. Thus, ethical criteria for the issue of corruption must be developed in such a way that their application should not be arbitrary but rather integrative to respecting the rule of law, the principle of justice and the dictates of practical and right reason. In the context of this academic work and in the approach to the issue of corruption, to emphasise the relevance of coercive law as a criterion is to delineate the fact that all manner of legislation against corrupt practices should be integrally applied, such that no part of the society would be treated as a group above the law. Such laws and policies must be reasonable and acceptable to the polity they are meant to guide, and they must carry the power of conviction so as to awaken the sense of trust in the masses, whose compliance to civil obedience stands as the focal point. In this way, the moral approach to the fight against corruption would qualify to be imperative, not just hypothetical. The imperative power must be convincing and trustworthy enough to ensure free and intrinsically based societal obedience and compliance. Recent studies, reflections and propositions in this direction would be further taken up by our research and given further attention within the cultural conventions surrounding our work settings.

1.6.8.1 Law and Intrinsic Motivation: Towards the Functionality of the Fight Against Corruption

Modern development studies and studies in ethics and morals recognise the persistent difficulties that arise when norms and laws appear to be only absolute juridical tools that impose legal coercions but predominantly through pure command, prohibitions and constant threats with punishments. However, in many cases, the mere juridical and legal application of the law has been known to the predominantly subsidiary (*nebensächlich*) to the actual motivation towards obedience to laws and norms. This would prove why, even in states, jurisdictions and sovereignties that have tried pure concrete legal and penal measures in the fight against corruption and other social crimes, the tendency to incessant ethical and moral deviations persists as a *negative behavioural constant*. Nigeria is a good example.

In the above line of thought, and examining the opinion of Baurmann, the functional basis of law and societal norms is predominantly dependent on the level of intrinsic motivations that the law itself awakens in the members of a particular society. As a basic principle and in the context of corruption, Baurmann stated that without a convincing convergence of social norms and morals, which must be a concrete and intrinsic part of the convictions and orientations towards actions among the members of society, all legal instruments in the fight against corruption would remain mere toothless weapons (cf. Ibid., p. 167). Therefore, a viable ethical approach to the issue of corruption must examine the relationship between the coerciveness of laws and norms and the extent to which the masses are intrinsically motivated and convinced about the attitude to the rule of law, both from the state and its public officials. A developing ethical model—*Die Theorie der positiven Generalprävention*—within the German-speaking areas of Europe and similar discussions about Law and Intrinsic Motivation within the anglophone intellectual space would offer relevant study elements to find the ethical *missing link* in the fight against corruption in the Nigerian case.

1.6.8.2 The Theory of *Generalprävention*: Mere Penal Code versus Intrinsic Moral Conviction in Approach to Corruption

The theory of *Generalprävention*—general prevention or integrated prevention strategy—is a recent form of approach in ethics and morals, with regard to examining the place of laws or social norms as relevant means for awakening moral orientations and consciousness in a particular society, instead of the mere application of penal instruments of law for the conviction for and penalisation of offences and offenders. The main ideas behind the theory place emphasis on the development of a new attitude towards a society's understanding of the law and penal justice, such that law and penal codes would not just be seen as imposed external means of the clampdown on social situations of moral failures, but rather as means of strengthening both social and moral influences on the masses. Furthermore, instead of such legal instruments to be looked upon as mere media of *prohibitions* for anticipated noncompliance or the unwillingness to obey the norms within the polity, they should be seen rather as sources (not weapons) of inducement for social consensus towards the acknowledgement and acceptance of norms. Again, instead of being looked upon as weapons against individual or collective freedom and autonomy, they should rather be seen as injunctions for better and convincing development of the individual sense of responsibility and self-determination (Baurmann 1994). Within the Anglophone Schools of Thought, the discussion is all about the need not just to read and observe the letters of the laws and norms but, more importantly, the relevance of understanding the spirit of (behind) the laws and norms.

The development of this theory has led to the differentiation between absolute and relative theories of criminal law and justice. The absolute theory is built on the principle of "*punitur quia peccatum est*", which means an application of a punitive measure purely because an offence has been committed. The approach in the absolute theory, therefore, follows two objectives for applying the law, namely (a) the *Talion principle* (the principle of retribution), which aims at establishing a balance, especially in a criminal assault, between the harm or injury done to a victim and the punitive measure that is awaiting the perpetrator himself or herself. An ancient form of expression in line with the Talion principle is an eye for an eye, a tooth for a tooth; (b) the *Atonement principle*, which aims at evoking punitive measures to attest that restitution and reconciliation have been carried out between the perpetrator and the legal system. The form and nature of the atonement, however, do not depend on the choice or free will of the perpetrator but rather on the inflictor of the punishment. This has become one of the criticisms levelled against the absolute theory.

On the other hand, within the relative theory, punitive measures follow the principle of "*non punitur quia peccatum est, sed ne peccatur*", meaning literally, it would be punished, not just because sin has been committed or the law is broken, but rather to prevent further breaking of the law. It describes the approach that focuses not just on applying punitive measures for atoning an offence but rather on how the punitive measure should work to achieve *prevention*. Within the relative theory, one finds the more favoured opinion that our reflections have chosen as a possible and applicable approach in the Nigerian context, namely the *Generalprävention*. Again, under the relative theory, one differentiates between the negative Generalprävention, which chooses its language of approach as *deterrence* of the general public, and the positive Generalprävention, which focuses on observing, evaluating and seeking to improve on the (positive) impacts, which the deterrent or punitive measures are expected to make in the general life and structure of a society. This is proposed to have three basic effects, namely (a) *Learning effect*, in the sense of a gradual opening up of disposition of the public towards legal compliance; (b) *Trusfulness effect*, in the sense of gradual building of trust in the public on the validity of a legal system, and (c) *Satisfaction effect*, in the sense of reassuring the general public that justice has been done not denied and the conflict between the perpetrator and rule of law resolved—this is known as *Integrationsprävention.* Thus, Integrationsprävention as a model of approach develops in both broader and narrower senses as represented below (Table 1.5).

Applied to the context of corruption in this academic work, it means that the moral approach to the deviant issues in any given society should not just focus on mere deterrent strategies seeking to enforce general conformity to laws and norms and, perhaps, punish offenders where and when it becomes necessary. The ethical and moral sense in the generalprävention model focuses more on discovering and developing strategies for awakening members of the society themselves to a willing

Table 1.5 Positive general prevention as *Integrationsprävention*

Broad sense	Narrow sense
Behaviour model	
Action based on the inner disposition of the individual	Action based on the belief in the legitimacy of law and lawgiver
Explanatory model	
Norms and laws binding through disposition	Acceptance of and compliance with the norm following conviction about the legitimate validity of the norm
Effect mechanism	
Learning process	Persuasion
Hypothesis	
Criminal law (norm) leads to the dispositional binding of norms by creating and maintaining freely applicable dispositions	Hypothesis I: Criminal law would lead to the acceptance of norms through thesymbolic manifestation of legal validity Hypothesis II: Criminal law leads to the acceptance of norms by securing legal trust viewed as (a) trust in compensatory justice (b) trust in norm enforcement

Source Adapted from the schema presented by Michael Baurmann in his "Vorüberlegungen zu einer empirischen Theorie der positive Generalprävention" (cf. Ibid., *Korruption, Recht und Moral*, p. 167). *Table format: Mine*

intrinsic commitment and responsibility towards social norms and, of course, through policies and institutions capable of promoting interests and non-compulsive acceptance of laws and norms of the society. Understood in this manner, therefore, retributive justice or the application of penal measures would assume the status of moral instruments of education and upbringing for the social and moral stability of the society. This general objective (Generalprävention) would be, at least to a predominant extent, to pave the way for behavioural and attitudinal conformity of the social participators of a given society who would and should feel intrinsically motivated to give themselves into such social contract and practice in solidarity. The reverse of this objective—*Abschreckungsprävention* (deterrence) (cf. Ibid., p. 171)—reduces the masses to mere passive objects with constant burning feelings of compulsion, constraint and oppression. No moral attitude may be possible without the freedom of the moral agent himself or herself. In every moral action, there is a will and this aspect of "willing" is a product of an intrinsic action and motivation that demand attention too.

In a broader sense, to view corruption in Africa generally and Nigeria in particular within the *generalprävention's* model of approach is to focus, on the one hand, on the question of the big moral gaps that exist between state legislations and the demand it places for conformity and compliance. On the other hand, it would mean a critical examination of how the anticipated mutual relationships (necessary connections) that should exist between forms of legislation and the moral statuses of the legislators themselves function. In other words, it is to ask the question of how legitimate and trustworthy the state, policymakers and enforcement agencies are with the administration of laws and norms. Here, we stand on the premise that there still exist wide gaps between laws and lawgivers themselves, especially in developing countries and economies of Africa, thus making the fight against corruption grossly trivial, ephemeral, questionable and unsustainable. This brings to the fore the question concerning the right dispositions (*Dispositionen*) and decisions (*Enscheidungen*) in situations that demand legal and moral conformity. The conditions for plausible norm conformity are not only dependent on the sociological factors requisite for norms and value orientations among human beings as *homo sociologicus* but rather a result of an added internal (intrinsic) disposition of the human person towards non-compulsive acceptance of the norms and values in question as an active, rational and autonomous actor—the *homo oeconomicus*. This combination leads to a more integrative model that would be more effective in any approach to the fight against corruption—the *integrationsprävention* (Integration of prevention or integrated prevention).

The exposition so far teaches that any functional normative process of a given society—in our own context, the law against corruption, anti-graft war—must constitute an integrative and endogenous part of the general life of that society so as to support actions that do not just focus on mere deterrence and sanctions. The supportive thesis here is that through both sanctions and gratification, laws would be able to create factors and induce actions that would form an essential part of the education, developmental and social life and process of a given society. Baurmann clarifies further that the societal normative process should possess attributes that have educational effects to be developed as socialisation factors that, together with other influences, could capably lead to an "inner bond" of the human person to social norms. To this effect, a state penal system is one of the factors that directly or indirectly contribute to the development of such dispositions to act and ensure their maintenance. From this point of view, criminal law should serve as part of a social environment that rewards certain personality structures and dispositions for positive actions through various forms of gratification while imposing sanctions and costs on others. This is a concrete and practical pathway to bring positive influence on longer-term personal development and learning processes.[11]

1.6.9 Integrationsprävention as a "Missing Link" in the Nigerian Fight Against Corruption

In his book, "Afrika Wird Armregiert oder Wie Man Afrika Wirklich Helfen Kann?", Seitz opined that corruption employs power relations and dependencies to destroy initiative and commitment where they would be required. Consequently, in both private and public spaces, projects are selected with the ulterior motive of how much "*corruption-money*" could be given or extorted. Such a statement about Africa is predominantly true in the Nigerian situation, where the political elites, who are the perpetrators of corrupt practices, frustrate every effort in the fight against corruption and bend laws and policies at will to suit and protect their crimes. Thus, the means and measures of the normative approach in the fight in Nigeria have remained trivial, arbitrary, selective and abusive. The cause of this is a form of conspiratorial *missing links* that are inbuilt into the socio-political system and which serve as effective weapons of manipulation in the hands of those in the corridors of power. These missing links are embedded in the lack of *consistency*, *predictability* and *invariance* in the general use and application of laws in the Nigerian system. Linking and defining these terms within our context would be insightful:

(a) *Consistency (Gleichmäßigkeit)*: The realm of the rule of law and criminal law or justice (Strafrecht) is deeply hijacked and manipulated in Nigeria. The level of inconsistency has always resulted in the multiplication of laws, policies and law enforcement institutions since every constituted or conscripted administration begins, at inception, to put in place varied new ways to use both protective and punitive laws to empower itself, pave the way for its strategy of corrupt practices and the looting of the national treasury, and of course, protect the so-called powerful or godfathers of the previous administration from being probed of their corruption and criminalities. Such inconsistencies in the dispensation of the rule of law have rubbished all efforts to make effective inroads in nipping

[11] The immunity clause is a constitutional clause enshrined in the Nigerian constitution which prohibits civil or legal proceedings against the principal public office holders in both federal and state levels—President or Vice-President, Governor or Deput-Governor—while the person is still in office (cf. Constitution of the Federal Republic of Nigeria, Chapter III, Part II, Section 308).

corruption in the bud within the polity. Consequently, many known corrupt statesmen and stateswomen still walk the streets untouched. Many among them even sit in the state and federal houses as law- and policymakers. The result is a total loss of confidence in their leaders among the populace.

(b) *Predictability (Berechenbarkeit)*: Effective law and criminal justice must be able to possess the character of being predictable. In other words, the normative powers associated with them must be made to work in such a way that the general polity should be able to predict situations that invoke gratification by law as well as situations that demand punishment by law. A citizen who has attained the age of reason should be able to reckon with existing laws, norms and rules as well as the consequences of his or her actions when he or she decides freely to act or freely neglect and omit an action. However, the Nigerian system suffers incessant legal deviance that portrays the whole polity as being infested with a deep *double-standard syndrome*, which invariable robs an average Nigerian person of any interest or disposition to care for the rule of law or norms of behaviour. An effective negative slogan to this effect in Nigeria is: Since we cannot beat them, we must join them.

(c) *Invariance (Invarianz, Beständigkeit)*: This concept shares some characteristics with the theme of consistency in the sense that their presence in a given society helps guard against arbitrariness in applying the law. Equality before the law is the invariance mantra in any functional polity. However, the variance of law or criminal justice in Nigeria with regard to the fight against corruption has awoken many criticisms within domestic and international socio-political spaces. This creates the *untouchable* syndrome where the rich and powerful elites could go scot-free no matter the extent of the crime committed, while the poor and social commoners pay high prices and serve jail terms. The absence of invariance as ethics of application of the law has made criminal justice or the penal code in Nigeria deeply controversial. To this end, it is to be noted that one of the most debated and persisting variant clauses within the Nigerian constitution is the *Immunity Clause* (Ibid., *Korruption, Recht und Moral*, p. 176), which has been grossly responsible for the provision of unnecessary cover for public officials and their corrupt practices. In the end, the amoral variance of the law and its application remains a bane of any ethical approach and can never guarantee any non-compulsive disposition of the masses towards laws and norms.

Conclusively, therefore, our research is convinced that rediscovery of and reinvestment in the effort to reintroduce these missing links in Nigeria's mainstream criminal justice system remains a prerequisite for a workable approach to the issue of corruption. It would help to introduce transparency and restore trust in the people and the system. The next sub-topic proposes a "three-thesis" strategy, all informed by the idea of the generalprävention model, which could unite to serve as a *control-apparatus* for any anticipated law or norm capable of awakening the non-compulsive disposition of the masses.

1.6.9.1 Applied Normative Ethical Approach: The Positive Generalprävention as the Applicable Moral Antithesis to the *Missing-Link* Theses

Our ethical evaluation of the Nigerian situation with regard to corruption has revealed both concrete facts about its existence and the complex networks of systems and structures that support and facilitate its persistence. The greatest bane in the fight has been identified to be the weak and inconsistent legal structure hijacked by the political elites and class and made an instrument of manipulation at their own will. Thus, the social cost, in the Nigerian context, has been the observable unwillingness of the citizens to ascribe any meaning again to the general system nor develop interest anymore towards conformity to the rule of law. Hence, Nigeria's current political and economic spaces, within the period of our research, present a jungle where corrupt practices are not just sporadically carried out but rather where its varied forms are gradually becoming a cancerous part of everyday concrete life. A time bomb may be awaiting the general system and structure since the so-called future generation is generically growing up with it.

Therefore, we propose an *ethical recovery agenda* in the fight against corruption. This should be an agenda in *restructuring* the general legal system and institution and concrete efforts to *de-corrupt* the penal code. The mode of the restructuring should and must be education-based, first, to take as a basis the integrative minimal knowledge of laws, norms and their application within the general learning process and upbringing. Secondly, leadership should be enthroned in the administration of law, emphasising equality before the law and not compromising the rule of law. This would mean a concrete positive contribution to convincing the body politic that no double standard approach could qualify as justice to the rule of law or motivate the free disposition of the masses to obey the rule of law. A restructured agenda of approach would, in line with the opinion of Baurmann, redefine the criminal law institution as an institution that not only externally enforces legal behaviour by threatening punishment but rather an institution that also promotes the intrinsic motivation towards voluntarily norm conformity and thus promotes the social morality of a society (Ibid.). Three normative principles would stand to assure the anticipated legal and moral plausibility as represented and defined below (Fig. 1.4).

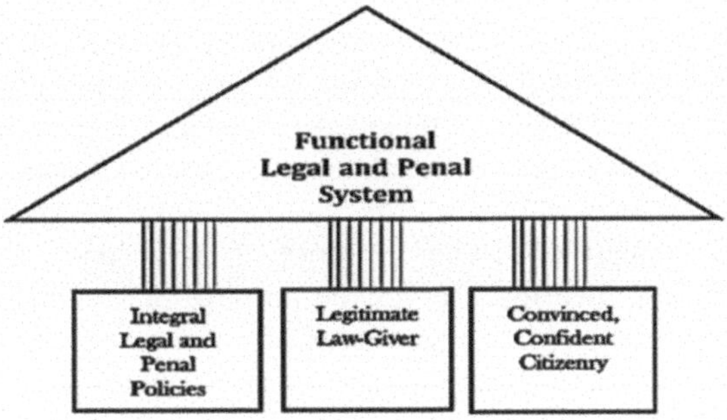

Fig. 1.4 The three pillars for a functional legal and penal institution in the fight against corruption in Nigeria. Pillar 1: integral legal and penal policies; Pillar 2: legitimate lawgiver; Pillar 3: convinced, confident citizenry. *Source* Mine (Nwosu 2020b). *Illustration mine*

We propose this three-pillar-approach as an urgent ethical pre-condition for a functional and effective legal order in the fight against corruption in Nigeria. Hence the following final theses:

Integration Prevention Thesis (Integrationsprävention)

To achieve an ethical recovery in the Nigerian penal code against corruption, the system must concentrate on practical and functional measures that must work to debase corruption from being attractive to the perpetrators and getting integrated almost as a way of life within a given polity. The higher the risk associated with corrupt behaviour in a society forms part of the general awareness and learning process of the masses, the less it would be worthwhile for certain groups of people to continue business as usual. The distributive application of the consequences of such moral failures can indirectly contribute to the development of a genuinely intrinsic commitment to dutiful and lawful action (cf. Ibid., p. 177). Therefore, concrete penal actions to ensure deterrence are still a missing link within the Nigerian system, added to the arbitrary form of selecting and punishing offenders. Such arbitrariness leads to contempt of law and the violation of the principle of equality. Thus, unless transparency is guaranteed through the emphasis on equality before the law with regard to corruption, there is bound to be an increase in the character traits that propagate and promote its integration as part of the daily practice and behaviour within the citizenry.

Legitimacy Thesis (Legitimitätsthese)

This second thesis places emphasis on the importance of the influence of a lawmaker or lawgiver in the awakening of the non-compulsive disposition of the masses towards legal and penal justice. Both the social status and moral character of a social and political figure provide critical requisites for moral judgement on the part of the body politic expected to engage what the lawgiver has legislated with the free disposition of the will and compliance to the rule of law. One of the greatest blows that the structure has done to the Nigerian penal administration in the fight against corruption is the illegitimacy of the persons and institutions that are always co-opted in the fight. The hijacked political and party system, which has witnessed an extended period of *changing-the-baton* from one corrupt group to the other, has been trapped in the celebration of this illegitimacy as *rulership* agenda. Thus, known and wanted criminals in both national and international spaces occupy the seats illegitimately as decision-makers since they are never elected by popular votes but selected by the manipulations of their corrupt bedfellows and political godfathers. Their rulership agenda, therefore, serves the purpose of bending the law to suit them and weakening every known apparatus of governance that may work to expose and punish their crimes. The legitimacy thesis would argue for more participation of the public and transparency organisations in penal administration. This would help to ensure transparency in the administrative process and to convince the body politic about the legitimacy of and compliance with the actor upon whom the mandate of using and applying the law has been placed. Baurmann defends the opinion that the legitimacy thesis remains powerful workable ethics in the fight against corruption even in situations where subcultural norms exist among a group of people that share a common interest and thus seem to be tolerant or permissive of some practices that may be viewed outright as corrupt or unacceptable in other cultures and worldviews (cf. Ibid., p. 178).

Confidence-Based Thesis (Vertrauensthese)

In the process of the proposed ethical recovery in the Nigerian system, the confidence-based thesis proposes the need for concrete proof and the practice of transparency with regard to effective retribution practices within the legal order as the only basis for building confidence in the citizenry. The expected proofs are the applied concrete forms of criminal justice, which in justice punish evil-doers and reward compliance with the rule of law. This is taken as society's fundamental objective towards retributive justice. The effectiveness of the confidence-based thesis builds confidence within the citizenry such that even potential evildoers are considered well-informed about the full consequences of breaking the law (Ibid.). Expected here is that proofs of concrete and practical retributions serve not just as deterrence, but rather they form essential parts of the learning and developmental processes in shaping attitudes and awakening the non-compulsive disposition of the citizenry towards the rule of law. Nigeria's fight against corruption, however, suffers the deliberate failure of the system to administer due punishments to known offenders of the law. It rather practises *selective retribution* where the *elite-offenders* are treated as "sacred cows" before the law, while the rest of the citizens are judged with a completely different standard and yet within the same polity, where the constitution holds everyone as equal and inviolable. Until this situation is revolutionised, Baurmann's assessment remains unequivocal in the Nigerian case, namely that there would be "zero" possibility of building trust in a legal system or gaining the voluntary willingness of the masses to conform with norms, especially in any known polity, where a majority of the criminals get away with their crimes and the very organs responsible for the administration of justice gulps itself as well in corruption, arbitrariness and practice of inequality.

Evaluation

Our journey into Nigeria's general structure as a sovereign state reveals both strengths and weaknesses. Strength, in the sense that it possesses so much potential capable enough to drive development and empowerment. Weakness, in the sense that the operating system possesses detracting factors that inhibit progress and institute a form of social struggle where the so-called powerful elite groups dominate and control the mainstream structure of the state's politics and economy. Our research recognises the factors responsible for such oppressive dominance as man-made, institutional and inbuilt. In the next chapter, we will discuss these factors under a uniting concept, namely "structures of sin"—structures that are dysfunctional, destabilising, detracting, oppressive and exploitative.

References

Andam KS, Hyacinth E, Oboh V, Pauw K, Thurlow J (2020) Estimating the economic costs of COVID-19 in Nigeria. NSSP Working Paper 63. International Food Policy Research Institute (IFPRI), Washington, DC, p 7. Available via https://doi.org/10.2499/p15738coll2.133846. Accessed February 4, 2021

Baurmann M (1994) Vorüberlegungen zu einer empirischen Theorie der positiven Generalprävention. In: GA, pp. 68–384. Available via https://www.goltdammers-archiv.de/

Bauermann M (2005) Korruption, Recht und Moral. In: von Alemann U (ed), Dimensionen politischer Korruption: Beiträge zum Stand der Internationalen Forschung, Politische Vierteljahresschrift, Sonderheft 35/2005. VS Verlag, Wiesbaden, pp 162–182

Emefiena E, Agudiegwu M (2015) Partisan politics in multi-ethnic Nigeria: democracy for national disintegration. Research on Humanities and Social Sciences 5(22):55, 53–63. Available via https://core.ac.uk/download/pdf/234674789.pdf. Accessed February 16, 2019

Falana F (2018) Restructuring Nigeria cannot guarantee unity and political stability. In: Vanguard Nigeria, Nigerian Daily, September 6. Available via https://www.vanguardngr.com. Accessed May 12, 2022

Fofack H (2015) The idea of economic development: views from Africa. In: Monga C, Lin JY (eds) The Oxford handbook of Africa and Economics, Vol I, Content and Concepts. 2015. Oxford University Press, United Kingdom, pp 271–272

Gelb A, Meyer CJ, Ramachandran V (2015) Development as diffusion: manufacturing productivity and Africa's missing middle. In: Monga C, Lin JY (eds) The Oxford handbook of Africa and economics, Vol I, Content and Concepts. Oxford University Press, United Kingdom, pp 115–116

Graf WD (1989) The Nigerian state: political economy, state class, and political system in the post-colonial era. Pearson Education, London, p 29

Harriman T (2006) Is there a future for democracy in Nigeria? In: Paper presented at the Department of International Development, Oxford University London, 5. June, p 2. Available via www.oregan.politic.ox.ac.uk/. Accessed December 6, 2021

Ike OF, Onyia C (2018) Ethics in higher education: foundation for sustainable development. Globethics, Geneva, p 479

Johnston M (2005) Syndromes of corruption: wealth, power and democracy. Cambridge University Press, London, p 12

Joseph R (2015) Inclusive growth and developmental governance: the next African frontiers. In: Monga C, Lin JY (eds) The Oxford handbook of Africa and economics, Vol I, Content and Concepts. 2015. Oxford University Press, United Kingdom, p 224

Kieh GK, Agbese PO (2021) From politics back to the barracks in Nigeria: a theoretical exploration. J Peace Res 30(4):409–426. Available via http://www.jstor.org/stable/424484. Accessed February 4, 2021

Korff W (2009) Handbuch der Wirtschaftsethik. Berlin University Press, Berlin, p 310

Kurer O (2015) Definitions of corruption. In: Heywood PM (ed) Routledge Handbook of political corruption. Routledge, London, pp 30–41

Meredith M (2011) The fate of Africa: a history of the continent since independence. Public Affairs, New York, p 581

Nwabueze B (2019) Our match to constitutional democracy. The 1989 Guardian Lecture, Lagos, July 24. In: Ojo E (ed) The military and the challenge of democratic consolidation in Nigeria:

positive scepticism and negative optimism, Journal of Military and Strategic Studies, JMSS 15(10):9–37. Available via https://jmss.org/article/view/58136. Accessed October 23, 2019
Nwodo JN (2021) Federal government making things difficult for leaders to douse youth restiveness in South-East. In This Day, Nigerian Daily, November 15. Available via https://www.thisdaylive.com. Accessed May 12, 2022
Nwosu CF (2016) Ethical approach to abuse of common good and human rights in Nigeria: the polity, oil, violence and Niger delta. AV Akademikerverlag, Berlin, p 8
Nwosu CF (2020a) Boko Haram: a threat to peace or a legal agitation? the political ethical implications. Module—political ethics. Seminar Paper Submitted as Requirement for Master of Arts Degree in Applied Ethics, Westfälische Wilhelms University, Münster, Germany, p 8
Nwosu CF (2020b) The socio-economic and political implications of unbridled corruption in the Nigerian system: an ethical approach. Unpublished Master's Thesis in Applied Ethics. Westfälische-Wilhelm University, Münster, p 5
Odey JO (2003) After the madness called election. Snaap Press, Enugu, pp 20, 25
Ojo EO, Adebayo PF (2008) The politics of states', local governments' creation and Nigeria's search for geo-political balancing. African Res Rev 2(3):334, 334–367
Osaghae E (1998) Nigeria since independence, crippled giant. University Press, London, p 1
Petersen M, Wehrmann D (2015) Geopolitics.In: InterAmerican wiki: terms, concept, critical perspectives. Available via https://uni-bielefeld.de/einrichtungen/cias/wiki/g/geopolitics.xml/. Accessed February 4, 2021
Rawls J (2001) Justice as fairness (as edited by Erin Kelly). Harvard University Press, Massachusetts, pp 42–43
Siep (2018) Ludwig Ewiger Friede und gerechter Krieg in der politischen Philosophie der Neuzeit. In: Althoff K, Meier T et al (eds) Frieden: Theorien, Bilder, Strategien von der Antike bis zur Gegenwart. Sandstein Verlag, Dresden, pp 256–273
Smith DJ (2015) The contradictions of corruption in Nigeria. In: Heywood PM (ed) Routledge handbook of political corruption. Routledge, London, pp 56–66
Smith M (2016) Boko Haram: inside Nigeria's unholy war. I. B. Tauris, London, p 9
Stokke O (1970) Nigeria: an introduction to the politics, economy and social setting of modern Nigeria. The Scandinavian Institute of African Studies, Sweden, Uppsala, p 15
Stahl C, Kassa S, Baez-Camargo C (2017) EARF corruption, social norms and behaviour in East Africa: drivers of petty corruption and anti-corruption interventions in developing countries. a semi-systemic review. UK Department of International Development (DFID) Basel Institute of Governance, Switzerland, p 5. Available via https://assets.publishing.service.gov.uk/media/599acae840f0b679518fb548/Drivers_of_Petty_Corruption_and_Anti-Corruption_Interventions_in_Develop....pdf
Stockmann R, Menzel U, Nuscheler F (2010) Entwicklungspolitik. Theorien – Probleme – Strategien. Oldenbourg Wissenschaftsverlag, München, p 221
Tetzlaff R (2018) Afrika: Eine Einführung in Geschichte, Politik und Gesellschaft. Springer, Wiesbaden, p 251
The Federal Republic of Nigeria (2018) National forest reference emission level (FREL) for the federal republic of Nigeria. Published by the Federal Department of Forestry and Federal Ministry of Environment, Dec., p 11. Available via https://redd.unfccc.int/files/nigeria_national_frel_modified_revised__for_posting.pdf. Accessed October 28, 2021
Toyo E (2006) Thirty-five theses on corruption. The Constitution: A J Constitut Develop; Lagos, Nigeria: Centre for Constitutionalism and Demilitarisation, CENCOD 6(4):1–14. Available via https://doi.org/10.10520/AJA15955753_84
Uzoigwe GN (1998) The Niger Committee of 1898: Lord Selbourne's report. J Hist Soc Nigeria IV(3):467, 467–476
Warren ME (2015) The meaning of corruption in democracies. In: Heywood PM (ed) Routledge handbook of political corruption. Routledge, London, pp 42–55

Chapter 2
Socio-Political and Theological Focus of the Research Work

Abstract Human societies are not just amorphous. They have structures that can be approached at the macro, meso, and micro levels. The relatively harmonious relationship within these levels determines how functional or dysfunctional a society could be. An attempt to view the Nigerian state at these structural levels paints a picture of a polity beclouded, on the one hand, with conflict of ideology. On the other hand, however, it possesses the potential for resilience and growth. The meaning associated with the concept of '*Structural Sins*' helps to identify what constitutes structural weaknesses in the Nigerian system and which contributes to keeping it underproductive. The reflection juxtaposes proof of structural dysfunctions within the polity with evidence of their effects in disrupting human capacity building (capability approach) for authentic freedom and development. The persistent decades of struggle to stabilise the energy sector of Nigeria and the records of induced leadership and institutional failures provide the fieldwork and the empirical specimen in proving the presence of such structural cancer.

Keywords Social structure · Structural sin · Polity · Conflict of ideology · Capability approach · Energy sector · Nigeria

2.1 The Social Structures: Socio-Economic, Political and Ecological Nigerian Geo-Polity

Our research explores the structural patterns upon which Nigeria as a nation is built. This exploration would examine the interplay among established political, economic, social and ecological institutions, which mark a nation as developed, developing or underdeveloped; stable or marred with instability; progressive or retrogressive, etc. The social structures of Nigeria as a nation-state define her type of ideology towards general human development and the general plans for a sustainable future.

C. F. Nwosu, *Systemic Remediation*, Environmental Humanities: Transformation, Governance, Ethics, Law 4, https://doi.org/10.1007/978-3-031-61909-0_2

2.1.1 The General Concept of a Structure

A Structure is both a pattern and an entity. A pattern in the sense of a structural guide intended to form a designed and unified whole, which benefits effectively and efficiently from the structured parts. On the other hand, it is an entity in the sense that a structure, when well built, reserves the ability to resist external debilitating forces, hence, independent as a formidable whole. In other words, the more co-related and lubricated a structural pattern is, the greater its chances for optimal efficiency and survival—political, social, economic, or ecological. The structure generally and in its dual classification (material and abstract) always retains some basic characteristics that define its nature, functionality and qualities, namely interrelation, formative attitude, stability and the ability to withstand stress and strain. Thus, the business dictionary defines structure as,

> […] Construction or framework of identifiable elements (components, entities, factors, members, parts, steps etc.), which give form and stability, and resist stress and strains. Structures have defined boundaries within which (1) each element is physically or functionally connected to the other elements, and (2) the elements themselves and their interrelationships are taken to be either fixed (permanent) or changing only occasionally or slowly. (Monash University 2019)

For Glatzer, the two most important characteristics of a structure are pattern and orderliness (*Muster und Regelmäßigkeiten*). In relation to human society, structural patterns are analysed in the form of social structures, which play important roles in the general interactions and relationships involving individuals living in a particular society and other extensive structural patterns of interdependence. Thus, Glatzer describes a structure as an arrangement or uniting together, not arbitrarily, of a majority of entities in a manner that such a union gives rise to orderliness or pattern. Consequently, one expects that what arises or occurs at one part or position can also reoccur in another part or position (Glatzer 1969). This definition, therefore, delineates a type of relational lubrication and interdependence within a structure, which makes for unity among the existing parts.

Based on this relational character of a particular structural pattern, this research examines Nigeria's structure in its macro, meso- and micro-scales to determine the level of stability or instability within the system, bearing in mind also that structure is an important issue for management. In social structure, roles, statuses and other related factors play combined roles within the general network of human interaction. In other words, structure defines the relationship existing among entities or groups. In a social structure, for example, one explores the patterns of behaviour among participants in a social system, their relation to one another, as well as the institutionalised norms or cognitive frameworks that guide and influence the actions of the actors within such a social system—Nigeria taken as one.

2.1.2 Social Structures and the Three Levels of Classification

Social or societal structures are predominantly viewed and studied within interactive and behavioural scales, which help to define the basic relationships among humans, institutions, organisations, policies, economies and conventions. The three known primary concepts in this regard are the macro, the meso- and the micro-scales.

2.1.2.1 Macro-scale Social Structure

The macro-scale social structure reflects a cross-section of patterns that provide the fundament of any organised social union. Such a union contains patterns of social institutions, which unite to animate a society. In sociology, the macro-scale represents an "[…] overarching patterns of social relations that lie outside one's circle of intimates and acquaintances. Macrostructures include classes, bureaucracies and power systems such as patriarchy" (Furze 2015). Thus, in popular politics, one of the widely debated macrostructures is *patriarchy*, "[…] the traditional system of economic and political inequality between women and men in most societies" (Ibid.). With reference to the generality of the society, the macro social structure gives an encompassing picture of interrelated institutions upon which a given society operates and functions, bearing in mind that the flexibility of the flow of interaction within a structural pattern would invariably determine its readiness to absorb change, implement workable policies and sustain development. Our research, therefore, would examine, in the course of its varied studies and expositions, the macro-scale upon which Nigeria is built, of course, within her claims about being a Nation-State that practises democracy.

2.1.2.2 Meso-scale Social Structure

The meso-scale discusses the relationship between individuals as subjects in a social structure and the available institutions of both macro- and micro-scales. It is concerned with the societal ties between and among individuals and organisations. Thus, it focuses on such themes as race, ethnicity, gender, age, income and organisation, as well as some adverse geographical factors that affect a community. When examined within the Nigerian State, the aforementioned elements unite to present a multiplicity, making Nigeria one of the most complex meso-scale structural patterns on the continent of Africa. The meso-scale examines the endogenous responsibilities that individuals in a given society have and are bound to fulfil within a network of related institutions. In other words, the meso-scale social structure directly affects the individual and the links they share with one another based on values, interests or experiences.

2.1.2.3 Micro-scale Social Structure

The micro-scale deals with the normative aspects of a social structure. It is the level of the "ought" in the social system. That means it searches for the ideals that are normatively relevant to functional social structures. At this level, norms are examined and laid out as guiding principles that influence the life and behaviour of both individuals and institutions. At this level, we examine,

> [...] the everyday interactions we have with each other in the forms of norms and customs. We can see it present in the way patterned institutionalized relationships shape our interactions within certain institutions like family and education, and it is present in the way institutionalized ideas about race, gender, and sexuality shape what we expect from others, how we expect to be seen by them, and how we interact together. (Crossman 2020)

In the micro-scale social structure, therefore, norms and principles are put in place to assist the functioning and organisation of social order. Such norms and principles must be coercive to impact the relevant results within the polity in question.

2.2 The Three Levels of Social Structures and the Nigerian Polity

Within its framework as a structural entity, Nigeria remains a dynamic developing nation with variegated odds and cogs in the wheel of development. The timeline of its institutional, social and infrastructural development shows both conflicts and harmony at the same time. Its structural, infrastructural and ideological development remain threatened by social, political and economic fluctuations and instabilities, which, however, are characteristics of all institutions within the general process of development. The expected integration of the social structure within the polity still experiences staggering tensions and frictions, and the United States Agency for International Development, USAID observes,

> Although Nigeria has the largest national economy in sub-Saharan Africa, it must overcome significant development and governance challenges to fulfil its potential. With an estimated population of 214 million, of which approximately 96 million people live on less than $1.90 per day, Nigeria is home to the world's largest population living in extreme poverty. USAID works in Nigeria to promote broad, inclusive economic growth, food and nutrition security, a healthier, better educated population, accountable, inclusive, and responsive governance, gender equality, access to renewable energy, and stability, including humanitarian assistance, in areas devastated by extremist violence [....]. (United States Agency for International Development, USAID, Nigeria 2021)

An examination, therefore, of the present situation and conditions within the polity, as they concern Nigeria's social and institutional policies, would help to make our focus more precise. Our analysis would focus on what the Nigerian structure looks like, its characteristics and how it affects both governance and the quality of life within the polity.

2.2.1 The Macro-Nigerian Society

The macro-Nigerian society is rather large and multifarious. With an actual population of over 200 million people, the country commands a deeply diversified macro society, giving rise to the defining elements of the macrostructure, namely social groups and organisations, institutions and states. The functionality, coordination and dynamics of integration within these elements present Nigeria as a whole with macro parts in need of proper and redefined direction.

Constant references to a seemingly ill-planned fusion of the parts that gave rise to the macrostructure known today as Nigeria regularly point to a known fact that the macro-Nigerian society of today remains shaky, superficial and poorly productive. Paul Eric noted that the "[…] aftermath of colonialism in Nigeria has affected Nigeria's socio-political development, especially in the false marriage with the sobriquet—amalgamation, where different entities were fused to live as one [….]" (Eric 2016). The tone of the foregoing assertion indicates compulsion, and within the concept of compulsion, one finds inherent tensions between rivals and among groups. Such tensions have been the bane of the Nigerian macro-scale structure, and the conflict has always led to structural destabilisation and instability. The manifestations are social groups that constantly build ideologies with rivalry undertones/underpin; the existence of organisations formed with oppositional and sometimes egoistic blueprints of approach; creation of institutions or groups that only promote the unhealthy practice of the survival of the fittest; the purported federating structures (states) remaining deeply immersed in politics developed and tied blindly within ethnic borderlines, etc. These tensions breed havoc in the entire macro-Nigerian structure. One of the founding fathers of the Yoruba extraction did not mince words but observed that "The entity called Nigeria was a mere geographical expression; there is no country called Nigeria because he happened to be first an Egba man, then a Yoruba man, before being a Nigerian." (Ibid., p. 67).

Recent appraisals of the macro-Nigerian State always appeal to the opinion that Nigerian macro institutions develop only along ethnic and religious lines. This situation leaves the country in continuous, unhealthy and frictional combats that hamper sustainable development.

2.2.2 The Meso-Nigerian Society

The key concept within the meso-scale or structure is relationship or interaction. In the meso-Nigerian society, we examine the nature and level of relationship existing between the polity, the people or individuals and the existing macro and micro-institutions that play their way into development. It is logical to say that functional macro and micro socio-political structures would invariably create the platform for a functional relationship and interaction between the structures and agents. In other words, the meso-scale exists in such a manner that it functions towards enhancing efficiency *en route* to a healthy interaction with the agents—human and institutional.

The relationship provided in the meso-scale is likened to the modern concept of corporate social responsibility, which places importance on a healthy and sustainable manner of interaction aimed at the benefit of all participating agents in a relationship, be it social, economic, political, ecological, infrastructural, ideological etc. The relationship has to be predominantly mutual so as to enlarge and expand the choices, capabilities and functioning of the agents in action (United Nations Development Programme (UNDP) 1990).

However, the concrete facts in the Nigerian situation show that the meso-scale structure has failed to form a healthy interaction due to incessant tensions within the macro and micro levels. In some cases, no atom of relationship or interaction exists since the required development structures are either totally absent or comatose through corrupt intents. In this type of scenario, the Nigerian meso-scale presents a broken interaction that stagnates progress, efficiency, innovation and initiative, lending credence to some opinion polls that an average vibrant Nigerian at home remains a failure. Yusuf wrote about this broken interaction in his article on why Nigerians succeed better abroad in the realm of academics. He opined:

> Nigerian institutions operate at isolated realms that disconnect them from the society that they exist in. When students graduate from these institutions, they are thrust out into a "foreign world" like a fish out of water and therefore find it difficult to operate. Institutions must always connect with the society so that the transition from tertiary institution to the workplace is seamless. Both school and society will suffer if students are not exposed to the practical society. (Omotayo 2015)

In the Nigerian experience, this connection suffers gross neglect due to clueless forms of governance, which invariably give rise to deficient policies that thwart Nigeria's capacity to offer sustainable development to her teeming population. The situation leads to a loss of orientation and perspective on the part of the citizenry, especially the youths whose endowed potentials are made helpless victims of man-made greed and exploits. The Nigerian case has never been the unavailability of resources. It is rather the manifest irresponsibility of a few who sit on the policy bank of a nation-state but fail to deliver functional leadership capable of revamping the structure entrusted to their care.

The meso-Nigerian society, therefore, evaluates the grounds for a proper "rule of the game", and such meso level defines the informal dimension of the general social structure since it is the home of the ethos, which connects the individuals and the institutional structures that go into development. Such ethos plays a fundamental role in human development and, as De Muro et al. put it,

> [...] (it is) characterised first of all by social norms, conventions, moral values, religious beliefs, traditions and other behavioural norms that have passed the test of the historical time and that determine the individual's behaviour as well as organisations in pursuit of their aims. These norms are institutions if they become 'standard of appropriate behaviour for actors with given identity'. (De Muro and Tridico 2008)

However, where the formal dimension is deficient or absent in some cases, the informal dimension suffers grievous lack and stagnation. The Nigeria of our research is deeply defective at this level.

2.2.3 *The Micro-Nigerian Society*

The microstructure is fundamentally the level of the more intimate social relationships that bring people face-to-face in varied interactions. To this level belong, families, friendships, work associates, etc. (cf. Ibid., Sociology in Today's World, p. 7). The legal and ethical approach to this level would be concerned with the normative standards that exist and guide such interactions within the microstructures. The microstructure is the realm where rules and regulations are needed among actors for responsive interaction. In line with this, the Nigerian micro society is to be regulated by norms and values originating from her as a democratic entity. Consequently, her micro (normative) structure is spelt out in her most current constitution as amended and promulgated. But the constant question in the Nigerian case remains: How coercive, functional and practicable is the application of the dictates of the constitution within the polarised Nigerian society, where the macro institutions suffer constant Balkanisation and the mesostructure remains trivial, and in some cases, suffer total lack of operational vitality?

The Nigerian micro-institution has long shown a recipe for anarchy and lawlessness. Archives and parliaments flourish with piles and piles of books on norms and regulations, but application stumbles always on the hard rock of inertia, manipulation and favouritism. Confirming this anomaly concerning Nigeria's abuse of norms and rule of law, Ilo observed:

> There must be punishment that is not just in paper. The reverse has been the case in Nigeria for a long time. Criminals, miscreants and misfits walk around the country even when it is obvious to people that such persons should be behind bars. Sometimes government does not even hold itself accountable. (Ilo 2012)

In line with the above-mentioned observation, we would propose an overhaul as a premise for further and deeper work on the Nigerian micro-scale. The cardinal point remains that Nigeria's normative institutions as a nation are weak, non-coercive, corrupt and, above all, abusive to human dignity and rights.

2.3 The State of the Nation: Appraising the Current Nigerian Social Structure

At the time of writing this work, Nigeria had witnessed over sixty years of experience in managing her affairs as a nation-state after her independence from its former colonial master, Britain. Its existing *state of the nation,* which our research tools met, would be examined under the socio-political, socio-economic, socio-cultural, socio-religious and socio-ecological structures and institutions.

2.3.1 The Socio-Political Structure

Nigeria's socio-political structure is a democracy. As a colonial state, it first took up the parliamentary system after independence, following the footsteps of the colonial master Britain, but later remodelled the system after the American system, thereby establishing the executive, legislative and judiciary arms of government. This socio-political structure has been a child of varied forms of plagues and vicissitudes, some of which are incessant political instability in the forms of military coup *d'état* and the consequent enthronement of a military government; incessant change of government and the consequent inability to consolidate policies; a devastating three-year civil war; political corruption, arson and electoral malpractices, etc. The six geo-political zones remain a political creation deeply tied to both mega and minor ethnic lines, which constantly affect and influence political decisions and elective offices, thereby hitting negatively hard on any concept of true democracy in Nigeria.

Furthermore, the nature of the party system typical to Nigeria presents an age-long grip on power by political elites designated as "godfathers" in Nigerian parlance but who constitute an oligarchy coated with the pretence of democracy. Consequently, their political wills are the "grand finale" and need not be challenged. The old leaders, mainly former military rulers, return themselves strategically in civilian apparel, while the new political, social and economic innovators, which Nigeria possesses in abundance, are constantly kept out of the circle of power. In other words, the Nigerian political scene is complex; it is as tribal as it is trivial, vindictive and hijacked by a few.

2.3.2 The Socio-Economic Structure

The socio-economic Nigerian structure presents an economy that has gradually degenerated into a mono-economy—Oil production. Nigeria leads Africa in oil production and ranks thirteenth in the world. The analysis of its economic history in chapter one shows that its economic journey did not begin with the story of oil. Nigeria had a strong agrarian economy. However, since the discovery of crude oil in a commercial quantity, her economic vibrancy has become drastically one-directional. Consequently, the sense of economic diversification was long neglected and played down. This has had and is still having constant grievous effects and debilitating influences on the general economy, putting sporadic pressures on it. It has sometimes led to recessions and Structural Adjustment Programmes, SAP (Pettinger 2019).[1]

The first structural adjustment programme was introduced in Nigeria in 1986 as an urgent response to an existing fact that "[…] Because over 90% of Nigeria's export earnings are from oil, growth in agriculture and manufacturing could offset little of the large drop in purchasing power that resulted from the collapse of oil export revenues […]" (World Bank Group 1994). Thus, the over-dependence on the *petro-dollar* has robbed Nigeria of the real integrated economic strategy needed to strike a healthy balance between external and internal revenues. Consequently, Nigeria has fallen deep into becoming a consumer country, with incommensurate efforts to revitalise local economies and maximise indigenous production. The present Nigerian economic structure is remote from the required integration of the top available five drivers of an economy, namely agriculture, oil and gas, trade, information and communication, and manufacturing. An independent scientific study reveals,

> The Nigerian economy still suffers inadequate diversification in the wake of the first commercial oil production in the late 1950s and the collapse of the nascent manufacturing sector from the mid-1980s onwards. For example, since that time car assembly in Nigeria has ceased and the tire industry has collapsed. The sporadic availability of electricity still forces manufacturers to deploy diesel generators for reliable electricity supply. As consequence this induces uncompetitive electricity costs and frequently leads to factory closings. (Deutsche Gesellschaft für Internationale Zusammenarbeit 2015a)

Despite the attempt, especially in 2013, to re-base the Nigerian GDP and by acknowledging the fact that Nigeria has the largest developing economy in Africa at the moment, according to a recent economic report (cf. Ibid.), the economy still fails to address poverty and development at the grassroots. The recent Brookings Institution's report published under the title "Re: Growth Spikes (5.01%), But People are Hungry?" indicates that Nigeria has overtaken India as the country inhabiting the greatest number of people living below the poverty level, with "about 87 million people in extreme poverty, compared with India's 73 million. What is more, extreme poverty in Nigeria is growing by six people every minute, while poverty in India continues to fall". (Brookings Institution Report 2018). In other words, the present Nigerian socio-economic structure needs revolution, revaluation and restructuring. Our reflections aim to proffer the "why" of such socio-economic deficiency.

2.3.3 *The Socio-Cultural Structure*

The geo-political entity, Nigeria, is socio-culturally diverse and multilingual. This makes its socio-cultural structure complex, with so many governance difficulties. The socio-cultural structure denotes,

---

[1] Structural Adjustment refers to a set of economic policies often introduced as a condition for gaining a loan from the IMF. Structural adjustment policies usually involve a combination of free-market policies such as privatisation, fiscal austerity, free trade and deregulation.

> [...] all elements, conditions and influences which shape the personality of an individual and potentially affect his attitude, disposition, behaviour, decisions and activities. Such elements include beliefs, values, attitudes, habits, forms of behaviour and life styles of persons as developed from cultural, religious, educational and social conditioning [...] These elements are learned and are shared by a society and transmitted from generation to generation within that society. (Akpor-Robaro 2012)

The Nigerian experience presents, therefore, rich socio-cultural dynamics prevalent in setting ethos and values in conflict with one another, sometimes leading to crises of identity, social tensions, unhealthy frictions and confrontations at the national levels and, of course, with conflict-infested ethnic lines. The socio-cultural Nigerian structure of our research harbours varied socio-cultural groups clamouring for self-rule, independence and full access to resource control. Such manner of clamour and often the confrontation with the federal character of Nigeria as a geo-political entity have always resulted in grievous conflicts of interests, intimidation, tribal, inter-tribal and civil wars, flight and displacement of people, bloodshed and loss of lives and properties. Nigeria has a documented history of such socio-cultural conflicts. At the time of writing this work, this struggle for socio-cultural identity and self-actualisation rages in the different parts of the ever-disputed Nigeria's federal structure, with the deepest impacts being felt through the Boko Haram in the North-East, Fulani-Herdsmen escapade in the North-Central, the Niger Delta emancipation struggle in the South-South, the IPOB-Biafra revolution for self-actualisation in the South-East, as well as the Oodua People's Congress in the South-West.

The federal character ascribed to the Nigerian structure as a polity, therefore, is a political "ought-to-be"—a utopia—to address its multi-socio-cultural dynamics. However, federation (or the alternative model in the minds of so many, "confederation") remains a developing ideology for bridging incessant socio-cultural conflicts and crashes in Nigeria. This academic work unconditionally confronts the already tensed-up reality of Nigeria's socio-cultural instability, which threatens her aspirations towards nation-building as well as all other expected matrixes for sustainable development.

2.3.4 The Socio-Religious Structure

Like many other human races, Africans, in general, are deeply cultural and religious, and to describe this passionate religious experience in Africa, Mbiti gave a classic presentation of Africans in relation to religion and belief thus:

> Africans are notoriously religious, and each people has its own religious system with a set of beliefs and practices. Religion permeates into all the departments of life so fully that it is not easy or possible always to isolate it. A study of these religious systems is, therefore, ultimately a study of the peoples themselves in all the complexities of both traditional and modern life. (Mbiti 1969)

The Nigerian geo-polity is densely networked in her socio-religious structure, with Islam dominant in the North and Christianity dominant in the South. Followership of indigenous traditional religions in terms of rituals and rites has declined over the years due to contact with the Islamic Arab world, the coming of the missionaries and the influence of colonialism. However, even in the practice of the various newly found religions, the Africans, in general, and Nigeria, in particular, show varied peculiarities, which stem from what is indigenous to them as a people. Within the northern and southern zones, the sporadic presence of followers of both religions, Islam and Christianity, is a recognisable social factor due to internal migration and other socio-economic factors.

Furthermore, both religions have different sectarian denominations within their confines that identify themselves with one form of teaching and belief or another. In some extreme cases, such sects or denominations take their practices and beliefs to the level of radical fanaticism or fundamentalism. Nigeria possesses enough of such experiences. In other words, the socio-religious structure of Nigeria has "brutal" records of conflicts, religious wars and bloodshed, sectarian violence, denominational and interdenominational disputes and the subsequent break-ups, denouncement, abandonment and re-grouping. In the Nigerian case, one witnesses the proliferation of religious interest groups alongside a Nigerian society that shows deep ethical and moral bankruptcy. One wonders then what the purpose of religion is and why a nation-state like Nigeria is almost drifting away despite its socio-religious institutions, where each claims to be the moral voice and conscience of the people. However, the reality is that the realm of religion, church, faith or denomination witnesses deception and exploitation within the current Nigerian polity. This is due to the failed system of governance, which has made the masses prey to all approaches to seeking and waiting for miracles, even for the slightest challenges of life, the solution of which lies within the human person's capability to think. The abuse within the socio-religious structure of Nigeria has unfortunately given rise to varied forms of false defensive mechanisms of so many of her citizenry, who have lost every sense of self-determination in confronting and facing our challenges squarely. Consequently, many average Nigerians have developed the attitude of attributing every daily life challenge to the work of an esoteric evil machination that must be cast out through bickering in prayer, crusades and exorcism. However, it must be noted that the foregoing evaluation of the socio-religious structure in Nigeria is never to be understood as casting aspersion to faith, religion and churches in Nigeria. On the contrary, it is meant to draw attention to our various structural defects, which invariably keep Nigeria stagnated in progress and development. Until they are nipped in the bud, sustainable development benefits will continue to elude the acclaimed "giant of Africa".

2.3.5 The Socio-Ecological Structure

Discourse on Nigeria's socio-ecological structure examines the *Anthropocene* issue. Using the concept within the socio-ecological structure in Nigeria is an attempt to examine the recent age of human activities in Nigeria and the impact they have on the ecosystem. Thus, the socio-ecological structure of Nigeria presents a geo-polity blessed with the gifts of nature, but unfortunately, with so many bruises due to the unguarded activities of man. Popular opinions, both national and international, show unanimity in decrying the extent to which human activities negatively impact the Nigerian ecology. In the Nigerian case, such negative impacts are seen in the forms of over-dependence on and improperly regulated forms of crude oil exploration. Consequently, oil exploration has degenerated into mere devastation and gradual destruction of the natural capitals. They are seen in the ecological woes, environmental degradation and aquatic poisoning due to incessant oil spills in Nigeria. They manifest in the uncontrolled carbonation of the Nigerian air due to power-generating sets emissions and other forms of usage, in the persistent poor canalisation in almost all parts of the country, which unleashes havoc in many communities through erosion and other related hazards. Other related issues are the attacks on biodiversity and man-made desertification, etc. With the reality of such states of emergency in the Nigerian case, it must be noted with reference to a previous research work, that,

> The Ecosystem is a build-up of interaction between the biotic and abiotic components of nature. In other words, it is the relationship between the living and non-living elements of nature interacting as a community. Human life, therefore, is intrinsically tied to this eco-world in a symbiotic way and in this symbiosis, the relationship becomes that of 'give and take'. The dependence on the provisions of the ecosystem automatically places on man the responsibility to maintain this source of provision – the ecosystem (itself). (Nwosu 2016a)

All references to the socio-ecological structure in Nigeria always present the outcome of the scientific research with demands for an urgent change of attitudes and approaches towards nature, the environment and the general ecology. In the opinion of this work, therefore, the failure to reflect the right attitude and approach in policies that have ambient on the environment in the Nigerian system would be identified as a structural defect—a structural sin. Simply put, Nigeria has various written socio-ecological and environmental laws to checkmate excesses of extreme Anthropocene. However, these laws are abused and stagnated due to embedded socio-economic corruption, which has deadly grips on the Nigerian ecology through national and international players.

Evaluation: The Nigerian Case of Structural and Institutional Weaknesses

After over sixty years of accepting her independence from British colonial rule, it is painful that a nation like Nigeria still marks time in the lower echelon of the development indices in general. Structural, infrastructural and basic institutional weaknesses have long accounted for the major setbacks that threaten the nation's corporate existence. The 2018 Human Development Statistical Update, as shown in the appendix, reveals a disappointing account of Nigeria's structural failures and neglect, which invariably have adversely affected her human development index. According to the informative data, countries are classified under the evaluative categories of (a) Very High Human Development, (b) High Human Development, (c) Medium Human Development and (e) Low Human Development. A country like Nigeria, with rich human and natural resources, ranks 157th with reference to the general parameter of development in a globe with 195 member countries and territories. Nigeria takes the

6th position within the category of countries with low human development structures (United Nations Development Programme (UNDP) 2018). This can only be an indication of a fundamental dysfunction that needs to be nipped in the bud. Commenting on such apparent humiliation and shame for a country like Nigeria, Saidu opined:

> No economy has ever grown to the age of maturity if the fundamental body structures of development parameters are under the tutelage of misguided personalities, because bad leadership and economic development are inversely related just as poor management and development are mutually exclusive [....] The leadership environment in Nigeria has (only) become a money-making venture and an avenue for easy stealing of national wealth with political protection [....]. (Saidu 2015)

The weak Nigerian structures at the macro, meso and micro levels have always given the nation heavy bruises because of incessant crumbling and collapse. Institutions that were built during colonial rule tended to last longer than post-independent policies; established corporations for a more distributive and more stable economy either dilapidated or went extinct; local economic institutions were sacrificed in preference to the mega and multinational networks for crude oil production. The result is that a nation that had *agro institutions* as its economic mainstream at one time now stands the risk of being unable to produce and feed its population. Manpower remains underused, talents are stagnated and youthful capabilities and initiatives remain underdeveloped.

It is exactly on this platform that we would set the tone of our journey in searching for the core causes of these gross *dysfunctions or defects* in the Nigerian polity. In other words, it would be a scientific effort to delineate the dysfunctional structures in the forms of policies, plans, decision-making, formulations of rules and regulations, practices, behaviours, thoughts and attitudes, as well as ideologies, which have worked or are still working against the principles of sustainable development, and invariably against an inclusive capacity building, in other words, the empowerment of the "Mmadụ"—the human person.

2.4 Internal Structural Weaknesses in the Nigerian System as "Structures of Sin"

In this sub-section, we will try to develop our central concept, "structures of sin", which serves as a unifying concept in examining the social, political, economic and ecological maladies plaguing Nigerian society. We will begin with a theological approach to respect the origin of the concept before getting down to applying it to real situations and states of affairs within the Nigerian polity.

2.4.1 The Concept "Structures of Sin": A Moral and Theological Explication

Theology, which focuses on studying the nature of the divine and the realm of divinity in contrast to humanity, would always, in human terms, be associated with such attributes as perfection, goodness, excellence, justice, love, etc. Thus, theology would study the purest attributes in relation to a *terminus ad quem* called God in the Christian world of belief. In other words, theology studies God as a being per excellence, a principle that is and remains the foundation and the vital force of all that exists. In the realm of such pure principles reside the pure virtues, which, as the theory of knowledge would prove, are the sources of all other abstractions that define humanity and the natural world. Nevertheless, although theology reflects on the supernatural, it does this not just in an abstract manner but rather seeks to bring the pure attributes to reflect on concrete religious experiences. In the context of our research, this religious experience or knowledge is the realm of the Christian doctrine—the Christian search for knowledge and applicable responses to the social and anthropological realities of human societies and existence. Therefore, appealing to theology would be an epistemic approach to the right principles of being and living. Theology would seek in this manner to emphasise the basic ideas and attributes of right reason in order to expose the implicit negativities which are consequences of human culpability, especially when the realms of ethics and morals are wounded and harmed by *privation* or absence of the good, of course, where the good has every right to exist.

In the light of the above common sense understanding of theology, the focus here would be first to understand sin as a wound dealt to nature when the right principles are deviated from, thus creating a moral gap that can only be compensated by appealing to the "good"—*restitution, reparation.* Therefore, we appeal to John Paul II's concept of "structural sin" to explain the draconian presence of the "privation of the good". In other words, to delineate the ethical and moral deviations within the Nigerian polity.

2.4.2 Understanding the General Theology of Sin: Hamartiology

The fact that a conscious feeling or manner of weakness (moral weakness) exists in human nature has led people of different epochs to seek interpretations, explanations and understanding of a purported "scar"—a wound associated with the nature of man. Thus, in the Christian worldview, "...the People of God in the Old Testament...tried to understand the pathos of human condition in the light of the "fall" narrated in Genesis...." (Catechism of the Catholic Church 1999). Search for forms of causal pathos has constantly shaped histories, narratives, writings and the human quest for knowledge. Even the New Testament Christian worldview, which is best encountered in the written Gospels and the teachings of the Father of the Church, began its quest with the account of the Temptations of Christ at the very beginning of his ministry on earth. Consequently, the account of Christ's encounter with the devil—or Satan, or the natural affinity to moral weakness—further exposes the New Testament's efforts to understand this undeniable part of human nature, which plays itself out in varied covert and overt forms.

Sin generally, in Christian parlance, is a breach and an omission. Christian Theology analyses this breach in action through a branch of theology called Hamartiology. Hamartiology thus studies sin, its origin, its effects on humanity and its result after death. The issue of original sin in connection with the biblical account of the cradle of life rooted in the creation and existence of Adam and Eve as well as the concept of habitual sin, which explains the human person's gullibility to weakness leading to actions that cause moral failures, form the basics of hamartiological interpretation of humanity and its moral existence. However, a deeper exploration of these two areas would be grossly outside the scope of our research. Our focus here would be on hermeneutics and etymology, which would rather offer an epistemic effort in understanding the words and the meaning associated with hamartiology.

Etymologically, hamartiology takes its root from two Greek words, "hamartia" and "logos". In its verb form, *hamartia* means to miss the mark, while *logos* defines terms like opinion, word, speech, reason or discourse—a discourse, opinion or reason about a given idea. Hamartiology literally becomes a discourse on "missing the mark". As a theological concept, it seeks to explain *why* we miss the mark, *how* we miss the mark and the *consequences* of missing the mark—all in relation to the Christian logos, God, and of course, God in relation to the natural (moral) order. Simply put, Christian theology understands sin as a breach of righteousness towards God and his ordinances. In other words, it would mean a breach in action or an omission of action in relation to God and his revealed ordinances that bind humanity in obedience to him. Therefore, departing from this bond would mean missing the mark on the redemptive or moral path that leads to righteousness and an authentic life. Basically, in theology as well as in theodicy, this proclivity towards missing the mark is viewed as an intrinsic weakness that receives the concept of "original sin", especially in Thomas Aquinas and later in St. Augustine, who separated it from the concept of a habit since it is a privation, simply put a lack or loss (Knight 2017). Privation, therefore, means a corresponding diminution of the "good", which is the intrinsic characteristic of all of nature. The *naturally good* builds up the entity upon which the action of privation happens, either by commission or omission of an agent that diminishes, detracts, imperils or corrupts the "good". In this way, privation retains within itself an embedded form of *imputation* in relation to the agent or a force(s) that militates against the good, thereby causing its diminution or corruption, defect or dysfunction.

However, whether the susceptibility to the human weakness that leads to missing the mark is by inheritance, imputation or one's own personal choices remains a vital issue for the theological exegetes to crack. Fundamental to our reflection would be the recognition of the fact that humanity, in general, has an embedded weakness, either by omission or commission, which has the capability of obstructing or neglecting the right principles of behaviour, thereby negating the good to enthrone evil or, using a more general concept, moral deviation. Thus, a holistic concept of sin acknowledges a vacuum that is being created whenever humanity, individually or collectively, chooses to miss the mark in the moral path towards the truth, the objective and the good of life. That means, where societies exist, laws and ordinances come in to checkmate and guide such human potent pushes towards the unacceptable—towards moral negativity.

2.4.3 The Hamartiological Trio: The "Why-How-Consequence" Approach

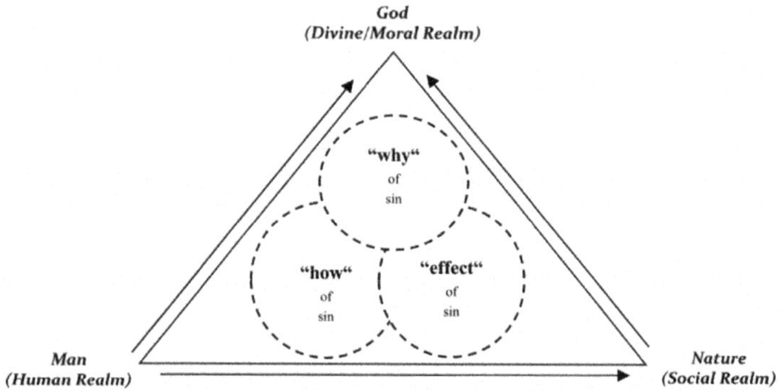

Illustration mine

The diagram above describes the three interdependent elements that intrinsically relate to any case of "sin" or socio-ethical aberration. The triangular relation shows the interconnectedness existent within the divine/moral realm, humanity and the social order. Thus, the *why-how-consequence* approach forms a trio towards a deeper understanding of hamartiology. Why does humanity miss the mark? The question becomes fundamental and can only appeal to humanity's real world for a valuable answer. The "why" of missing the mark presupposes an obvious possibility rooted in various art of weaknesses rooted in human daily life and experiences. John Paul II calls this weakness "the dark forces" (Paul 1984). This weakness manifests itself in every aspect of life with varied instinctive power of control over individuals. It influences concepts, ideas, interactions and groups. Consequently, the "why" of missing the mark (the "why" of sin) responds to both the conscious and unconscious, overt and covert phenomena of human inclinations, as well as affinity towards harming or deviating from the good. St. Paul tends to affirm this reality when he writes: "For I do not do the good I want, but the evil I do not want is what I keep on doing" (Rom. 7:19—RSV).

As an insight into the reality of this inherent weakness in humanity, Keltner of the University of California reported how he would always request his students to complete a statement like "Human nature is…" during most of his seminars. He would then later state emphatically that 70% of their responses refer to some form of selfishness, competition or aggression (*Dacher Keltner on the Evolutionary Roots of Compassion:* www.youtube.com). That means, when humanity is bent towards and influenced by such inclinations, as is evident in Keltner's experiment, the tendency would be a creation of conflicts and tensions which *private* the good and *enthrone* evil.

The "how" of missing the mark describes the reality of an "innate passion for deviating" within humanity, such that the natural order and existing principles of natural law receive, often and on, breaches that have the capacity to disrupt or distort. With such an innate passion for deviating, the "how" of missing the mark (sin) manifests characteristics like: "self-sufficiency instead of faith; self-will instead of submission; self-seeking instead of benevolence and self-righteousness instead of humility" (Strong 2017). All these characteristics combine to reveal humanity's inherent weaknesses and deviances. They constitute the combined negative powers and forces which make moral deviations happen. However, such innate inclinations and their negative influences are not "absolutes" in themselves. In other words, humanity and the human person (the moral agent) reserve the capacity and power to make choices—the ability to hold such inclinations in check.

The third aspect of the hamartiological trio deals with the "consequences" that ensue from *hamartia—missing the mark—Sin.* Elementary Christian doctrine would refer to the consequence of sin as meaning death. Thus, in this sense, a hamartiological consequence has an eschatological connotation, namely, death as punishment with hell fire at the end of life; death as damnation ascribed to the phenomenon of the loss of both the body and soul at the end of time, due to the break of union and relationship with the creator—God.

A broadened analysis and understanding of the consequences of sin also treats its social, psychological, physical and legal consequences. The socio-phenomenal level explains the social and physical impacts of sin. Here the consequence of sin becomes an introduction of a wound and hollow in the order of nature and natural law, creating a vacuum which nature abhors. When this is done, the order of nature and the *pure principles* like Justice, Love, Equity etc., become thwarted, giving rise to conflicts and tensions that wound the internal functions of the social system, thereby breeding structural dysfunctions, defects and incongruities. In this sense, we speak of social sins, which are fundamentally rooted and abstracted from personal sins. Mohandas Gandhi would write about seven such social sins thus: "[…] Politics without principles; wealth without work; pleasure without conscience; knowledge without character; commerce without morality; science without humanity; religion without sacrifice" (Gandhi 2013). Such social sins and their effects would be one of the virtual preoccupations of this academic work, whereby we seek to identify the ways through which they affect Nigerian society and its general polity.

The psychological consequences of *hamartia* are manifest, especially in the creation of insecurity, instability, suspicion, distrust, feelings of guilt and shame, as well as trauma. They usually assume varied forms while they act as wounds done to the humans and psyche, which may invariably lead to depression, disruptive behaviours, forced affinity to lethal and weird thoughts as well as attempts to execute the same. In other words, the psychological consequence introduces disorder in the operative general approach within a structural unit, namely within the human society. They unite to build up behavioural and emotional instabilities in the human agents—*ndị Mmadụ.*

The legal consequences of missing the mark (hamartia) have both causal and punitive aspects. Thus, the legal consequence, in its causal aspect, examines the injury done to the normative institutions of human society, the disruption of the natural order, as well as the instituted ethos and norms upon which a legal entity is built. The laws and ethics that guide human societies exist basically to checkmate breaches at this level. Thus, breaches at the causal levels are seen in the forms of social crimes, juvenile delinquencies, theft, burglary, murder, arson and manslaughter, misappropriation of public funds and properties, human rights violations, etc. On the other hand, the punitive aspect of the legal consequences results in the loss of basic rights and freedom on the part of the "agent" who misses the mark. Such loss of rights and freedom can be inflicted on an agent, be it human or institutional, through arrest, prosecution, detention or incarceration; through court martial with reference to the armed forces, or in some extreme cases, through direct capital punishment in the case of a culprit killed at a crime scene; etc. Such punitive consequences are meant to serve as deterrence, restitution, punishment, prevention, rehabilitation, admonition or, using a dogmatic theological concept, "the making of satisfaction for sin"—a sense of restitution or of mending what has been broken.

2.4.4 *Other (Anti-)Deontological Concepts Associated with Hamartia*

In ethics, deontology studies the moral duty considered requisite for actions viewed as morally acceptable and which carry in themselves the power to command imperatively. They are actions carried out just for the sake of duty and for the good they represent. In this regard, one talks about justice, the sense of common good, respect for life, human rights and dignity, love, equity and equality, etc. Acts contrary to duty are anti-deontological, and branches of human knowledge, especially moral philosophy and moral theology, use different but complementary concepts to articulate such anti-deontological moral deviations. Some of the concepts discussed here would lay a foundation for the development journey when dealing with the human acts that hijacked Nigerian society and kept the system under-productive.

2.4.4.1 Ungodliness and Lawlessness

Ungodliness and Lawlessness would define the attitudinal act of "missing the mark"—hamartia. While one is in relation to God, the other is in relation to humanity and, in general, nature. In the Christian understanding, sin would mean a departure from God and his ordinances made known in history through the prophets and brought to perfection in Jesus Christ. Acting in an ungodly manner means breaking the vertical relationship, which means in relation to God. In its varied forms, this causes wounds and breaches as well in the horizontal relationship, in other words, with humanity and nature. The Christian holy scripture gave a classical Judeo-cultural allegory of this breach in the vertical relationship, which at the same time has a grave consequence on the horizontal, with the narrative of humanity's experience at the building of the tower of Babel. Reflecting on this in his Apostolic Exhortation, John Paul II puts it in this form:

> Why did the ambitious project fail? Why did "the builders labour in vain?" They failed because they had set up as a sign and guarantee of the unity they desired, a work of their own hands alone and had forgotten the action of the Lord. They had attended only to the horizontal dimension of work and social life, forgetting the vertical dimension by which they would have been rooted in God, their creator and Lord, and would have been directed toward him as the ultimate goal of their progress. (Ibid., Reconciliatio et Penitentiae, no. 13)

Further, John Paul II would see so many similarities between the experience of Babel and the tragedies of humanity today: a world believed as a global village, yet the gaps between people and human institutions keep widening on a daily basis; a world that seemed to have bridged the gap in interpersonal communication, almost putting communication devices in every individual's hand and household, yet divisions and enmities are on their increase. In sum, in Ungodliness, what people embarked on as "a symbol and source of unity" plunged them into being more scattered than before, more divided in speech, more divided among themselves and increasingly incapable of consensus and agreement.

The analogy above should and must be strictly made only in our context of proving the reality of this "sense of breach" in the vertical relationship. When taken outside this context and further, one faces relevant criticisms and exegetical hitches with the yet-to-be-answered question about whether the division at Babel has more advantages than disadvantages to humanity. However, this would be outside the scope of our research.

2.4.4.2 Unjust Rebellion

Hamartia, as rebellion, presents an attitude of rejection. It is a culpable ignorance which keeps the victim trapped in refusal and disobedience towards the just order of reality. H. G. Wells, a then British writer, would see the rebel as a recalcitrant, especially when the rebel's intent for rebellion is directed to the just and moral order. Therefore, laws should be made and enforced for such a rebel since recalcitrance is a negative passion that manifests itself in an unwillingness to obey orders or follow the right principles of behaviour. Rebellion can be an attitude of the mind, which in varied cases becomes covert, especially at the socio-phenomenal level. The rebel, therefore, misses the mark both in relation to God/moral order and the natural order when the intent of rebellion goes against what is just, moral and ethical.

2.4.4.3 Treachery

Treachery is an attitude towards betrayal. It is an overt or covert attitude that implants or awakes in another person the painful feeling of having been deceived, betrayed, tricked, disrespected, abused, cheated upon etc. Hamartia as treachery deeply harms such moral concepts as trust, love and authentic relationships. In a socio-phenomenal realm, it breeds sycophants and creates both real and imaginary enemies, which generally keep the societal structure tensed, aimless and stagnant. Thus, in treachery, the perpetrator misses the mark by creating a vacuum for distrust in a relationship with the other that is supposed to be mutual, inclusive, purposeful and reciprocal. Thus, the treacherous person abuses the tenets of reciprocity, which should bring the partners mutual advantages of common interest instead of pain and fear.

2.4.4.4 Covetousness

Covetousness is a type of greed or inordinate passion for obtaining what belongs to another. Such passion leads to swindling, forgery, manipulation, lies and perjury, or, in some extreme cases, murder. The burning passion here misses the mark and injures another's rights to property and possession in general. Covetousness has no positive connotation. It creates an injury, violates or breaches a system, and it is both an infringement and an infraction. Covetousness, therefore, is a form of idolatry, which has the material things of nature as the exclusively targeted objects.

2.4.4.5 Debauchery

Debauchery is excessive indulgence in pleasurable acts, which gives rise to situations like alcoholism, bestiality, drug addiction, etc. It amounts to an abuse of freedom and insults human rationality. Debauchery has a co-relation of effects and could be social, political, economic, etc. When debauchery is socio-political, it leads to the abuse of power, money and material gains. It leads the political debaucher to use his position of office to abuse and exploit the people he or she is supposed to lead. Social debauchery harms decency and moderation, while economic debauchery makes corruption, theft and misappropriation look like acceptable social conventions.

2.4.5 Structures of Sin: A Brief Reflection on the Use and Application of the Term

There are fundamental ideas associated with the concept, *Structures of Sin.* First, the reference to structures refers to social and experiential phenomena within the human society and realm of knowledge. Second, is that the reference to "sin" is not just confined or limited to the biblical account of "human fall" and its eschatological consequence—death. Sin, as a concept and in our context, has a relational character between an "anthropological object"—an individual, a human person or a group of persons, and a social phenomenon (a system), an organisation or other forms of social structures. The two realms in this context, namely the anthropological and social, remain inseparable in understanding the concept, *structures of sin.* Thus, in relation to the focus of our research, namely the human person and the social situations surrounding his or her life, this core concept would be dealt with as a social and anthropological specimen.

With the above-mentioned focus in mind, the anthropological character of the concept of sin recognises as fundament the natural attribute or ability of the human being (person) "to err"—attitude to making mistakes, weakness and inclination towards deviation, etc. Joachim Wiemeyer refers to this as *Irrtumsfähigkeit* (capacity for error) (Wiemeyer 2015). This ability manifests in factual errors that originate first in individual persons but at the same time prevail against known values, norms, ethos or conventions, etc. In reference to the human person and his or her actions, this attitude would often be referred to as moral misconduct or malpractice, otherwise known as bad conscience in theological language. The factual content of this human weakness, especially in the anthropological realm and following the opinion of Wiemeyer is a combination of factors stemming from spite, striving for power, greed, etc., predominantly when people want to realise their own interests, goals and advantages at the expense of other people, but possibly also at the expense of nature itself. They try to dominate people, oppress them, exploit them or otherwise use them for their own purposes. If they resist, they are willing to kill, injure, torture or psychologically torment people (Ibid.).

Furthermore, because the human person possesses both natural inclinations, namely, the propensity for the *morally good* and the propensity for the *morally bad*, both anthropological inclinations elicit, as a common feature, concrete social consequence (*gesellschaftliche Konsequenzen, Joachim Wiemeyer*). With the reality of social consequences, we, therefore, refer to established structures and institutions, in other words, the social structures which constitute the direct object(s) of the consequences of this individual attitude to err, to deviate (*der Hang zum Bösen, Joachim Wiemeyer*)—in companies, financial industry, social organisations, governmental and political institutions, as well as in the church. Nevertheless, while this individual negative attitude in isolated cases, may not qualify enough to constitute a structure, Wiemeyer spoke here about "constellation", whereby the individual negative inclinations, especially if they are not discovered, sanctioned and/or controlled, can solidify and spread to others. Thus, such cases of constellations, in other words, consolidation or solidification of the human propensity for evil, elicit predominantly adverse effects in any system they are found, and therefore, they build up structures of sin—structures that are dysfunctional to or within a system. To this effect, John Paul II spoke in the plural in reference to this *constellation network* when he referred to structures of sin as being rooted in the concrete acts of individuals who introduce them, which would eventually grow stronger and spread. Their individual presence unifies and relates with one another in emitting hydra-headed characteristics that attack the system or structure. The National Programme on Technology Enhanced Learning, NPTEL listed such general characteristics, namely (a) They always have injurious consequences for all sections of society; (b) They always take the form of deviations from the ideal situation or from the values of a society; (c) They have a predominantly common basis of origin; (d) They are predominantly social and political; (e) They enthrone pathological social conditions; (f) They are intertwined and interconnected (constellation) and (g) They require collective effort or approach to be able to proffer solutions to them.[2]

Again, the influence of these negative constellations on a social system or structure has been known to determine a system's progress or retrogress directly, in other words, its development or backwardness. However, where functional measures keep the constellations in check, such a system's development process would be proportional to its norms and rule of law. On the contrary, where a system becomes badly infested with uncontrolled negative moral and structural constellations, the development process would be dysfunctional to the dictates of norms and the rule of law. To this effect, one speaks about such structural constellations being obstacles to progress, development, innovation, etc. Thus, the structures of sins considered as obstacles mean that they are violations of the core values and conventions behind a given system or structure. The control over such obstacles in a given system, according to Wiemeyer, is both a duty and a prerequisite. This is because known workable societies have been so designed that they possess the means to control and curb people's propensity for evil through moral values, social control, economic incentives (e.g. against laziness) and social institutions such as criminal law (cf. Ibid.).

2.5 The Social Teachings in Relation to Structures of Sin

The Church's social doctrine contains bodies of teachings under varied themes, through which the church makes its presence in human society more concrete, effective and participative. These teachings carry principles of objective truths with moral obligations that are not just mere human designs but are rooted in the principles of right reasoning and natural law. These teachings are predominantly handed down

[2] *As adapted from the National Programme on Technology Enhanced Learning, NPTEL. Humanities and Social Sciences: Issues and Problems. Joint Initiative of IITs and IISc, Founded by MHRD. Available via https://archive.nptel.ac.in/courses/109/103/109103024/.

as Encyclicals, Apostolic Exhortations, Messages, Homilies, etc. In these teachings, the (catholic) church bares its mind unequivocally on the social issues of our time, through in-depth and inspirational discourses on diversified areas and issues affecting society, human dignity and rights, justice, political situations, the conditions of the poor in our world, issues of morals with obligations and responsibilities that follow them, issues affecting the general life in the society, with critical regard to human science and technological developments, etc. Two of such social documents would help to trace the originality and further analysis of the concept of *structural sins*, namely *Reconciliato et Penitentiae* and *Populorum Progressio*, both issued by John Paul II.

2.5.1 The Originality of the Concept "Structures of Sin" in John Paul II

John Paull II, now St. John Paul II, goes down the annals of history as a pastor and pontiff of outstanding character and deep faith. His philosophical and theological views have given both the twentieth and our twenty-first centuries invaluable teachings on Christian humanism, social justice and doctrines of the faith.

As an ardent preacher of social justice, the originality of our concept, *structures of sin*, is attributed to his teachings, first in and through his Apostolic Exhortation, Reconciliatio et Penitentiae, and later reflected upon deeper in the encyclical Populorum Progressio—On the Progress of the People. An applied and reflective study of these two documents would help to set the platform upon which our journey of discovering the causes and effects of Nigeria's covert and overt structural defects would be laid.

2.5.2 Reconciliatio et Penitentiae: On Reconciliation and Penance in the Mission of the Church

At the time of writing, this post-synodal Apostolic Exhortation has, for more than thirty-eight years, borne witness to the truth, which it once communicated to our world on the eve of the year 1984. However, years after the issuance of this exhortation, the world that was addressed by its teachings and admonitions still marches on and remains buried in both sporadic and incessant social, political and economic confusions, bilateral divisions, sectarian violence, threats of wars and rumours of wars, etc. Perhaps most excruciating among these threats for our twenty-first century is the social and psychological plagues which are the results of incessant acts of terrorism and the never-ending widening gap between the rich and poor. The exhortation took up this pertinent situation of the world from the very first statements and never minced words in naming it: "A Shattered World". Thus, John Paul observed:

> [...] divisions are seen (in our world) in the relationship between individuals and groups, and also at the level of larger groups: nations against nations and blocs of opposing countries in a headlong quest for domination. At the root of this alienation, it is not hard to discern conflicts which, instead of being resolved through dialogue, grow more acute in confrontation and opposition. (Ibid., Reconciliatio et Penitentiae, no. 2)

The above words of the Pope, concerning the unfortunate presence of divisions and the consequent in-built alienations, prepared from the on-set the very foundation upon which the idea of a worldly structure that is infested with man-made socio-phenomenal maladies, in other words, man-made dysfunctions, structural defects or structural sins would be argued upon and discussed. These defects are as personal as they are social. Such structural impasses eat deep into the general social phenomena, giving rise to various societal ills identified with the total loss of respect for basic human rights and the persistent neglect of human worth and dignity, even at a time in our history when our world and society have seen and developed powerful and convincing rhetoric on issues of human rights, freedom and dignity. Everywhere in our world seems punctuated with threats against freedom and peace, seen in concrete forms in racial, cultural and religious discrimination; unjustified socio-cultural repression and the so-called practice of cleansings; lethal forms of violence, misguided interests and misappropriation of resources in the arms race and atomic weapons stockpiles. All of these happen alongside apparent abject poverty and death due to hunger, malnutrition, lack of proper access to clean water, healthcare, etc. (cf. Ibid.).

Tracing, therefore, the root cause of such a world trapped in structural disunity, the exhortation calls for recognition and acknowledgement of the tragedy of man and the concrete circumstances of sinful humanity. Such recognition would be a journey to self-discovery, the purpose of which is to bring humanity to the conscious reality of its failed responsibility seen in the "personal sins", which are intrinsically tied and concretised in "social sins". Giving a Christian view of the need for this journey to self-discovery, the exhortation teaches:

> In the concrete circumstances of sinful humanity, in which there can be no conversion without the acknowledgement of one's own sin, the church's ministry of reconciliation intervenes in each individual case with a precise penitential purpose. That is, the church's ministry intervenes in order to bring the person to the "Knowledge of Self" – in the words of St. Catharine of Siena – to the rejection of evil, to the re-establishment of friendship with God, to a new interior ordering to a fresh ecclesial conversion. (Ibid., no. 13)

The journey into this true self-discovery exposes the truth about human nature in relation to sin, which is seen as a product of human freedom but is deeply connected to what the exhortation calls the dark forces. The dark forces, therefore, are "[...] forces at work which place it beyond the merely human, in the border area where man's conscience, will and sensitivity [...]" (Ibid., no. 14) are infested. Such contact creates a relationship and a union between the person and other social factors, which unite to form structural dysfunctions that weaken a whole system. First, it is always a personal act deeply connected to conditions, incitements and influences from numerous and powerful factors (cf. Ibid., no. 16), but they are not completely reduced to these external factors so as to deny the role of the human subject as an accomplice.

Much relevant to us is the end product of this interrelationship located in the concept of *social sin,* which has consequences both on the individual and the general structure at large. The concept "structures of sin", in the opinion of John Paul II, has its roots in social sin and such social sin, embedded in the general structure, carries with it three general characteristics, namely: (a) it has an effect or consequence on others; (b) it breaches the natural law of love for the neighbour; (c) it affects relationship, justice and freedom in the human communities (Ibid.). Understanding, therefore, structures of sin in John Paul II's exhortation relates the concept to all breaches of the natural law of love, conceived and perpetrated by the individual but which invariably has consequences outside of him, such that it affects others and the community in general. These socio-structural sins, according to the exhortation, include, among other things,

> [...] sin against love of neighbour; [...] sin against justice in interpersonal relationships, committed either by the individual against the community or by the community against the individual [...]; sin against the rights of the human person beginning with the right to and including the life of the unborn or against a person's physical integrity [...]; sin against other's freedom to believe in God and adore him; [...] sin against the dignity and honour of one's neighbour [...]; sin against the common good and its exigencies in relation to the whole broad spectrum of the rights and duties of citizens; sins of commission or omission, on the part of political, economic or trade union leaders, who though in a position to do so, do not work diligently and wisely for the improvement and transformation of the society according to the requirements and potential of the given historic moment. (Ibid.)

These sinful and dysfunctional structures weaken the system. They create internal tensions and strife that make authentic development either lacking in content and application or become a total failure. That they exist in almost all human societies is already a fact a posteriori—based on experience. Further effects and the danger they pose to the general structure become deeply troubling when the actors in the socio-phenomenal scenes portray the attitude of total indifference, engulfed in the passion for persisting and continuing the status quo, thereby showing direct or indirect disregard for the dictates of moral conscience. This leads to a loss of the sense of sin, which Pius XII called "the sin of our century" (Pius 1946). The exhortation teaches the following:

> Nevertheless, it happens not infrequently in history, for more or less lengthy periods and under the influence of many different factors, that the moral conscience of many people becomes seriously clouded; (and the root causes and sources are identified in): …a rejection of any reference to the transcendent in the name of the individual's aspiration to personal independence; from acceptance of ethical models imposed by general consensus and behaviour, even when condemned by the individual conscience; from the tragic social and economic conditions that oppress a great part of humanity, causing a tendency to see errors and faults only in the context of society; finally and especially, from the obscurity of the notion of God's fatherhood and dominion over man's life. (Ibid., Reconciliation et Penitentiae, no. 18)

It means, therefore, that a society's structure or system that is bereft of citizenry with optimal value for the moral conscience must invariably find itself trapped in chains of structural social sins or dysfunctions which, as a result, remain the products of three internal dysfunctional ideologies of modernity namely: (a) Secularism, an ideology "[...] completely centred on the cult of action and production, and caught up in the heady enthusiasm of consumerism and pleasure-seeking, unconcerned with the danger of losing one's soul" (Ibid.); (b) the wrongful evaluation of certain findings of the human sciences with certain affirmations of psychology, leading to a total rejection of the feeling of guilt or admittance of shortcoming within the realm of the *human act*; and (c) historical relativism, an "[...] overthrowing and downfall of moral values" (Ibid.), leading to a relativising of the moral norm, and consequently denying its values (cf. Ibid.). In all of these, the dictates of the moral conscience remain thwarted and violated. The actors in the forms of persons, policies or institutions persist in the discoveries of new ways and strategies to perpetuate further and consolidate the status quo while at the same time rejecting any ideology or philosophy that imputes responsibility or culpability on the human person.

2.5.3 Solicitudo rei Socialis: On the Twentieth Anniversary of Populorum Progressio

This document should serve as one of the primary sources of our research. It was written and issued in 1987 by Pope John Paul II in commemoration of Paul VI's Populorum Progressio, issued in 1967. The name, Solicitudo Rei Socialis, would be translated as *social anxiety*. Thus, the document came as a social response from the church to the growing anxieties threatening our modern world, the modern economies, socio-political institutions and the global market, as well as how such challenges affect the *development of peoples* everywhere in the globe. Recalling the different themes that were dealt with in Paul VI's Populorum Progressio and with a focus on calling the world's attention to the need to observe the "signs of our time", the encyclical laid out from the on-set its purpose and destination thus:

> [...] the awareness of the duty of the church as an expert in humanity – to scrutinise the signs of the times and to interpret them in the light of the Gospel; the awareness, equally profound, of her mission of "service", a mission distinct from the function of the state, even when she is concerned with people's concrete situation; the reference to the notorious inequalities in the situations of those same people; the confirmation of the council's teaching, a faithful echo of the centuries' old tradition of the church regarding the "universal purpose of goods"; the appreciation of the culture and technological civilisation which contribute to human liberation, without failing to recognise their limits; finally, on the specific theme of development, which is precisely the theme of the encyclical, the insistence on the "most serious duty" incumbent on the more developed nations "to help the developing countries. (Paul 1987)

Viewing the anxieties expressed in these themes with all sense of justification, the encyclical, therefore, hit the nail on the head in setting out its originality by focusing on the need to understand the true meaning of development. This very approach would later lead to the analysis and discovery of the various structural imbalances (structures of sin) that are cogs in the wheel of development in general. In the words of the encyclical, "[...] True development cannot consist in the mere accumulation of wealth and in the greater availability of goods and services if this is gained at the expense of the development of the masses, and without due consideration for the social, cultural and spiritual dimensions of the human beings" (Ibid., no. 9).

However, almost everywhere, in developed, developing and underdeveloped countries, the individual, as well as groups of individuals, find themselves under heavy pangs of structures that are intrinsically inhibiting and oppressive. In the words of the encyclical, such inhibiting and oppressive structures are forms of modern imperialism found in "[…] a world which is divided into blocs, sustained by rigid ideologies and in which instead of interdependence and solidarity, different forms of imperialism hold sway […]" (Ibid.). The result is the subjection of our worldly structure to the negative pressures of varied forms of man-made "structural sins" that make integral development difficult and, in varied concrete cases, stagnant and illusory. Giving a practical and simple description of the constituents of these structural sins, the encyclical teaches that "[…] (they are) the sum total of negative factors working against a true awareness of the universal common good, and the need to further it, (they unite with) the impression of creating, in persons and institutions, an obstacle which is difficult to overcome" (Ibid., no. 36).

The above expression brings to the fore the complexity of such a social phenomenon wherever it is found. A repeated and observable consequence is the blatant disregard of the dictates of the moral conscience in the life and the activities of the perpetrators, such that the sense of sin and the acknowledgement of human imputation remain denied, trivialised and rejected. The actors on the scene act like impostors who perpetrate evils that they have designed and taken as absolute values for themselves. When such evils are entrenched in the life of a particular system, infest the entire structure and keep it weighed down with morally bankrupt ideological deviations, which are predominantly rooted in personal sins and behavioural patterns of the perpetrators. The encyclical elaborates further:

> […] (in this form) it is not out of place to speak of "structures of sin", which, as I stated in my Apostolic Exhortation, Reconciliatio et Paenitentiae, are rooted in personal sin, and thus always linked to the concrete acts of individuals who introduce these structures, consolidate them and make them difficult to remove. And thus, they grow stronger, spread, and become the source of other sins, and so influence people's behaviour. (Ibid.)

The complexity and the difficulties involved in dealing with such structures cannot be underestimated, especially when one looks at the challenges that confront our world today. It has often become an axiom in the social, political and economic world that when one cannot beat them, he or she has to join them. A simple mode of expression that gives credence to the amount of influence such structures can exert on the polity, especially when such structures are negative and anti-developmental. As concrete phenomena, the encyclical spoke about such dysfunctions and defects as being selfish, short-sighted, mistaken and misguided, thereby leading to imprudent economic decisions (cf. Ibid.).

As driving forces within these *sin-infested structures,* the encyclical pointed out two fundamental inordinate passions that consume the actors on the scene, thereby beclouding and weakening the moral conscience, namely: the desire for profit and the thirst for power, which tend to hijack a given system. They often become detractive weapons for those human attitudes that seek achievement "at any price" (Ibid., no. 37). In the explicit words of the encyclical: "If certain forms of modern "imperialism" were considered in the light of these moral criteria, we would see that hidden behind certain decisions, apparently inspired only by economics and politics, are real forms of idolatry: of money, ideology, class, technology" (Ibid.).

Our present work, therefore, is bent on identifying these structural sins which threaten the general interdependence within a given structural and social existence. They constitute a detractive system which works to destroy the core societal networks upon which the fundaments of integral development ought to be built. In this regard, our research proposes that an authentic viable weapon against all forms of detractive man-made social structures would be an investment in both ideological and moral solidarity. This must be pushed on to become a virtue as well as a moral value upon which social attitudes should be built. This should be an ethical way to do away with the structures of sin and chart a sustainable course for peace and integral development which, at the same time, ensures the maximisation of capabilities in *Mmadụ*, the human person, the Imago Dei. According to the encyclical, the Christian word for the initiation of this battle is "conversion"—a conversion that leads to the right attitude of self-discovery and the interdependent attitude to value the moral conscience. The conversion needed, in other words, the change of attitude advocated, must be far from being shallow, as it is often observable in our world. The feeling of conversion and the resultant solidarity with one another should not be, according to the encyclical, "[...] a feeling of vague compassion or shallow distress at the misfortunes of so many people (the world over and, in particular, my country Nigeria) both near and far. On the contrary, it is (and should be) a firm and persevering determination to commit oneself to the common good; that is to say to the good of all and of each individual, because we are all really responsible for all" (Ibid., 38).

Our two primary sources from the social teachings of the church demonstrate, not only in idealistic terms but also in concrete terms, the known experiences and the virulence of structural sins in the general system of our world today and, by application, within the specimen of our research—Nigeria. Since the field of study within which our research is done has a normative character, we would then begin by examining the place and the limits of law in our approach to the concept in order to ascertain the availability of instruments or criteria needed for combatting such dysfunctional structures.

2.5.4 The Place and Limits of Law in the Normative Approach to the Concept "Structures of Sin": Interpretation in the Light of John Paul II

An instrument of work in combating these structural sins remains a total conversion, as the encyclical pointed out. But the question remains: How does this come about? To what extent is such an attitude realisable within the entire social structure? However, to answer these questions, the role that law, as a normative principle, plays in society invariably takes centre stage since these laws have boundaries that are not just merely conceptual but essentially prudential (Breen 2008). In other words, they deal with structures that have the human person as a focus—an individual human person different from a robot or other inanimate objects, a possessor of moral conscience.

But employing law as a criterion in eliminating the structural sins appears to be limited in approach "…because it cannot reach the recesses of the human heart, law is severely limited in bringing about true social transformation" (Ibid.). It then means that there is a need to find an approach where the law ends and makes way for a true society and association. This brings us, therefore, nearer to the truth that only in a true and deepest change of heart and attitude lies the purest criterion for combating the structures of sin and paving the way for integral development, with its consequent solidarity of and for all. John Breen presented this image again in his analysis of John Paul II's concepts using the narrative about an innate reflective conversation, which made the difference between Zosima and Mikhail. Zosima was a former officer in the military who dated a woman who later married another man. Disappointed and deeply angered about the woman's attitude and decision, Zosima resorted to insulting the "breadwinner" publicly, which led to the two agreeing to a duel with pistols. But the bloodshed would later be averted with the simple attitude of asking for forgiveness. Mikhail, the partner in this conversation, represents a personality and character who, as an opponent, struck a deal that averted deadly violence, not with the mere application of the dictates of law but with attitude and commitment to personal conversion rooted in personal conviction, and stemming from his previous personal act and experience of shedding the blood of a young woman who once rejected his marriage proposal. However, more relevant to us in assessing the true criterion for proper combat against structural sins deeply entrenched in the society is the convincing rhetoric embedded in the words and convictions of Mikhail. He opined:

> […] In order to change the world, man's way of thinking must be changed. Thus, there can be no brotherhood of men before all men become each other's brothers. There is no science, no order based on the pursuit of material gain, that will enable men to share their goods fairly and to respect each other's rights. There will never be enough to satisfy everyone; men will always be envious of their neighbours and will always destroy one another [….] When heaven on earth will come about, I can only promise you that it will come without fail, but first the period of man's isolation must come to an end. (Ibid., p. 326)

The word "isolation" in the discussion strikes a resounding chord of deep meaning in this journey of discovering the criterion for the fight against structural sins. The demand for the liberation of the *self* from such an isolated world leaves solidarity, which emanates from a true and convinced change of heart and attitude, as the only option and weapon of combat. Solidarity becomes then "[…] the antithesis of the self-consuming isolation that Mikhail in the above discussion criticises. It reflects a readiness to […] lose oneself for the sake of the other instead of exploiting him, and to "serve him" instead of oppressing him for one's own advantage" (Ibid., p. 331).

The attitude of exploitation and oppression, therefore, is identified as the causal factor behind the enthronement of the sinful and dysfunctional structures in a social system. Such an attitude is identified as the ultimate source of injustices that afflict the world since it originates from the wrongful appeal to absolute human freedom, which has its root in the individual. When such structures are present, their origin remains traced back to individual sins, which unite to build up the dysfunctional parts that eventually challenge and, in most cases, destroy the optimal functioning of the whole. Now, how has this raised concerns for the Nigerian polity?

2.6 Structures of Sin with Reference to the Concept of Sustainability

A sustainable society, system or structure is contra-intuitive where a constellation of social ills hijacks the administration of its institutions. Thus, the challenges posed by the structures of sin within a particular polity are invariably the challenges to its paths towards sustainable development. This is remarkable in our focus on delineating the obstacles and suggesting pathways towards remediation and recovery. Therefore, an unhealthy system, in other words, an unsustainable system, requires a concrete social agenda as a remedy since the dysfunctional structures that keep it unhealthy are not just arbitrary. They are rather planned and constellated factors of detractions. To this effect, Vogt sees any remedial agenda as a process towards *a new social contract* (Vogt 2013). Our research would discuss this contract as *a new socio-cultural orientation* needed in the approach to Africa, particularly Nigeria.

However, a planned agenda has invariably set objectives. In sustainability discourses, developers, innovators and policymakers unite to find strategic, ethical principles that would help confront the various domestic and international structures that have their roots in individual persons' or groups of persons' tendencies to impose their inordinate desires over and above institutional and societal values. Vogt names seven representative areas within the social, economic and ecological areas as the varied platforms where the powerful influences of structural sins concentrate and which provide the sustainability discourses only the conceptual framework for the development of ethical strategies in approach to the general problems (cf. Ibid., pp. 110–111). This is represented below (Table 2.1).

The above-presented framework goes into the modern discussions on sustainable development processes and procedures, and to date—it must be emphasised—our world has yet to lay claims to the fact that it has developed a finite approach to sustainability and society in general. We believe that the various past and present efforts in the discussions must be seen as an ongoing process in responding to situations and challenges of the human world based on epochs, nature and the development of human knowledge. Thus, our research remains one among varied contributions to the general discussion.

Table 2.1 Conceptual framework of the core concerns of sustainability questions

1	The disputes about how much natural capital should be allowed through technical innovation and which concept of nature is required (our research work discussed this dispute under the *Eco-Eco dichotomy*)
2	The question about the social dimension of sustainability and thus about the connection between poverty, poverty reduction and nature conservation
3	The interpretation of the so-called three-pillar concept of sustainability, in the ethical-conceptual sense, cannot consist in the postulate of equal rights for the economy, ecology, and social affairs but in the claim of cross-sectional politics and networking
4	The feasibility of the motive for the hard core of sustainability, which is analysed under the keyword "ecological economy" as a new model of prosperity
5	The connections between energy, work and capital as strategic key issues of a sustainable economic policy
6	Philosophical and critical appraisal of the contemporary growth drug as new measures of progress and basis of sustainability
7	Fighting structural blockades to social participation in the open search process of sustainability or to understanding the term from a political-theoretical perspective

Source Adaptated from Vogt, M. Das Prinzip Nachhaltigkeit (Ibid.). ***Illustration mine***

2.7 Structures of Sin and the Nigerian Polity

The Nigerian polity, taken as a specimen of our research and reflections, presents Nigeria as a socio-political entity. In other words, a group of people that represents a collective unit within which they identify themselves. As a political entity, Nigeria is viewed as a nation, a people and a unit with concrete ontological institutions that build up such a system. Nevertheless, the question about the availability or deficiency of human development institutions necessary for its integral development would be appraised. However, to understand the concept, "polity", the *Fachhochschule Nordwestschweiz* noted,

> Polity umfasst die Form oder Struktur des Politischen und bezieht sich auf institutionelle Aspekte. Der Fokus ruht auf den verfassungsmäßigen politischen Strukturen und Ordnungen einer Gesellschaft (Regierungssysteme, Parlamente, politische Parteien, internationale Organisationen, Interessenverbände etc.) und auf deren Rechtsordnung (Verfassung, Gesetze, Verordnungen etc.) welche die Handlungsspielräume, Zuständigkeiten und Abläufe bei Auseinandersetzungen festlegen. Neben der institutionellen Ebene (Staatskunde) ist die normative Ebene (Wertvorstellungen) Absichten hinter Regelungen) angesprochen.(Leimgruber 2016)[3]

[3] Citation in English: "Polity comprises the form or structure of politics in relation to its institutional aspects. The focus rests on the constitutional political structures and organisations of a society (system of governance, parliament, political parties, international organisations, international relations, etc.) and on its legal order (Constitution, Laws, Enactments, etc.), which defines the scope, jurisdiction and procedure in both general and conflict situations. Besides the institutional level, one addresses as well about the normative level (the ideals) in view of regulations.".

Within the democratic Nigerian polity as a nation, our research evaluates her present structures with regard to observable "structural dysfunction"—the structural sins, which form the main theme of this research. Whether the present structure works or is being detracted from by organised human factors serving as negative weapons in the hands of the users would be the pertinent question to answer. In this regard, we further examine the general Nigerian structure in its socio-political, socio-economic, socio-infrastructural, socio-ecological, as well as socio-technological realms.

2.7.1 *Nigeria: The Socio-Politically Related Structures of Sin*

It took only barely seven years after Nigeria gained independence for it to record her first official and bloody political strife, which claimed lives and properties and prepared the stage in Nigeria for subsequent political mistrusts, ethnic and regional confrontations, religious and faith-based feuds, the blazing passion for revenge at every slightest provocation, as well as a devastating politics of greed and corruption enthroned by sectarian inclinations of northern and southern agendas.

2.7.1.1 In-Built Structural Neo-Colonialism

The colonial era of Africa, in general, and Nigeria, in particular, brought with it many socio-political and economic impacts. The influences remain indelible in the life and history of Africa. At the same time, there exist observable trends viewed as negative or disadvantaging, which colonialism left within the African soil and worldview. Some of such trends tend to keep many African politics and politicians seemingly enslaved and perpetually tied to foreign identities and ideologies, thereby keeping the development of a people shallow and depraved. This situation has earned Africa and, of course, the entire world the detractive case of *neo-colonialism.* Attributed to the famous African leader, Kwame Nkrumah of Ghana, neo-colonialism became an awareness concept and campaign which decried Africa's continuous dependence on foreign ideologies in dealing with her household matters, and which has proved with varied and numerous known cases, quasi-productive or totally detrimental to African institutions. For Nkrumah, it is the last stage of imperialism and finds its essence in bequeathing to States "[...] all the outward trappings of international sovereignty" (Nkrumah 1974). In reality, however, it keeps the economic and political policies imprisoned by heavy external influences (cf. Ibid.). It is, therefore, a system of indirect rule.

With reference to Nigeria, a neo-colonial Nigeria still reserves within its system an institutionalised structural defect that keeps development shallow and ineffective. Perhaps, the most deadly and detrimental is her virtual, and in many cases, total dependence on foreign actors in matters of commerce, production and industry. Consequently, Nigeria has degenerated into becoming an *import-economy* instead of an *export-participator* or, better still, a combination of the two. Nigeria is heavily dependent on exogenous economic policies and has a high rate of consumer-goods consumption. It has become so handicapped and powerless to produce even enough for its citizens. This structural sin has even led gradually to a loss of identity embedded in a feeling of inferiority, such that the average Nigerian person feels better and proud using a product coming from outside of his/her country than increasing her effort to support indigenous innovations and productions. A few indigenous companies and factories, which battle to liberate a trapped national economy, bury themselves in the forgery of identity so as to put their products across in the markets. As a result, the limited products manufactured locally receive rebels and tags of other foreign European, American and Asian countries as their countries of origin. This is a socio-political structure of sin which has deep economic consequences for a developing country like Nigeria. It is just one example, but it is endemic.

Our research identifies as well, but as a subjective impute, the existence of a dual form of neo-colonialism, namely: *Neo-Colonialism ad extra and Neo-Colonialism ad intra*. The situation we have already described above, where foreign "chains" hold captive the inner recesses of a nation, thereby bringing authentic and integral development crumbling on its knees, is an exhaustive definition of neo-colonialism ad extra (from outside). On the other hand, neo-colonialism ad intra (from within) represents an internal structural and ideological dysfunction, in other words, internally operative structural sin, being perpetrated by individuals or groups of individuals who use their socio-political positions for oppressive, dehumanising and anti-developmental activities. Contemporary Nigeria describes neo-colonialism *ad intra* as the realm of the *political cabals*. Talking about such cabals, the famous Nigerian Author Chinua Achebe made it clear in an interview:

> Following a series of crises that culminated in the bloody Nigeria-Biafra war, Nigeria found itself in the hands of military officers with very little vision for the nation or understanding of the modern world. A period of great decline and decadence set in, and continues to this day. The civilian leadership of the Second Republic continued almost blindly the mistakes of their predecessors. At that point in our history, the scale of corruption and ineptitude had increased exponentially, fuelled by the abundance of petro-dollars.... By the time the Third Republic arrived, we found ourselves in the grip of former military dictators turned 'democrats' with the same old mind set but now donning civilian clothes. So, Nigeria following the first republic has been ruled by the same cult of mediocrity – a deeply *corrupt cabal* – for at least forty years, recycling themselves in different guises and incarnations. They have then deeply corrupted the local business elites who are in turn often pawns of foreign business interests. (Achebe 2013)

In neo-colonialism ad intra, politics and political positions become the single and easy medium for material gains. In the main, the corporeal actors are not direct foreigners. They are (and more painfully) elected leaders who allow themselves to be used as instruments in the hands of both external and internal actors. Their only concern is personal and selfish interests and their political ideologies are only designed to loot, oppress and manipulate. In Nigeria, they receive, as peoples' representatives, the resource benefits due to their constituencies but convert them to personal gains and properties. In a deceptive strategy, they distribute occasionally bags of rice and other foodstuffs to the hungry and their poverty-stricken subjects in a show of ephemeral and disguised philanthropism. This is just to cover their theft and loot. Since they have no sustainable plan for the development of the community, their political mantra is only seen in the Afro-Igbo expression "*Onye Ube ya ruòlù, ya ràcháa*", which explains the phrase: "This is my opportunity (to loot); therefore, I have to use it to the maximum." In line with the foregoing analysis, neo-colonialism ad extra and ad intra are not myths in the Nigerian socio-political system. They build up one of the contemporary structural and ideological misgivings that threaten the existence and development of Nigeria as a Nation. Therefore, they are identified as structural sins that must be nipped in the bud.

2.7.1.2 The Bad Trend of Lucrative Politics: The Nigerian Misconstrued Expensive Democracy

Politics has become so deadly in Nigeria because it has been taken to the level where it has been made a lucrative avenue for inordinate gains. Although "lucrativeness" as a word does not have a direct negative connotation, the ideology, however, that brings about a lucrative venture has the power to influence when the questions concerning legitimacy and legality are the focus. A morally acceptable lucrative business is invariable with success or viability. A lucrative job or activity carries with it a high sense of corporate social responsibility. Even in politics, lucrative politics can only mean a political setting and ideology that brings about integral and people-oriented dividends. However, in the Nigerian context, the general approach to politics has become a structural sin that has turned Nigerian politics into a den of *hijackers* and *political jobbers* who employ and apply every means possible to attain a political position. Such politicians ascend their positions with no vision and no plan except the plan to loot and steal. A glimpse of an awaited Nigerian leader who would eventually break the jinx of Nigeria's negative approach to politics, thereby breaking free from the status quo, remains deeply elusive for Achebe. He opined, "[…] That elusive great Nigerian leader (who) is able to transcend our handicaps—corruption, ethnic bigotry, the celebration of mediocrity, indiscipline etc.—will only come when we make the process of electing leaders—through free and fair elections in a democracy—as flawless as possible, improving on each exercise as we evolve as a nation" (Ibid.).

The lucrative mentality attached to the general approach to Nigerian politics shows an indigenous off-shoot of what seems to be a general problem of leadership in Africa. Many African leaders of the past and even of varied present dispensations trailed and are still trailing the paths to negative power clinch, which unfortunately has led many and is still leading many to power addiction and masked personality. Describing such victims as "power mongers", Oforka writes,

> The status quo of power mongering made many of the sub-Saharan African leaders to get drunk with power and lose focus. It was like influenza that raged through the independent young states. Caplan regrettably notes: 'It is Africa's curse that only a handful of African rulers have been able to resist wallowing in a full-blown cult of personality.' They were sacrificing anything on their way to increase and maintain power. (Oforka 2015)

The Nigerian situation shows a constant political feud that has always turned out to be a parody, dashing all hopes, expectations and democratic values to the floor of designed deceptive political intrigues. Examined and proven qualifications for political positions in Nigeria are forcibly rendered irrelevant today. Political positions now are business sites where the *power mongers* take their turn in looting. A few who show a guided understanding of what real political leadership should mean always face rejection, oppression and even the risk of elimination. This unfortunate socio-political morass has created a structural sin that has weakened the true meaning of service, leadership, authority, discipline and the rule of law.

The reality of the whooping cost of running over-crowded governance in Nigeria keeps the international community always in bewilderment. The greater percentage of the annual budget would always go into the running of government and the maintenance of the over-populated network of her politicians. Eventually, when the so-called "national cake" for structural and infrastructural development is divided—*the petro-dollar*—the revenues still make their way into the private pockets of the so-called people's representatives (the corrupt politicians) who serve only their selfish interests and desires.

The bad influence of money and the related factor of institutionalised godfatherism remain the bane of Nigeria's political failure. Consequently, political positions are bought through all (negative) means available to the mongers. They are not won or merited. In other words, the electorate is cowed and fooled to the polls, laden with well-planned voting activities shrouded deeply in deception and disenfranchisement. The greater part of the masses in Nigeria are victims of political money-lobbying due to the situations of hunger and abject poverty. In other words, the power mongers who view politics as an existing lucrative ground for amassing wealth eventually resort to buying their votes so as to push through their inordinate ambitions. In the end, they keep the same disenfranchised victims disappointed, oppressed, exploited and dehumanised when they assume office. Furthermore, in a bid to achieve their selfish interest within their construed vision of politics, they dive deep into all manner of corrupt activities, taking all sorts of criminal steps to eliminate any person or thing they consider an obstacle or opposition on their way. Such structural sin has kept Nigeria in man-made persistent bondage.

2.7.1.3 Constant Political Strife and Instability

This is another in-built structural sin that has kept the nation called Nigeria unproductive and stagnant. The trend of events since the purported independence from colonial rule has kept Nigeria marking time amidst other progressive nations and economies graded at the same level as her at the inception of independence. Such instability has rendered development, ideologically and realistically, incongruent with the concrete social and economic situations of its citizens and with the expected prospects of sustainable development in our world. Fagbadebo observed this when he wrote,

> The Nigerian State is a victim of high-level corruption, bad governance, political instability and a cyclical legitimacy crisis. Consequently, national development is retarded and the political environment uncertain. The countries authoritarian leadership faced a legitimacy

> crisis, political intrigues, in an ethically-differentiated polity, where ethnic competition for resources drove much of the pervasive corruption and profligacy. (Omolulu 2007)

Although Nigeria is not alone in the endemic issue of political instability in Africa or the world generally, her case of instability constitutes one of the structural sins that have dealt her heavy blows in the timeline of her development as a nation-state. Ake, in his general observation of the African States and their general situations from the 1960s, when they started gaining independence, gives a true picture of the Nigerian case. For him, a cross-section of analysis of the African States and how they have fared since their purported freedom from the colonial rule presents "[…] stagnation, regression or worse. The tragic consequences of this are increasingly clear: a rising tide of poverty, decaying public utilities and infrastructures, social tensions and political turmoil, and now, a premonition of an inevitable drive into conflict and violence" (Ake 1995).

That this structural sin has bred stagnation, regression and other social crises in the Nigerian story is a self-evident truth. Nigeria's political intrigues and feud took centre stage only three years after independence when the military struck. This reached its peak with the unfortunate period described as *the thirteen-year unbroken military rule*. It was and has always been the drama of the survival of the fittest, deeply rooted in politics driven along ethnic lines: the Nigerian North versus South dichotomy; supremacy tussle among the major ethnic groups; political crises that have religious undertones; politics of resource control and power devolution; unjustified cost of governance and the consequent agitation among the masses; the politics of petro-dollar and the consequent ecological issues; the politically "super-rich" amidst the suppressed and marginalised daily workers, etc. These confrontational and radically tensed-up situations leave, within the system and structure, a level of instability that has made the socio-political Nigerian environment incessantly charged with social, political and ethnic, as well as religiously sponsored violence, and they have remained cancerous in the general system. Omololu captures this situation by citing Kew (2006) thus:

> The giant (Nigeria) was brought to its knees by 20 years of brutal and corrupt military rule, which left a legacy of executive dominance and a political corruption in the hands of Nigeria's so-called "god-fathers" – powerful political bosses sitting atop vast patronage networks who view the government primarily through the lens of their own personal enrichment. (Ibid., Corruption, Governance and Political Instability in Nigeria: A Dysfunctional Conundrum, p. 30)

The struggles, therefore, to combat these anomalies have created unending political and social squabbles that have for long plunged the Nigerian nation into ideological, structural and system bankruptcy, leading to stagnation of and, in many cases, the total absence of purposeful development policy.

2.7.1.4 Political and Ministerial "Mis/Dis-appointment"

Political and ministerial appointments in Nigeria bear the mark of "mis-appointment" or, better expressed, disappointment. The cases of nepotism and favouritism remain cogs in the wheel of the general political, ministerial and personnel management. The qualified are denied their rightful positions and the privilege to serve, while the unqualified and the unmerited infest the key professional and political positions into which they have no competence. The whole situation got more complex due to the Nigerian socio-political and ideological failure to put in place corresponding institutions and create job opportunities that would be able to respond to the teeming population of graduates, their professional fields and qualifications. In consequence, there arose the situation where graduates and job seekers give in to whatever comes their way, even outside their individual competencies, provided they get themselves fixed somewhere. The result is and remains a form of political and ministerial disorientation which has led to a general disappointment in the need and effort towards integral and sustainable development in Nigeria. A civil engineer, on a fateful day, ascends to a position as a bank worker or its manager; a secondary school drop-out receives the portfolio as a personal assistant, just for the fact that a relation of his or her has taken up a political appointment; a medical doctor takes up a ministerial post as minister of works and housing, while an architect becomes the sports minister, etc. At the time of writing, Nigeria's Minister of Sports was Solomon Dalung, who graduated as a lawyer and once worked in the legal department at the Prison Headquarters, Abuja. His approach and show of ignorance in his political post speak volumes of the mess into which Nigeria has found herself. His choice of words in public spaces and his inability to make coordinated statements leave Nigerians disappointed. This remains an example of ministerial and political disorientation.

Although a mere academic certificate or qualification may not necessarily guarantee optimal or successful handling of a political and ministerial post, however, respect and valuation of areas of specialisation and competence remain invaluable in a socio-political responsibility. Found deeply immersed in the horror of *mis-appointments* and disappointments in the Nigerian political environment of the 80s, one of the Nigerian dailies was reported to have described the general political setting as *the government of the idiots, by the idiots and for the idiots.*

The consequences of the above-examined Nigerian situation manifest in an overt institutionalised love for mediocrity, and any Nation-State that settles and feels comfortable only amongst mere political jobbers would be condemned to persistent waste of her life and rich resources, be it economic and human, in a man-made orgy of cynicism. Political post in Nigerian society, therefore, has deep-rooted structural sin, which is and remains the aftermath of a "hijack" from self-appointed elites, who have long held the country captive. They come into political office, loot the office and step aside after they must have fielded their likes again into the same position so that the looting galore would continue. The whole structure, therefore, becomes tied up in planned lop-sidedness that continues to set the different sections and sectors of the general polity against one another. Thus, as a structural sin, the so-called king-makers thrive in the distasteful attitude of *imposition,* which infringes on the citizens' rights and freedom to chart their course and determine their political destiny. Because it is consciously designed and executed, one expects nothing other than the persistent practice of a corrupt status quo which has continued making progress and development stillborn in Nigeria since independence. The system repeats itself with impunity. Trapped in the foregoing situations, therefore, the golden chances for new and innovative ways of thinking, new ideologies, new sustainable policies, etc., become improbable for a nation like Nigeria that has the resources it would take to stamp its feet strong in the global space.

2.7.2 Socio-Economically Related Structures of Sin: Nigeria @ Over 60 Years After Independence, Yet with Persistent Youth Unemployment Indices

The twenty-first-century scientific determinant measure for the progress and development of a country on the global scene is now a relevant eye-opener. Progress and development do not depend anymore on foreign reserves nor on the number of rich persons a country has; neither does progress depend on the huge billions and trillions of dollars voted annually as state or national budget. The modern scientific approach examines the GDP per capita of each country in order to determine the effects and impacts of both macro and microeconomic institutions in relation to the general socio-political settings in the concrete life situations of the citizenry. In this sense, when a country prides itself on sequential and sumptuous budget records running in billions and trillions and at the same time witnesses a scandalous rise in unemployment of her citizens, then it is an indication of a structure that is sick, mismanaged, misguided and disoriented.

The unavailability of the relevant institutions which should offer and provide job markets for the skilled and less skilled workers, as well as train the unskilled ones, has contributed grossly and negatively to the constant rise in youth unemployment in Nigeria. While the number continues to be staggering on a daily basis, many ask relevant questions and wonder what generally should define the education and learning process in Nigeria as being relevant. Many Nigerian graduates and skilled workers today roam the streets and wander in the dark clouds of hopelessness and confusion triggered by a dysfunctional system that has dashed their hopes, aspirations and visions to the ground.

Deeply embedded in the above-mentioned situation is a structural sin described as *visionless politicking*, which takes its root from socio-political and economic selfishness and ignorance, added to the neglect of the dictates of modern moral pathways to sustainable political economy. In such a situation, one sees the overall political misappropriation of infrastructural relevance, love of mediocrity and a clear manifestation of indifference in the effort to guard and strengthen the human ecology.

Many of the affected youths carry the scar of failure inflicted on them by man-made systems and negligence. Conversely, when the slightest opportunity is offered to such affected citizens elsewhere, they excel and break the bounds because they now meet opportunities and systems that help them re-write and transform imposed failures into successes. Between 1999 and 2021, the unemployment rate in Nigeria rose on arithmetic progression without experiencing any effective policy change to stem the tide. The table below represents recent independent research on this rise, with the observation that intermittent growth occurred in adult and youth unemployment rates, accounting for 9.71% and 19.67% of the monitored statistics, respectively. It means that Nigeria must grapple with a percentage average between 24 and 29% of its unemployed citizens. A recently published independent report with particular reference to youth unemployment opined:

> In 2022, the estimated youth unemployment rate in Nigeria was at almost 13.44%. According to the source, the data are estimates from the International Labour Organization, an agency of the United Nations developing policies to set labor standards. The youth unemployment rate refers to the percentage of the unemployed in the age group of 15 to 24 years as compared to the total labor force. Youth unemployment rates are often higher than overall unemployment rates, which is true in Nigeria as well: the general rate of unemployment was approximately six percent in 2018. One reason for this contrast is that many of the youth under age 24 are studying full-time and are unavailable for work due to this. (Statista 2022)

To understand, therefore, the case of unemployment as an embedded structural defect in the general polity, we must understand the fact that unemployment as a concept defines a situation whereby the availability of a proportion of a labour force meets a limited, or in many cases, a total absence of the labour market. In the Nigerian case, the causes and effects are so intertwined that both the provider and the receiver share substantial levels of blame, which concurrently impute attitudinal neglect and failure on each of the opposite camps. To the government and policymakers, it is an attitudinal indifference to the basics of sustainable development. To the working class, it is the persistent lack of a non-violent revolutionary will to confront and say an effective "No" to man-made systems and policies that betray justice, equity and the common good. For the renowned Nigerian Laureate Prof. Wole Soyinka, the man will always die in all who keep silent in the face of tyranny (Soyinka 1976).[4]

For many well-intentioned Nigerians, including the present researcher, the unfortunate story of unemployment in Nigeria can never be told without bringing to the limelight the entrenched cult of corruption among its leaders at the national level, which, as a consequence, drowns the country in an ocean of economic and development woes. For Ojukwu, the *oil boom* days of Nigeria's economy remain in the annals of history the wasted years and contributed disappointingly to the present rate of unemployment in the country. This is because the architects of the socio-political scene of those economically fruitful days suffered political and ideological blindness, which led their different administrations into clueless forms of governance (Ojukwu 1989).

---

[4] The above expression the "man dies", is used in reference to Wole Soyinka's literal prose 'The Man Died' – An account of the arrest and the consequent 22-month incarceration he experienced during the regime of former Nigerian Military Head of State, General Yakubu Gowon.

The structural sin of unemployment in the Nigerian polity must be approached a posteriori. That means there must be the need to reinvigorate and reinvest in the richness of its human resources. The needed institutions and infrastructures must receive now a workable and sustainable policy, which would be able to counter the wastage of human labour and provide the job market for the teeming number of her youths who are unjustly being forced down to the level of social liabilities. Yet, the persistent absence of this determination to impact the necessary changes remains a socio-political and socio-economic structural defect.

2.7.2.1 A Geometrically Dichotomised Socio-Economic Polity: The Rich and Poor Saga

Equality and equity are two socio-phenomenal themes that characterise any approach in the discourse on the right and justifiable place of individuals in a general network of interaction and participation in human society. Both themes play a very important role in discussing the reality of the conspicuous gaps between the so-called advantaged "rich" and the unfortunate "poor" within our world and societies of today. In a broader sense, the issue encompasses the global reality of dividing the world into blocs, with sociological and ideological ascriptions like developed, developing, underdeveloped, rich, poor, privileged, and/or under-privileged worlds; or the first, second or third (perhaps fourth) worlds and societies. These ascriptions have helped to constantly deepen the feeling of inferiority or superiority in a diversified world, where such diversity should not, under normal circumstances, undermine any group, race or colour and ought not to be superimposed or set one against the other. In functional societies, diversity should rather be appreciated, coordinated and used as a means of complementarity and enrichment.

It is, therefore, the reality and practice of such divisions which brings about an in-built structural sin cancerous in the Nigerian system, namely the rich and poor saga. The existence of such a gap means simply that equity is thwarted while equality is unjustly politicised. As a social sin systematically entrenched in the general structure, the ratio of its growth and virulence runs in a geometrical progression in the Nigerian case. Most conspicuous is the reality of the rich getting richer while the poor sink deeper into excruciating penury. The builders of such oppressive networks stop, as a matter of fact, at nothing in keeping their general network functional, even if it entails the elimination of the same helpless victims of their criminal oppression who are always at the receiving end—the *Mmadụ*, the human person.

When talking about equality and equity in the discussion about the rich and poor saga within the Nigerian context, one needs to understand the basic characteristics of the two concepts in question. Equality in its socio-political and socio-economic connotation does not need to be absolute, in so far as the concept of ability and capability within humanity remains broad, unique and diversified. In giving attention, therefore, to the existing problem of the growing gap between the rich and the poor, attention would turn mainly to the requirements for functional "social equality", which should focus on members of a specific social group(s) expected to share justifiable common status, respect for one another, promote freedom of speech and equal rights, as well as have common access to the goods and services available for the common good. In other words, a determinant factor in the assessment of functional social equality remains the possibility of every individual having "equal opportunities" to the needed goods and services, basically judged by individual abilities, interests and capabilities. It means the absolute determination to respect differences while integrating the individual chances to participate in and contribute to building a justifiable whole.

Any abuse, therefore, of the tenets of social equality would constantly create structural tensions and disruptions that forcefully balkanise individuals as well as groups of individuals into sociological, ideological and psychological inequalities that create imbalances among different parts and sectors of a given society, setting one against the other. To avoid this, certain norms and action principles must come into play, namely:

> [...] the priority of meeting the basic needs of the poor and the importance of increasing the level of participation by all members of the society within the economic life of (any) nation. (Such) norms establish a strong presumption against extreme inequality of income and wealth as long as there are poor, hungry, and homeless people in our midst [...] Extreme inequalities are detrimental to the development of social solidarity and community. (United States Catholic Bishops 2009)

Equity, on the other hand, embodies the concept of social justice and demands the establishment of the principles of fairness, impartiality and justice towards all; it demands distributive and inclusive access to the necessities of life and an equitable share of responsibilities in dealing with the resources of nature. As a pre-requisite to bridging the widening gap between the rich and poor, the Brundtland Report opined, "Poverty is not only an evil in itself, but sustainable development requires meeting the basic needs of all and extending to all the opportunity to fulfil their aspirations for a better life [...] (there is the need) of meeting essential growth for nations in which the majority are poor, (by assuring) that these poor get their fair share of the resources required to sustain that growth" (United Nations 1987).

In equity, justice is to be applied according to the dictates of natural law and human rights, and every approach to the distribution of natural resources must be free from bias and favouritism. In other words, such an approach must respect the rights that are inherent to the human person by virtue of human nature, made coherent and comprehensible through human reason. Conclusively, therefore, discourses on equity and equality bring together both the political philosophers' views about them as "egalitarianism" or the American similar approach to the concepts understood as "liberalism". However, central to both is the unitive idea of distributive justice, which stands against all forms of discrimination, impartiality and impoverishment, as well as against all manner of exploitation and oppression with regard to resources and material goods. Where equity and equality are neglected, in other words, where distributive justice in a political economy is downplayed, one awaits the forceful enthronement of a violent class struggle which always creates the two opposing camps of the *haves* (the rich) and the *have-nots* (the poor), setting one against the other. Such a conflict is entrenched deep in the polity of the structural entity called Nigeria.

The rich and poor saga in the Nigerian context, therefore, is a structural, social, ideological, economic and political *sin*. The oppressive gap it creates between the two opposite camps widens with impunity even in twenty-first-century Nigeria. This happens alongside Nigeria's intimidating population and its dynamic human resources. The general scene presents the stark reality of hunger amidst affluence and an abundance of resources, corruptly appropriated by a few who see them as their birthright. While the rich few count their millions daily, the hapless poor sink deeper and deeper into unspeakable yet avoidable penury, consciously created and consolidated by a camp of political money mongers who profit from such socio-economic disorientation and imbalance. The cry of the poor masses remains a waste of time and tears because they (the exploiters) usually lock themselves away in mansions, castles and Government Reserved Areas (GRAs), surrounded by great walls and security forces which they must create for themselves and their ill-gotten wealth for the fear of the unknown.

Such a great divide plays itself out in all facets of Nigeria's life as a Nation. As an illustration, the field of Education is rotten due to the widening gap between the rich and poor. This has brought the standard of education crumbling to its knees in Nigeria. The reason is simple. If the rich and, at the same time, the politicians are only concerned about sponsoring their kids and kin to obtain quality education elsewhere, of course, where it is better organised and structured, the only option left, therefore, is the total neglect of and poor investment on what is indigenous. The poor have no other option than to grapple with what they see as being available, no matter how dilapidating and dehumanising it may be. It is indeed a structural abuse that played out in the American society of the 50s and 60s during the time of Martin Luther King Jr. and Malcolm X; a structural sin which blackmailed the British Policy during Apartheid in South Africa; a socio-structural separatism which implicitly played into Hitler's excesses and caused the lives of millions of Jews, etc. At the time of our research, the events in Nigeria show no prospects of bridging the gap and the perpetrators are not convinced that a veil of ignorance has blindfolded their sense of moral judgement. The Nigerian situation depicts a socio-political scene where inequality is celebrated alongside *resource-affluence* that better equips the country to practise an inclusive economy for the benefit of all.

The gap between the rich and poor, generally in Africa and particularly in Nigeria, is yawning. It has a history that is relatively viewed from the vantage point of the different states that make up Africa. For example, the then president of South Africa and a victim among other victims of the evils of apartheid in the same country, gave an articulate description of the rich and poor saga, epitomised by organised oppression inflicted on the black majority by the rich-white majority, thus:

> [...] an African child is born in an Africans-only hospital, taken home in an Africans-only bus, lives in an Africans-only area and attends Africans-only school, if he attends schools at all. When he grows up, he can hold Africans-only jobs, rent a house in Africans-only townships, ride Africans-only trains, and be stopped at any time of the day or night and be ordered to produce a pass, failing which he will be arrested and thrown in jail. His life is

> circumscribed by racist laws and regulations, which cripple his growth, dim his potential and stunt his life. These racist laws include the regulation of the 1913 Natives Land Act, which ultimately deprived blacks of 87% of the territory in the land of their birth; the Urban Areas Act of 1923, which created a teeming number of African slums and helped to create cheap labour to white industries; and the Colour Bar Act of 1926, which banned Africans from practising skilled trades. (Teferra 2000)

In Nigeria, the rich and poor saga takes up a description as being *extreme*, which means a degeneration to the lowest rung of poverty calibration that humans can ever imagine (cf. Ibid., The Bleeding Continent: How Africa Became Impoverished and Why it Remains Poor, p. 2). Defining extreme poverty, Jeffrey Sachs writes: "Extreme poverty exists when families are unable to meet basic needs and survival. This means they are chronically hungry, cannot access health care services and lack the amenities of safe drinking water and sanitation. They cannot afford education for some, or all of the children and they also lack rudimentary shelter, clothing, shoes, etc." (Ibid.).

Affirming the above statement as true for Nigeria, the present Nigerian social structure offers no remedy to the yawning gap that widens uncontrollably between the *haves* and *have-nots*, both socio-economically and socio-politically. The societal picture is like a den of extremely poor masses whose economic lives hang in a balance and who, on a daily basis, are being molested by "[...] the choices they face, the constraints they grapple with and the challenges they meet (all within the general jungle of socio-political and socio-economic exploitation)" (Seligson and Passe-Smith 2013). Such a man-made gap becomes an organised corrupt structure built for a strategised impoverishment of millions of innocent citizens while at the same time creating wealth for a privileged minority. It is not just a lack of money. It is rather the stark reality of being forced down to the situation of having only a few real choices to make amidst streams of possible options that are attainable yet are made exploitatively difficult for realisation by imposed factors. It is the enslavement and suppression of the capability towards authentic development, an organised crime against humanity—a structural sin.

2.7.2.2 Lost Sense of Macroeconomics

The Nigerian macroeconomic sector is a shattered structure in terms of performance, organisation, behaviour and decision-making. The Nigerian state has always been viewed as a leading potential whenever the focus of the international community goes to Africa. However, the dreamed potentials, especially with regard to macroeconomics, suffer neglect and mismanagement. In terms of performance, Nigeria's macroeconomic output falls below expectations. Her naturally endowed capabilities and resources suffer abuse due to the lack of the necessary institutions and infrastructures that should aid optimal output. The general structure presents a weak network of resources that are being mismanaged and maladministered due to the incompetence of those entrusted with such responsibility. Deeply entrenched within the system is a culture of negligence seen in endemic socio-economic corrupt practices. Thus, the general output in the behaviour of this sector remains shaky and fluctuates to the detriment of both her national and international socio-economic status. In all these, a fundamental problem takes its root from her continuous experience of corrupt, weak and colluded decision-making machinery, which serves so many vested interests and which keeps the economy perpetually weak and non-performing.

Perhaps the deadliest blow to Nigeria's macroeconomic sector remains the additive policy of mono-economic culture which has put Nigeria in the satire of history as a nation held on the international scene as the giant in the production of crude oil and other liquefied natural gases, but which remains trapped in the mess of disoriented macroeconomic structure at both national and international levels. The World Bank Economic Review on Nigeria stated clearly:

> Given the high dependency of Nigeria on oil revenues, the recent sharp decline in oil prices has given rise to major challenges in the form of external imbalance, steep falls in government revenues, and slower economic growth. In contrast to the period of 2008–2009, Nigeria no longer has a large fiscal reserve to buffer government budgets from the revenue shortfalls. As a result, distributions to federal and state government budgets declined in normal terms by 39% in the first half of 2015 relative to the same period of 2014. Federal and state governments have slashed capital spending, and a number of states have struggled just to pay salaries to civil servants and service domestic debt obligations [...] Nigeria will need to create the fiscal space needed to provide the public services essential for maintaining and accelerating progress in job creating and poverty reduction. However, it will be unable to rely on oil revenue as in the past [...] there is potential for boosting non-oil revenue generation, and investors stand ready to bring significant resources to the country if they receive signals of commitment by the new government to policy directions and regulations consistent with strong private-sector-led growth. (World Bank Group 2015a)

The above-cited review and report indicate an observable lack of viability and vitality in a nation's economy. In the Nigerian case, namely being trapped in a mono-economic culture (revenue from crude products), this has kept every other potential for neo-investment and re-investment in non-oil resource sectors inert. In other words, being a macroeconomy that has completely lost the sense of dynamics and diversification, it reserves, therefore, within itself and the entire system, a structural dysfunction tied to the weakness and insensitivity of those entrusted with the responsibility of designing and applying the policy.

The general problem lies in Nigeria's lack of political will to design a viable macroeconomy that would be able to respond to the demands of sustained socio-economic life, both nationally and internationally. The lack of the needed political will is both a cause and a dysfunction. The World Development Report of 2017 spoke about this lack of political will, using the crafted film "Amka and the Three Golden Rules" to drive home the embedded structural weakness within the general system. The film has been screened to tell the story of Mongolia and its constant lack of firm political will to curb its unmanageable levels of indebtedness. The climax came when Amka, in whose personality, action and behaviour the State of Mongolia was personified, should and must, as a matter of urgency, seek refuge in the personality of his uncle (the right decision-maker) who must proffer a solution and teach him the three golden rules. Such rules, as a principle, ought to be followed strictly if any act of reconstruction is to be achieved. In this line of thought, the World Development

Report reiterates: "[…] unless the interests of the powerful groups in society were to change (and follow the rules), the commitment to a policy of fiscal prudence would continue to fail and the county would remain in debt" (World Bank Group 2017). Lack of fiscal prudence and political will, therefore, become identified as structural dysfunctions that keep the macroeconomic sector in Nigeria moribund. They are socio-economic structural sins that must be done away with.

The fiscal space of Nigeria's macroeconomy is exploited, badly administered and high-jacked by the vested interests of policymakers who structurally manipulate the general socio-economic sector for their own personal gains. Consequently, the problems posed by poverty and economic inequality have remained persistent and intertwined. In other words, the inability to design and put forward a pathway to fiscal sustainability keeps the macroeconomic sector weak and predominantly under-productive. To tackle this, the varied inbuilt structural microeconomic ineptitude must be replaced by neo-sustainable fiscal rules, which must be prudently applied.

The known ineptitude is a structural sin with the character of lopsided benefits that serve and enrich the few alongside mega cases of excruciating penury within the vast majority. At the time of this research, agitation for a total overhaul and restructuring of the whole macroeconomic sector is heard and considered overdue from almost all known sections and corners of the country. This has set opinions, governmental and non-governmental organisations, as well as economic and political institutions in Nigeria against one another. The benefiting few see the situation as a challenge and a threat to their well-designed economic slavery and exploitation imposed on the rest of the masses. Consequently, for the agents of such an oppressive agenda, those who put up revolutionary pressures are viewed and condemned as people with oppositional strategies to destabilise the country. To push forward their evil and exploitative agenda, therefore, a new campaign slogan designed by them is echoed in the general socio-political and socio-economic Nigerian atmosphere, namely, *that the unity of Nigeria as a country is non-negotiable.* In other words, any agenda for a restructured political economy for Nigeria hits the hard rock of ineptitude if ever desired and pursued. In a most recent assessment of the state of the nation, with an attempt to proffer a way forward, the document *Nigeria Vision 20:2020* gave an appreciable evaluation of the observable content and challenges of Nigeria's economy and its micro-institutions thus: the undeniable realities of poor and decaying infrastructure; epileptic power supply; weak fiscal and monetary policy coordination; fiscal dominance; pervasive rent-seeking behaviour by private and public agents, including corruption; weak institutions and regulatory deficit; policy reversals and lack of follow-through; inordinate dependence on the oil sector for government revenue/expenditure; disconnect between the financial sector and the real sector; exchange rate instability; insecurity of lives and property (The Federal Republic of Nigeria 2009).

In general, the political will fails, both on the part of the decision-makers and on the part of the citizens themselves, and as the document emphasised, the situation requires an obligatory renewed political commitment and determination. The *"man dies"* (Wole Soyinka) in so many citizens of Nigeria as a country and would only require courage and determination to say *"enough!"* to such structural sin. Ibe made a good articulation of our situation thus:

> We are not only too shy (afraid) of confronting the fundamentals, we are too sacred of facing real realities. Out national mindset is corrupt, false, fraudulent, opportunistic and potently predatory. Our political structure is fraud. To call Nigeria a nation is a lie. To describe Nigeria as a federation of states is patent falsehood. To describe the civil service as the service arm or executing agency of government is blasphemy against the truth, for the civil service is more an organised crime syndicate than anything else. The political infrastructure is systematically corrupt and patently deformed. (Ibe 2016)

In such a deformed state of affairs, therefore, fiscal policies and other decision-making ideologies designed by corrupt policymakers serve only as avenues to perpetrate stealing and looting of the resources meant for the overall sustenance of the country's socio-economic life. The consequence is that sustainable people-oriented policies remain grossly illusory for Nigeria. Such structural sin keeps governance in Nigeria exclusive, not inclusive, and exclusive governance is nothing but an organised form of exploitation and scam.

2.7.3 Infrastructural-Related Structures of Sin: The General Problem of Economically Related Facilities in Nigeria

Facilities that are economically related play major roles in the general life and development of a given geo-political entity. Socio-economically related facilities represent the generality of infrastructure that goes into the mainstream of economic activities of a nation, and which helps to ensure the good functioning of a society. They contribute to optimising the quality of economic output of the entire socio-political and economic structure. These infrastructural amenities are social, cultural, financial, natural, technological and human. They all unite to build "[…] a network of transport, communication and public (social) services—all functioning as a system or as a set of interrelated and mutually beneficial services provided for the improvement of the general well-being of the population" (Olaseni and Alade 2012). Scientific study and research have established close interconnectedness between economic growth and infrastructural development. Thus, the presence of relevant infrastructural facilities remains a precursor to stable and effective economic development. The general socio-economic infrastructure is examined in terms of it being either physical (roads, major dam and canal works for irrigation and drainage, and other transport projects like urban and interurban railways, urban transport, seaports and waterways and airports) or social (education, health, recreation, housing, etc.).

An observable culture of negligence is like cancer in the Nigerian system, even in the twenty-first century. The general infrastructure presents a scene of horror and dilapidation arising from years and decades of utter neglect and mismanagement. Worst still is the practice of voting huge sums of money for the maintenance and upgrade of such projects, only for the budgeted fiscal resources to later find their way into the private pockets and bank accounts of the corrupt politicians. Contracts worth millions and billions of Naira had been awarded in the past and many of them have never been executed, nor has anyone accounted for the embezzled fund. The embezzlers move about freely and even control the nation's state of affairs. Some sit tight at the national assembly as lawmakers and decision-makers. For them, the loots are part of their own national patrimony hewed out of the "national cake" called Nigeria, and as citizens, they have the right to it. The means of participating in the sharing, whether foul or fair, is not a matter of importance.

The culture of infrastructural negligence has cost Nigeria fortunes. So many infrastructures left by the colonial regime would later have to battle an attitude of negligence which eroded many of these institutions and forced them to wither away. Regulating and preserving such properties, in other words, the infrastructure of a nation forms one of the key functions of any political power as theorised in John Locke's Second Treatise of Government. It is the power of civil society to protect and preserve the welfare of the citizenry. Any amount of negligence would reduce the affected government to the proverbial fool who sits complacently on the same branch of a tree for which he invests time and energy cutting down. In this regard, we assert without equivocation that the general conditions of social facilities in Nigeria always tell the story of organisational and structural bankruptcy. They are often treated as *no-man's* property because they often fall under the responsibility of varied irresponsible regimes. Nigeria's political history has shown that such social amenities are constantly exploited as avenues to loot and steal money since the money meant for maintenance disappears into private banks and pockets. The affected areas litter the whole country, and they are not difficult to identify, even for a first-time visitor to a nation like Nigeria. They are the epitome of man-made socio-economic depletion. Some concrete examples would help to drive this point home.

2.7.3.1 Energy Sector

The social facility of power and electricity in Nigeria is a disgrace in the history and experiences of the nations within Africa. Decades after her independence, the energy and power sector of the economy is nothing to write home about. The generating capacity of the power sector, in a nation of over 200 million people population-wise, trails far behind several countries and nations that either equal Nigeria in population or are far behind in comparison. Countries like Libya and Morocco, which inhabit only 3.5% and 18.8% of Nigeria's total population, respectively, share the same capacity of household power output as Nigeria (Nwosu 2016b). That is excruciatingly disproportional.

Increments in the megawatts and the improvement of the output capacity have continued to feature in political rhetoric, only to be buried in organised inaction afterwards, leaving the masses always with the false assumption that something has been or would be done or would be initiated as policy-response. At the time of conducting our research, the power minister claims a total installation of 12,000 MW capacity of electricity generation in Nigeria but regrets that the general output and supply remain disappointingly 6000 MW due to, according to the minister, vandalism, obsolete equipment, poor planning and corrupt practices of the sector (Fashola 2017). Inbuilt corruption and mismanagement build up the structural defects, which have continued to hold the sector captive, thereby keeping the generation, transmission and distribution of power in this sector stillborn. This sector would be the focus of the fieldwork to be undertaken by our research.

2.7.3.2 Health Sector

Nigeria's health institutions are often described as glorified mortuaries. This is due to the hazards they pose to the citizens and the general health environment. Nigeria has no organised common healthcare system and as such, healthcare services pose many challenges to both the entire Nigerian society and particularly to millions of individual homes and families. Nigeria's health sector suffers a structural sin which is often blamed on the leadership's misplacement of priorities. As a primary and necessary policy to be embraced and harnessed, the utter negligence of this sector by Nigeria's policymakers becomes one of the banes of Nigerian politics, leading to uncountable situations where so many of her citizens still meet their death untimely or forced to live the rest of their lives incapacitated, due to health challenges that could have been averted. On the other hand, apart from the general dormancy in policy implementation, the healthcare sector suffers chronic health infrastructural challenges, depletion in manpower as well as below-capacity output due to obsolete healthcare equipment and services which could no longer confront modern health challenges—all due to the structural dysfunctions rooted in poor funding and corrupt expending of health institutions' budgets.

This structural sin of misplaced priority has left the entire health sector and services in drought, and the decision-makers are not concerned since they themselves have unhindered access to healthcare services elsewhere outside the country. This they do when, how and wherever they want it, even at the expense of the *stolen taxpayers' money* that would never be accounted for. Oforka noted correctly,

> It is estimated in 2013 that the capital flight for medical tourism in Nigeria alone stood at 250 billion naira ($ 1.25 billion) annually. The rest of the population that cannot afford the luxury of travelling abroad in the face of such illnesses that defy the sub-Saharan African healthcare environment have the singular option of succumbing to untimely death. Most affected by this condition are women and children. (Ibid., The Bleeding Continent – How Africa Became Impoverished and Why it Remains Poor, p. 15)

This appalling situation is found within the same nation with a constituted ministry trusted with the healthcare sector, namely the Federal Ministry of Health. An excerpt from a document of the National Health Act and the parallel inaction on the part of the same ministry reveals a structure that is sick and deplorable. Stating the functions of the Federal Ministry of Health, the Health Act in question makes it clear that the Federal Ministry of Health shall: ensure the development of national health policy and issue guidelines for its implementation; collaborate with the states and local governments to ensure the appropriate mechanisms are set up for the implementation of national health policy; collaborate with national health departments in other countries and international agencies; promote adherence to norms and standards for the training of human resources for health [...]; conduct and facilitate health systems research in the planning, evaluation and management of health services; ensure the provision of tertiary and specialised hospital services [...]; promote the availability of good quality, safe and affordable essential drugs, medical commodities, hygienic food and water; etc. (The Federal Republic of Nigeria 2014).

The above-defined functions, unfortunately, stumble at the heavy rock of negligence since the Nigerian health sector offers no hope to the less privileged citizens or the common men and women on the street. The whole picture is pitiable, as revealed by the World Health Organisation, which ranked Nigeria 187th among the 191 member States of the organisation. The consequences become obvious, namely: high infant and maternal mortality; the prevalence of varied forms of diseases (many of them preventable), many of which degenerate with the slightest chance into epidemics; deficiency and malfunctioned growth and bodily development, especially in children, due to poor or total lack of supportive health care; etc. The whole situation resulted in a detrimental fall in life expectancy to 48 years in Nigeria. Thus, by the year 2000, the dream of affordable *health-for-all* ideology, which successive regimes and administrations sometimes took as a *mantra*, sunk into oblivion (Adefolaju 2014). Statistically, the ratio of the Patient-Medical-Practitioner relationship within the most populous country in Africa, Nigeria, is insulting. The WHO's report of 2006 gave Nigeria away as having the lowest statistics among all other countries in sub-Sahara Africa with the following description: "[...] 0.3 physicians per 1000 persons, 1.7 hospital beds per 1000 persons, 1.7 Nurses, 0.02 Dentists, 0.05 pharmacists, 0.91 community health workers and 1.7 Midwives, per 1000 persons respectively (WHO 2006; Ogbolu 2007)" (Ibid.).

The above statistics reveal that the number of health personnel in Nigeria has suffered and is still suffering emigration drought due to the mass exodus of many to Europe and America, of course, in search of opportunities that offer them better chances of survival and development in the fields. All these deficiencies manifest themselves in both structural and infrastructural defects, which Akindutire articulated thus:

> a) Inadequate coverage: Many Nigerians still have no access to modern health care services. Rural communities and the urban poor are also not well served; b) The orientation of the services is inappropriate with the disproportionately high investment on curative services to the detriment of preventive services; c) The management of the services often shows major weakness resulting in waste and inefficient targets and goals; d) The lack of basic statistics is a major constraint at all stages of planning, monitoring and evaluation of health services; e) The financial resources allocated to the health services, especially to some priority areas, are inadequate to permit them to function effectively; and f) The basic infrastructure and logistic supports are defective owing to inadequate maintenance of buildings, medical equipment and vehicles, unreliable supply of water and electricity, and the poor management of drugs, vaccines and supplies system. (Ibid., p. 153)

2.7.3.3 Educational Sector

In the educational sector, a historical and terrible blow was done to the Nigerian educational system the moment the corrupt Nigerian political economy appropriated all educational institutions as marketplaces for money making. Thus, with the government takeover of schools from the original owner, the Religious Mission, the quality of education was forced to dwindle both in general academics and in the areas of morals. A seed of structural discord was then systematically sown since the training of character became clearly denigrated and, by instalments, severed from mere knowledge acquisition. This monstrous seed found its entrance into the Nigerian polity first, with the Edict of 31st Dec. 1970 declaring the takeover of schools from the voluntary agencies in the then East Central State by the Nigerian government (which later was extended to the whole country), led by the Administrator of the defunct East Central State, Dr. Ukpabi Asika. Consequently, in the new government-owned schools, education began to witness a blow in moral and character formations, as well as in general academic output. The time for moral and religious instruction was gradually reduced to only 45 min per week. In some schools, it was totally obliterated. What more? The temporal dimension of education was succinctly severed from the psychical and ethical. As a result, attention became gradually directed to the *divinisation* of the material and the devaluation of the spiritual. But the human person remains a product of both the "psyche" and the "soma" and any attempt to introduce disequilibrium would predominantly land in moral and character imbalances (cf. Ibid., Ethical Approach to Abuse of Common Good and Human Rights in Nigeria, The Polity, Oil, Violence and Niger Delta, p. 41).

The educational infrastructure in Nigeria has kept experiencing records of man-made culpable abandonment. The amount of money that goes into the educational sector in the history of Nigeria's fiscal policy and budget shows a nation in love with mediocrity and a half-backed citizenry. The state of so many educational infrastructures in the country, be it primary, secondary or tertiary, leaves the mind with so many questions, one of which is: What will eventually become the products of such anomalies of visionless policymakers? The opinion and assessment of Oyeyinka (2011) give credence to this:

> [...] university education in Nigeria has experienced considerable decline in quality over the last two decades or so, owing to a confluence of factors acting in tandem. They include episodic and uncertain political-policy environments that led to declining support from governments [...]; declining investment in teaching and research facilities resulted in poor products in graduates and low employment opportunities and diminishing value of earned income. The declining quality of education is largely a result of continuous budget cuts (since 1980) together with rapid increases in enrolment rates. This made the financing of education recurrent costs more difficult. Public expenditure on education, generally, declined from 6% in 1980 to 0.65% in 1995. (Oyelaran-Oyeyinka 2011)

Because of the man-made structural dysfunction, the Nigerian educational sector has been in the history of Africa and the world over as a nation associated with incessant teachers' and lecturers' strikes, violent student riots and often the consequent brutality of the government as a response to riots and unrest. The average Nigerian child is supposed to have unhindered access to formal education, given the enormity of Nigeria's economic resources, but in Nigeria, the reverse is the case. Families are forced excruciatingly to grapple with the greed of those entrusted with the provision of quality education for the training of future minds and leaders of the country but who are heartlessly bent on keeping the sector corrupt and visionless. In such a man-made scenario, many children and youths are deprived almost permanently of their rights to good and meaningful academic life, and as such, remain unwillingly and helplessly lost out in a global village within which they have full participation rights. The hollow and horror that have been created in this sector breed many consequences that are detrimental to the entire society, namely: youth crime and delinquencies; rise in youth unemployment; weak and vulnerable socio-economic institutions due to lack of expertise; mass exodus of the youths to other countries in search of better educational opportunities, in other words, a senseless loss of available and potential manpower for the Nigerian nation; death of interest and motivation in the lives of the younger minds in the face of witnessing so many so-called graduates who roam the nation's streets, towns and suburbs, jobless and with shaky orientation and battered perspective; the disdain of the Nigerian academic certificates in the wider societies, even in some of the African countries, among whom Nigeria prides herself to be a giant; the general loss of a sustainable citizenry in the face of man-made educational oppression, which, as a consequence, stifles every opportunity to break free and make the best out of the numerous opportunities open to the citizenry. It is really a bemoaned sector in the general life of Nigeria as a geo-political entity.

2.7.3.4 Oil and Gas Sector—The Main Economic "*Gold Mine*"

Despite its status as an oil-producing giant in Africa, Nigeria experiences structural and infrastructural malfunction in this sector. This sector of Nigeria's economy has been reckoned as the most corrupt and the root of all other aspects of corrupt practices in Nigeria. One of the greatest themes for constant debate in this sector has remained the question of why and how the infrastructural machination of this sector has been exploited, neglected and forced to perform below standard. Refineries in Nigeria, manned by multinational companies in collaboration with the corrupt Nigerian government, have for decades failed to prospect this sector optimally so as to attain the expected standard within the global community. Unfortunately, the activities of the multinational partners have been hijacked and they remain under the control of the same corrupt actors in the Nigerian political economy. Again, because the multinational companies are accomplices, they have become strange bedfellows to the Nigerian economy. The consequences intersperse a whole lot of areas in the general life of Nigeria as a nation, namely: poor and exploitative exploration methods that have bequeathed ecological woes to the country; constant and huge losses of revenues due to poor infrastructural maintenance policy and, in this context, the poorly maintained and technologically obsolete refineries and oil power plants; the total devastation of the local economy, especially in the areas where exploration and refining equipment are located; multiple counts and cases of health hazards resulting from leakages, spills and other aspects of environmental pollution and contamination, etc. By 2020, Nigeria had plans to increase its oil production capacity to 4 million bpd (barrel per day). However, to date, the realities and conditions of the infrastructure in this sector already show political and economic projections, the success of which would be threatened by so many flaws and malpractices.

Cut in the web of these infrastructural delinquencies within the hub of its economic mainstream, the oil and gas sector, Nigeria is often made a laughingstock both in the national and international forums. Thus, amidst her affluent natural resources like oil and gas, the citizens themselves face, on a constant basis, the reality of grappling with the scarcity and high cost of such commodities due to inbuilt structural and infrastructural negligence. Of course, a former Nigerian Minister of Petroleum Resources, who, at the time of our investigation and research, was facing indictment for corrupt practices, embezzlement and misappropriation of public funds as well as accusations of playing accomplice to so many corrupt deals, was indeed right in expressing shock at the condition of Nigeria's oil and gas infrastructures. For the Minister, the whole situation presents a type of complication that reflects both the extent of negligence and the consequent confusion in plans and policies on tackling the existing horror. For her, the level of dilapidation is appalling,

> [...] To get a replacement for the equipment was not possible, because they were obsolete [...] For over 20 years the equipment was not changed or maintained [...] It was clear that contracts had been given out by previous governments, but the people who took the contract did not know what they were doing because of the state the refineries were at the time. (Madueke 2015)

With this situation, the Nigerian oil and gas infrastructure has been described as the worst in Africa. It is a typical example of inbuilt structural defects constituting the bane of Nigeria's development and progress.

2.7.3.5 Environment and Ecological Sector

In Nigeria, one is confronted with environmental hazards and pollution, not only in the cases of crude oil exploration and exploitation. The reality of illegal structures, poor housing systems and total absence or inadequate functioning of the drainage systems and canals, especially in semi-urban and suburban areas, always depicts the bane of Nigeria's infrastructural negligence. Illegal structures litter the Nigerian environment here and there. Ironically and in the Nigerian corrupt system, most of the sites for those illegalities are often sold to their occupants by corrupt cabals who use their positions of office to defraud innocent citizens. In most cases, they issue their victim fake documents that have no legal bindings, luring them into believing that such documents and issuances are authentic. The occupants themselves are either innocent or ignorant victims of planned illegal deals, or accomplices and beneficiaries of nepotism, with a strong affinity to the administrative powers that be. The proceeds are often appropriated and embezzled by the individual perpetrators or groups of perpetrators, while the real and unfortunate victims of their escapades are condemned to face stark defenceless losses or the outright destruction of their means of livelihood, for which they were lured into appropriating and possessing.

In a similar analysis, even when everything is legally procured, the lack of proper siting plays itself out in poor planning, which leaves the environment littered with structures and housing infrastructure that are sub-standard and geographically ill-positioned as well as aesthetically fractured. The incessant cases of collapsed buildings or buildings that must be pulled down later for contravening an original plan attest to this fact. Sub-standardisation and the habit of cutting cuts with set-down rules and regulations have been the real problems. Adeoye opined,

> Studies have shown that the urban housing in Nigeria is in a deplorable condition. Almost 75% of the dwelling units in Nigeria's urban centres are substandard and the dwellings are sited in slums (Onibokun, 18; Wahab et al, 19; Olotuah, 20; Jagun, 21). These result from combined effects of natural ageing of the building, lack of maintenance and neglect, wrong use of the buildings, poor sanitation in the disposal of sewage and solid waste, wrong development of land, and increasing deterioration of the natural landscape. (Adeoye 2016)

One of the devastating effects of such structural dysfunction has been poor sanitation, especially in tickly populated residential areas, which at the same time causes the breed and the never-ending story of mosquitoes and malaria in Nigeria. Again, there are the catastrophic effects of diseases and epidemics in the affected communities and beyond when they break out. These are all due to stagnant dirty waters, the poor or total absence of waste management systems, unplanned or manipulated geo-structural architectural activities and a lost sense of daily and interactive ecology. As it stands now, Nigeria has a long way to go in making the eighth target of the Sustainable Development Goals a reality in its life as a nation, namely, *Ensuring the availability and sustainability management of water and sanitation for all.*

Linked closely to the issue of poor sanitation is the extinction of the water supply system in Nigeria. Before the Nigerian independence (1 October 1960) and some years after, the society could, to an extent, talk about a system and corporation of water supply, which remains one of the basic needs of life in general. Present Nigeria inhabits only abandoned remnants of a defunct Nigeria Water Cooperation which, unfortunately and disappointingly too, has been left to go extinct by successive governments, regimes and policymakers. Many of the infrastructures have now seen decades of administrative inertia and, as such, have suffered dilapidation and rust. Typical of course of the Nigerian system, much of the archived information about the development and architectural plans of this corporation has been lost. So many of the underground networking and connecting linkages have been destroyed due to road constructions and other activities that paid no attention to such infrastructural and socio-economic edifices.

Today, the masses have been forced to face the challenges of what, under normal circumstances, should be a need and necessity to be made available by the government. The richer part of the citizenry has resorted to the provision of the borehole water supply system for themselves and perhaps for extended family members. Some larger communities try to pull efforts and resources together to realise a water supply system for the community, but numerous experiences show that most such projects fail due to the general attitude given to collective properties and interests. Such infrastructures suffer nonchalance and neglect. These challenges are felt more in the rural and suburban areas.

At the time of our research, Nigeria adopted in the year 2000 a National Water Supply and Sanitation Policy, entrusted to the three constitutional levels of government in Nigeria, namely the Federal Government, State Government and Local Government. However, a World Bank Assessment Report gave a situation report:

> A National Water Supply and Sanitation Policy has recently been adopted (January 2000). This has many positive points, but some weaknesses and inconsistencies. In particular, this policy makes the supply of adequate water supply and sanitation a right of all Nigerians, and gives responsibility to the three tiers of government, the private sector and the beneficiary. It recognises water as an economic good and the need to run water supplies as businesses, identifies the need for reform and for private sector participation, recognises the special needs of women and the poor, and the need to link improved sanitation with water supply. At the same time, it promotes unachievable targets for coverage, and recommends free water for the poor. There are inconsistencies in stated targets level of services, and sanitation policy is insufficiently developed; there are conflicts, for example, with other government actions regarding sanitation in other ministries. (World Bank Group 2000)

The above analysis exposes again the structural crack that impairs policy actualisation in the Nigerian system. Some of the states in Nigeria, with regard to the adoption of the Water Supply and Sanitation Policy, adopted State Rural Water Supply and Sanitation Agencies as well, charged mainly with implementing UNICEF programmes. During our research, many such projects were dormant. The seemingly functional ones have production capacities that do not meet the average demands of the State's population.

2.7.4 *Techno-Infrastructural Structures of Sin*

Mechanisation and machination have translated into necessary scientific and technological languages, which must never be ignored if ever any definitive evaluation of the world of today would be exhaustively done. Apart from the ethical questions and debates surrounding the avowed excesses of modern science and technology, such advances and innovations of human history have imparted so many positive strands in the society, particularly on human development.

However, in the Nigerian case, there is a failure of technological development and innovative culture, perhaps due to the general aftermath of colonialism, neocolonialism and imperialism experienced in almost all the African states. This deep experience impacts Africa's, and in particular, Nigeria's lack of the political will to enculturate colonisation and integrate globalisation. The level of Nigeria's dependence and enslavement in matters of technological infrastructures has enthroned a general culture of disdain for anything that is indigenous. Every product must be made in China, and worst still is the impunity with which these products come into the country in sub-standard forms and qualities. This situation has imparted so negatively on the general attitude of the citizenry. Consequently, the majority of the people have developed a burning passion for only "cheap articles", a strong disdain and, in some extreme cases, total rejection of anything indigenous, incessant fraud, forgery and piracy of indigenous products by designating them with foreign identities so as to have easier access to the markets, etc. The Manufacturers Association of Nigeria (MAN) has been ardent in bemoaning the Nigerian government's failure to address the techno-infrastructural dysfunctions within the system, which invariably stalls sustainable local industries and creates safe havens for exploitative foreign influences and dominance within the Nigerian economy. The concrete and most challenging example is China. The failure to create a sustainable and innovative environment to address this problem is identified as the structural sin which, in the Nigerian case, is intentionally planted within the system, especially by the corrupt actors who make fortunes out of it. They constitute themselves as mainstream importers and market price regulators, thus rendering the real sector moribund and isolated.

One of the most disastrous negative impacts that the techno-innovative infrastructural defect has brought to Nigeria is the crumbling and near extinction of the *agro institutions*. A country that once excelled in the agrarian economy has now failed woefully in mechanisation, empowerment and revitalisation of this sector due to a maddening grip on a mono-economy, summed up in "oil exploration and exploitation". The consequence of this neglect and failure is the painful draught in the generating capacity and the performance of the Internally Generated Revenue (IGR) being experienced in almost all the federating units. At the time of our research, only one federating unit, Lagos State, among the other thirty-six states, could boast of a functional IGR that could independently service the state. Other federating units write varied histories of persistent debts and non-performances in mainstream sectors.

With regard to ecology and techno-infrastructural structures, Nigeria has a history of ecological woes that have kept her on the international watch list over the years, which evokes an unceasing outcry against such man-made structural sin. The environmental woes seen in the Niger Delta, the menace of erosion seen in different parts of her geo-political zones due to poor canalisation as well as inappropriate disposal of refuse, which leads to obstructions and blockages of waterways, the challenging increase of greenhouse gas as the inevitable consequence of excessive use of power-generating sets for both domestic and commercial purposes—all these have for so long portrayed Nigeria as a nation on the crossroads.

As a general analysis, greenhouse gas emissions have been recorded as one of the major causes of the rise in climate change and global warming. The effects have clearly proved to be not only atmospheric. The observable health challenges remain a growing concern, especially in developing countries and economies, one of which is Nigeria. Although scientific study is still gaining more prominence today, so as to delineate and determine the empirical pieces of evidence bearing on the extent of damages already caused and the challenges which the future holds, the rate of unguarded emissions and the disinterestedness of policymakers in Nigeria to monitor emission-proliferation, speak volumes about our distasteful attitude of neglect and nonchalance. To our research, the greenhouse gas in question here is the *Anthropogenic Human-Induced Greenhouse Gas*. It contrasts with the *Earth's Greenhouse*

Gases, which, under normal circumstances and going with ecological principles, do not constitute a danger in themselves or to the general environment. It means, therefore, that Anthropogenic Greenhouse Gases are products of human activities that have maintained an increasing turn since the inception of the Industrial Revolution (about 1750) and the consequent rise in scientific and technological innovations. Thus, it remains the question of striking a balance between the acknowledgement of the trends, processes and systems of global human and economic development while at the same time invoking and applying the ethical principles that keep excesses in check. In the case at hand, the anthropocentric excesses have been projected (if allowed to move at their present pace) to adversely impact the ecosystem, biodiversity and livelihood of people worldwide (Wikipedia 2019), worst still, in developing countries.

The Nigerian case is a combination of factors. Apart from the known challenges of her oil and gas sector, which is not a monopoly of Nigeria, but rather the challenging contingencies shared by oil and gas producing areas all over the world, the instability of the power-generating sector has reduced power generation in Nigeria to private and individual venture. Consequently, the culture of generating sets excessively and recklessly employed for commercial, private and domestic purposes has been injected into the socio-cultural and economic system and mentality to the tune of madness in Nigeria. The present situation heightens to an alarming degree on a daily basis. The consequences have led to fatalities in some cases. It has become an economic paradox since the economy of the service provider in the "Gen-Set" technology thrives and prospers, however, at the expense and exploitation of the consumer-agent like Nigeria, which faces a future that is economically, ecologically and socio-biologically opaque. The danger and effects of the increase in greenhouse gases in the Nigerian case, therefore, manifest not only in the larger environmental challenges but also have heavy domestic concerns. Writing in "Generators: Providing Powers and Taking Lives", Osasona remarked,

> At least 60 million Nigerians own generator sets and more than N3.5 trillion is spent yearly on fuelling these sets. When one considers that 82.1 per cent of the 28.9 million households in Nigeria do not have access to electricity from the national grid and that Nigeria's electricity consumption per capita is the lowest in Africa, then the imperative of generators to public and private life in Nigeria becomes compelling [...] (and) [...] Tragically, Nigeria has no comprehensive policy framework regulating the use of generators or mitigating the environmental consequences of running generators in millions. Neither has policymakers given much thought to the threat that generator fumes pose to the productivity, health and overall wellbeing of the nation. (Osasona 2017)

The failure and the organised culpable indifference of Nigerian policymakers in this area, especially in the effort to provide institutions that would foster and validate energy-efficiency programmes, have been identified as inbuilt structural, socio-political and economic defects. The fieldwork undertaken during our research would provide quantitative analyses and statistics of this policy failure and the consequences ensuing from this sector of the economy. In this way, we would further expose the gullibility of the general Nigerian system to such man-made societal dysfunction.

2.8 Structures of Sin in the Nigerian Polity: A Capability Approach

To relate the whole concept of structural defect in the Nigerian system with the recent developmental and socio-economic theses associated with the "Capability Approach" is to evaluate the extent to which such ideological and institutional patterns could affect and influence the human agents within a given socio-political and socio-economic setting. Structural patterns of a society are the homes of the relevant institutions that go into the "inclusive development" of such society, of course, with the human agents at the centre of interaction. Consequently, therefore, a healthy structure would produce a healthy interaction. However, when the structural pattern that defines a society is defective—be it in the form of weak and visionless policies, selfish and incompetent political leaders, or lack of development institutions in general. It produces disorientation in the general system and impacts the human agents negatively.

2.8.1 Understanding Capability Approach

One of the recent breakthroughs that have been made in the study of human development in general and economic development, in particular, has been the "capability approach". It was originally developed by the Indian economist and philosopher Amartya Sen and subsequently elaborated in the works and writings of the philosopher Martha Nussbaum. The application has gained prominent entrance in development thinking, welfare economics, social policies and political philosophies of modern political thoughts (Robeyns 2019). The capability approach focuses on evaluating development that is basically inclusive and holistic; development concepts that do not isolate the human person and all he or she is endowed with to pursue a kind of life that dignifies the human person as an *autonomous being* (Immanuel Kant). For Ingrid Robeyns,

> The core characteristic of the capability approach is its focus on what people are effectively able to do and to be, that is, on their capabilities. This contrasts with philosophical approaches that concentrate on people's happiness or desire-fulfilment, or on theoretical and practical approaches that concentrate on income, expenditures, consumption or basic needs fulfilment. A focus on people's capabilities in the choice of development policies makes a profound theoretical difference, and leads to quite different policies compared to neo-liberalism and utilitarian policy prescriptions. (Ibid., p. 5)

The analysis above describes Sen's focus in the entirety of his work, namely, to handle any social evaluation in such a way that the central focus should bear on what people are able to do and be. It is an approach that should be based more on the quality of their lives, on activities and policies geared towards the removal of obstacles that militate against their authentic freedom to live the kind of life they find valuable, in other words, the kind of life that dignifies the human person. Two core normative claims deserve emphasis here, namely "[...] first, the claim that the freedom to achieve well-being is of primary moral importance, and second, that freedom to achieve well-being is to be understood in terms of people's capabilities, that is their real opportunities to do and be what they have reason to value" (Robeyns and Byskov 2021). For Sen, therefore, institutions and policies should make human capabilities the central focus for any development concept to qualify as being inclusive:

> The capability approach to a person's advantage is concerned with evaluating it in terms of his or her actual ability to achieve various valuable functionings as a part of living. The corresponding approach to social advantage – for aggregative appraisal as well as for the choice of institutions and policy – takes the set of individual capabilities as constituting an indispensable and central part of the relevant informational base of such evaluation. (Sen 1993)

Taken as a fundament in the capability approach is the complementarity between being and the end. While the "being" describes the aggregation of the different capabilities in a person, the "end" describes the attained *functionings* as a result of either the proper channelling of the capabilities or the mismanagement or abuse of the same. Therefore, capabilities are means that always have the open potential for either positive or negative ends. The end in question within the capability approach has intrinsic importance. In other words, it reserves within itself the ability to reveal the real functionings of well-being and development—attained or stagnated. The "means are only instrumental to reach the goal of increased well-being and development" (Ibid., The Capability Approach: An Interdisciplinary Introduction, p. 6).

Central to the capability approach and a necessary tool in analysing development is the need to evaluate well-being and development in terms of people's capabilities to trigger various *functionings* in their lives. In other words, capabilities in terms of the human persons' naturally endowed abilities which make them effective to "[...] undertake the actions and activities that they want to engage in and be whom they want to be" (Ibid.). In analysing human and economic development or development in general, therefore, the individual (the human person) is and remains a *functional possibility* with numerous open-ended capabilities that define his or her motivations and freedom in life. For Sen, the attained and consummated capabilities—the functionings—define authentic development, and such development is freedom. In other words, wherever the indices for true development—people-oriented development—fail, be it political, socio-cultural, socio-economic, etc., authentic freedom would be invariably jeopardised, while human potentials loaded in the individuals' capabilities to function become grossly stagnated.

The capability approach, therefore, is the people-oriented approach, which has, since 1990,

> [...] inspired a new way in which the United Nations Development Programme (UNDP) assesses Human Development (Sen, 1999). Hitherto, the focus was on Gross Domestic Product (GDP) and the human capital, suggesting that 'economic development was not human-centred and that, development was in practice inadequately conceived and operationalised as economic growth (Gasper, 2002). In this practice, human beings were/are seen as the means rather than the end of production (Sen, 1999). (Ntibagiriwa 2014)

However, Sen's ideas in this developmental approach can be traced substantively to Marxian and Aristotelian ideas that laid emphasis on the centrality of truly human functionings as a prerequisite for human flourishing, in other words, as fundaments for authentic development. Sen also acknowledged the relevance of Adams Smith and Karl Marx, whose concept of well-being requires the necessary inclusion of human activity and capability to function as determinant factors (cf. Ibid., p. 272). Utilising these backgrounds in his innovative thought, Sen acknowledged in his "Resources, Values and Development" the need for the individual's development freedom seen in Marx's quest for the replacement of the "domination of circumstances and chances over individuals by the domination of individuals over chances and circumstances" (Sen 1997). In other words, no policy that ever threatens the freedom and well-being of the individual should be acceptable if development should ever be authentic and sustainable. Policies, therefore, should have as first principles the maximisation and harnessing of the people's capabilities into viable ends (functionings) that necessitate well-being and development in the subject, the *Mmadụ*, the human person who always stands at the receiving end—in Sen's terminology, the "agent". Two critical words, therefore, are fundamental in Sen's capability approach, namely, *Functioning and Agency.*

2.8.1.1 Functioning and Agency

Since people reserve the ability to lead the lives they value and have reason to value, it means that in analysing development and freedom, the people must be seen as agents and the driving force. Thus, fundamental to Sen's capability approach are these two concepts, Functioning and Agency, which are both inclusive and complementary. For Sen, functionings are achievements. They are the physical as well as mental states of people's lives, which emanate from the *dialectics* of people's capabilities. Such functionings define the people's ability to participate in the life of their society and in a manner they themselves have reason to choose. These functionings are self-evident and are rooted in the principles of right reason. They are fundamentally influenced and driven by the human natural inclination to the basic needs of life.

It should be noted that Sen's original components in his assessment of human capability feature five basic concepts, namely: (a) the importance of real freedom in the assessment of a person's advantage; (b) individual differences in the ability to transform resources into valuable activities; (c) the multi-variate nature of activities giving rise to happiness; (d) a balance of materialistic and nonmaterialistic factors in evaluating human welfare; and (e) concern for the distribution of opportunities within society. These concepts would be further developed, culminating in the ten basic classifications of functionings in the capability approach. This was done in collaboration with the political philosopher Martha Nussbaum, the development economist Sudhir Anand and the economic theorist James Foster. Our focus and references would concentrate more on the closeness of thought patterns existing between Sen and Nussbaum, with sporadic highlights of their differences where necessary. Below, therefore, is a presentation of Sen's classification of functioning in an analytical comparison to the classification made by Martha Nussbaum (Table 2.2).

Table 2.2 represents predominantly a collaboration of ideas approached from unique backgrounds of knowledge, experiences and conventions. For Robeyns, fundamental to their differences,

Table 2.2 Amartya Sen's and Martha Nussbaum's versions of basic functionings

	Martha Nussbaum: Central human functional capabilities	Sen: Functionings
1. Life	Not dying prematurely	Avoiding escapable morbidity and premature mortality; longevity
2. Bodily health	Having good health, having opportunities for sexual satisfaction, being able to move from place to place	Being adequately nourished, nourishment, being in good health; being free from malaria; being well-sheltered
3. Bodily integrity	Ability to avoid unnecessary pain and non-beneficial pain, to have pleasurable experience	Move about; travelling
4. Senses, imagination and thought	Access to information, education, etc	Being literate, Cultural and Intellectual pursuits
5. Emotions	Ability to have attachment to things and persons outside ourselves	Being happy; being close to people one would like to see
6. Practical reason	Being able to form a conception of the good and engage in critical reflection, being able to seek employment and participation in political life	Taking part in the life of the community
7. Affiliation	Ability to live for and to others, concern for others, social interaction; capability for justice and friendship	Ability to entertain and visit friends; being close to people one would like to see; having self-respect, appear in public without shame
8. Play	Being able to laugh, play, enjoy recreational activities	Vacationing
9. Control over one's environment	Ability to live one's life and nobody else's (personal choice), freedom of association, integrity of personal property, being able to live one's own life in one's own surroundings and context	Ability to entertain and visit friends; vacationing; travelling; being employed; being decently clothed
10. Other species (further addition by Martha Nussbaum)	Concern for and in relation to animals, plants and the world of nature	Concern for and in relation to animals, plants and the world of nature

Source Ortrud Lessmann, 2008. (Adapted from Ortrud Lessmann 2008. In: Ntibagiriwa, S., Local Cultural Values and Projects of Economic Development: An Interpretation in the Light of the Capability Approach. 2016, Geneva: Globethics, pp. 28–30). Martha Nussbaum, as a developer of Sen's Capability Approach, added the concept, "other species" and gave further analyses of the basic functionings as an elaboration to the work (Katzer 2010)

> [...] is that Sen's work on the capability approach is closer to economic reasoning that Nussbaum's and more attuned to quantitative empirical applications and measurement. It lies closer to those fields and paradigms that are characterised by parsimonious, formal, non-narrative, and axiomatic modelling. (While) Nussbaum's work is much closer to traditions in the humanities, such as narrative approaches. Her work engages more with the power of narratives and poetic texts to better understand people's hopes, desires, aspirations, motivations and decisions. (Ibid., The Capability Approach: An Interdisciplinary Introduction, p. 7)

In Sen's opinion, these capabilities manifest themselves massively in the lives of individuals in the society, particularly in their feelings and activities in a participatory and interactive manner within their societies. With the range and complexity of participation and not being ignorant of the conflict of interests, the individuals become agents who must work within systems and methods that would be able to ensure individual freedom and development within the complex network of interaction and participation. Within this network, the capability approach asks the basic questions, as Robeyns articulated them:

> [...] it asks whether people are being healthy, and whether the resources necessary for this capability, such as clean water, access to medical doctors, protection from infections and diseases, and the basic knowledge on health issues, are present. It asks whether people are well-nourished, and whether the conditions for this capability, such as sufficient food supplies and food entitlements, are met. It asks whether people have access to a high quality of education, to real political participation, to community activities which support them to cope with struggles in daily life and which foster real friendships, to religions that console them and which can give them peace of mind. For some of these capabilities, the main input will be financial resources and economic production, but for others it can also be political practices, such as the effective guaranteeing and protection of freedom of thought, religion or political participation, or social or cultural practices, social structures, social institutions, public goods, social norms, traditions and habits. (Ibid., p. 24)

The whole network shows that there is inevitably a confrontation between the individual and the life and activities existing outside of him or her. Given the inevitability of tensions and conflicts between the individual and all other forces and interests existing outside of him or her, the attainability of the functionings would, therefore, remain utopic if a criterion is not defined so as to mediate the conflicts and tensions and, invariably, establish a common ground, where the individual could exist and function authentically within the collective interests and, at the same time, remain *an agent and not a patient.*

On the other hand, such criteria must help define principles of common interests, which the individual must learn to integrate and respect if the individual functionings are to be attainable. Both Sen and Nussbaum, in this regard, proposed different (independent) but complementary methods of achieving the functionings respectively, namely, *public debate* and *overlapping consensus.*

2.8.1.2 Public Debate Versus Overlapping Consensus: Towards a Congruence

Sen and Nussbaum proposed two different but complementary approaches to the question of how the functionings, which are results of well-harnessed capabilities, could be mirrored to success, given the tensions and interactions between the individual and the world existing outside of him or her. Thus, on the one hand, Sen, in his position, maintained that this method should be *Public Debate and Reflection.* In other words, a kind of democratic approach that would strike a balance between the individual person's freedom of choice and the general common good and welfare. In this sense, therefore, a debate involves participation and guarantees the freedom of expression and opinion. It offers a forum for reflection on how best to find common ground among conflicting possibilities. Sen's method, apart from emphasising the individual as the main determinant of what authentic development should describe and mean, does not, at the same time, lose sight of the larger influential factors outside of the individual, which play important roles in shaping the life of the individual himself or herself. Thus, appealing to public debate and reflection, in whichever form it should be organised, must never be extricated in the general development strategies if ever such strategies should be seen as sustainable.

Nussbaum, on the other hand, maintains the mainstream characteristics of the capability approach, by emphasising that the basic capabilities remain integral to a functional life, a life worth living and "**any life that lacks any one of these capabilities, no matter what else it has, will fall short of being a good human life**" (Nussbaum and Clover 1996). However, unlike Sen's option for public debate, Nussbaum proposed the method of arriving at the functionings to be *overlapping consensus.* The concept is originally rooted in John Rawls' idea in his "A Theory of Justice", where he opined that despite the apparent inconsistencies that are evident in the citizens' conceptions of justice, it would still be possible, through overlapping consensus, to arrive at principles of justice that would mediate the perceptible differences. The method, therefore, seeks "a conception by which people of differing comprehensive views can agree to live together in a political community" (Nussbaum 2001). In proposing this method, Nussbaum believes that, while the basic capabilities form the nucleus of functional life, individual approaches may suffer and be confronted with conflicts of interests which inevitably need to be mediated in order to achieve stability, progress and development. Thus, the consensus of ideas, interests, values and institutions needs to overlap so as to create a more complementary platform for the agents who have the right to lead the kind of life they individually have reason to value.

While not hanging strictly on the syntax of words associated with the two proposed methods, our research sees congruence within the ideas and approaches between Sen and Nussbaum, namely: Whether the approach to maximising the capabilities has to be publicly debated or the necessity of allowing some consensus to overlap become paramount, the mediating factor becomes rooted in the ideas of conformity, unification, agreement, cooperation, inclusion—simply put democratisation. In this sense, the perfect ground would have been set where agents in the society could responsibly, that means in freedom, pursue their basic aspirations and motivations, in other words, exercise their autonomy within divergent socio-political, socio-economic, cultural and religious spaces. In this regard, *our research takes off from the premise that the human agents in the Nigerian case are daily victims of heinous lack or total absence of consensus forum or public debate, thereby suffering man-made and organised stifling of the basic capable inclinations that yearn for autonomy and actualisation.* The unfortunate situation plays itself out in the persistent widening of the gap between the "haves" and the "have-nots"; the dehumanising inequality that is inbuilt into the social system; the practice of double-standard before the rule of law; the mortifying presence of abject citizenry in a land greatly endowed with mineral, natural and human resources—and the list goes on. It is evidence of a polity's failed politics of inclusion.

2.8.1.3 The Individual as the Autonomous Agent

In using public debate and reflection as a method towards achieving functionings or seeking the common ground, where consensus should and ought to overlap, the whole process culminates in a deeper level of participation which would inextricably unite opinions, peculiarities and interests. It is the participation of the moral agents—the human persons—who, at the same time, are embodiments of their individual autonomies. In this sense, using the concept of agency in the capability approach describes two basic characteristics that help to ground development processes on the human agents and on people's cultural values, namely the *ability to act* and the *respect for autonomy*.

Agency, in general, describes a factor that is intrinsic to the capability of acting or being in an activity. Applied to an individual human person, it describes the ability of a person to be capably and actively involved in shaping his or her destiny. It means that the individual is not and ought not to be a passive or dormant recipient of the consequences of decisions already made and concluded without the individual's involvement, which may sometimes conflict with the individual's worldview and freedom. In other words, decisions or policies that leave no room for complementarity, participation and dialogue. Therefore, "[...] lack of agency means not acting, being passive, patient or recipient, and eventually allowing others to act on one's behalf [...] (and this may arise because one is not allowed the opportunity to exercise such capability)" (Ibid., Local Cultural Values and Projects of Economic Development: An Interpretation in the Light of the Capability Approach, pp. 40–41). This lack of opportunity or, better put, the non-provision of opportunity for a viable agency forms the bone of contention in dealing with Nigeria's in-built structural dysfunctions or *structural sins*.

A lack in the agent, especially when the forces acting upon the agent come from without, makes the agent bereaved of autonomy. Such external circumstances could be "poverty, powerlessness, exploitation and a lack of those things, such as education, health, food, which could increase our spiritual and material capacity" (Ibid.). They are usually man-made and could be great stumbling blocks to the human being's natural inclination to go beyond himself or herself—self-transcendence—which invariably guarantees autonomy. Hence, another concept intrinsically connected to the term agency is the issue of autonomy.

Autonomy, in reference to a person's capabilities, is understood to mean "[...] the capacity to be one's own person, to live one's life according to reasons and motives that are taken as one's own and not the product of manipulative or distorting external forces" (Christman 2018). For Sen, this autonomy should not be understood as mere selfish interest or preference, which would rather influence agency and the attainment of functionings negatively. The autonomy of the agent should be inclusively exercised for the proper integration of the agent and other pressures and influences outside of him or her. In other words, autonomy, in this sense, is never inimically individualistic. It is rather interactive since "the options that a person has depend greatly on relations with others and on what the state and other institutions do, as well as those (individual) opportunities that are influenced by social circumstances and public policy" (Drèze and Sen 2002).

To evaluate agency and autonomy in the context of Nigeria is to focus on the various ways and manners it exposes itself as a nation with a failed human development index. The state of affairs gives the country away as a man-made haven for political and socio-economic *patients and victims* whose natural capabilities are, on a daily basis, being sacrificed at the altar of social and political misdeeds and recklessness. The natural autonomy of the agents in Nigeria remains in chains since both policy and polity remain the double victims of manipulative servitude, which brings development in general always on its knees, given the fact that development policies, if any, are grossly not *people oriented*. This is, in the Nigerian experience, a structural dysfunction and defect, a structural sin. Our research, therefore, is poised to propose a need for an alternative, namely: the need to revitalise the human capital in Nigeria, the need for empowerment and the need to train capabilities so as to properly integrate the autonomy of the individual agents with the massive network of the state's political and socio-economic life. This work believes that questions surrounding the issue of capabilities embody all vectors that should translate potentials into functionings, particularly through the proper harnessing of abilities, exploring of possibilities, creating opportunities, discovering accessibilities and maximising availabilities, all with regard to authentic development of the agents, the *Mmadụ* and, consequently, for the attainment of true freedom, in other words, development.

2.8.2 The Three Angles of Applying the Capability Approach to the Nigerian Situation

The research question uses the premise that the Nigerian approach to development, in general, is culpably deficient, and such deficiency manifests itself in almost all the spheres of her existence as a polity. Applying the sense behind the capability approach as a relevant tool of approach, therefore, is to make our general approach evidence-based and more practical. Our principle is to seek viable means of application that would be able to reflect the polity as concretely as the reality in Nigeria depicts and to explore grounds for making our research and propositions more relevant in the people's (agents') lives. Our research chooses to view these three angles as three steps to a *Rediscovery of the Agents' Autonomy* towards (a) choosing the life they have reason to value, (b) the development of capabilities that make them agents, not patients, (c) accessing development vectors that lead to authentic freedom and to which they have right.

2.8.2.1 Rediscovery of the Agents' Autonomy Towards Choosing the Life They Have Reason to Value

The whole issue of awareness is fundamental to the concept of identity. It asks the fundamental questions about the existence of a united group of people, their understanding of their corporate existence, beliefs and customs; their manner of relating with the world around them and the influence each has on the other; their uniqueness within the diverse world and societies around them etc. To trace and understand these fundaments play a significant role in the choice of life an agent, *Mmadụ,* chooses and has reason to value. Ntibaririrwa articulates this line of thought this way:

> First of all, the ability of people to lead the life they value starts by questions of self-definition, which, in turn, leads to self-awareness, and awareness of the context and the potentialities people have. Questions such as: Who are we as a people? How should we live given who we are, our circumstances and the circumstances of the world we live in? – are about people's identity and their unique way of living, of conception of the world around them, and the way of doing things. (Ibid., Local Cultural Values and Projects of Economic Development: An Interpretation in the Light of the Capability Approach, p. 34)

To understand the foregoing more succinctly, emphasis should be laid on people's culture and their perception of who they are. When such perception is absent or gravely distorted, the agent's autonomy, or, simply put, human freedom, becomes mutilated with false policy imageries that misrepresent all that development can be and mean. The general situation in Africa, and particularly in Nigeria, is the battle for a rediscovery of identity. In numerous cases, it is the fight against the feeling of a total loss of identity. This is partly, on the one hand, due to some excesses of colonisation and disappointingly, on the other hand, due to a persistent wave of neo-colonisation, which Kwame Nkrumah rightly identified as the *Last Stage of Imperialism*.

We would argue that there is an urgent need for a rediscovery of the autonomous self in Nigeria, both individually and collectively, if ever the matrices of development are to be actualised within the general polity. The crises of identity and the stark loss of autonomy are monstrosities that have left both institutions and ideologies clueless and dysfunctional. On the individual level, the situation leaves the agent disoriented and visionless, full of ideas but handicapped in means of application. This has led to a common description of many average Nigerians as failures. Failures, of course, but only within a system that enslaves and denies them their rightful opportunity towards capability development. Such a situation is simply the oppression of the independence of the human spirit. Ntibaririrwa summed it up thus, "In the process of development, capability as "the freedom of people to lead the lives they value and have reason to value allows people first to bring to their awareness those values that are the deep source of their aspiration" (Ibid., p. 38). It has been described as the secret behind Japan's economic success (Morishima 2016). The Nigerian ethos towards the development of capabilities is, in many cases, disoriented and structurally dysfunctional. Scholars argue that the general failure of development in the so-called developing countries of Africa, despite breakthroughs of modernisation or the most recent concept, globalisation, has been the failure to articulate the universal principles guiding modernisation or globalisation from the pivot of independent spirits and ethos. Nigeria is not exempt. Our research presents the "why" of the concrete situations in the Nigerian case and attempts to proffer some remedial steps.

2.8.2.2 Rediscovery of the Agents' Autonomy Towards the Development of Capabilities that Make Them Agents, Not Patients

In choosing the life one has reason to value, varied sets of vectors are brought into play. They serve as the motivational forces of empowerment. In the Capability Approach, they are the capabilities that define the agent, *Mmadụ*; they are the vectors of empowerment. Such capabilities are intrinsic and, as Nussbaum implied, "any life that lacks any one of these capabilities, no matter what else it has, will fall short of being a good human life" (Ibid., Women, Culture and Development: A Study in Human Capabilities, p. 185). When these vectors are made viable, the *Mmadụ,* the human person, takes his or her proper place as an active participator in shaping the society and the world around him or her and avoids being reduced to passivity in a world of interaction, given his or her natural kinetics for activity, not passivity. In other words, the individual in the context of development should take the place always as an *active agent* of development, not the *passive patient* of misery, exclusion, unemployment, intimidation and other socio-political and socio-economic maladies.

In the Nigerian case, one speaks of a lack or the denial of agency due to oppressive social and political anomalies that are inbuilt within the general ideology. The consequence is the growing number of human resources that feel molested and socio-economically and politically marginalised. Such numerous societal *patients* are, on a daily basis, forced to grapple with the reality of imposed poverty which further holds them prisoners in a country and system where they are all eligible to be freeborn. They feel deprived, dejected and disoriented. The argument of Alan Gewirth (2007), as paraphrased by Ntibagirirwa, is in this context ad rem, namely, "Focusing on the deprivations which poverty can cause, […] people do lose their capacity to act as moral agents because their freedom and well-being are undermined by their lack of the means of subsistence" (Ibid., Local Cultural Values and Projects of Economic Development: An Interpretation in the Light of the Capability Approach, p. 41).

Basically, every average human person, aside from the natural cases of incapacitation, is endowed with various capable vectors (capabilities) that can help secure the needed means of subsistence in life. These capable vectors can only be made viable when corresponding grounds of application (opportunities) are made possible, available and receptive. The Nigerian case is viewed by our research as the case of insufficient, or in most cases, the total denial of opportunities, especially the absence of institutions that are supposed to provide adequate platforms for human capabilities integration.

Trapped in the web of such denial, the natural inclination in the human person to transcend the self and confront the realities that be and make sense out of them remains stifled and stagnated. Consequently, the inbuilt powers and talents suffer inertia, leading to confusion of ethos and values, crises of identity and inauthenticity—in a more global convention and usage, it leads to underdevelopment or stunted progress. This is the case in Africa, and particularly Nigeria. Our research identifies this as the neo-colonial weapon of modern tyranny and slavery. Most colonial countries of Africa, Nigeria for one, suffer immensely from this deprivation which holds the agency captive due to the imposition of "[…] certain types of development or conditions (from without) that prevent them from defining policies of development as they understand it or as their reality dictates it" (Ibid.). The foregoing is true since agency predicates the ability of the agent concerned to be aware of its autonomy exercised critically as well as substantively (Ibid., pp. 42–43). Ntibagirirwa elaborates further,

> Agency presupposes that people who act are aware of their autonomy. According to Gasper (1997:298), autonomy can be understood from two points of view, namely, the critical and the substantive points of view. The critical perspective of autonomy has to do with the 'ability to form, not just adopt, one's own conception of the good'. Here autonomy has to do with the individual decision-making of various matters of one's society […] Although there may be certain aspects of the African culture that may need some caution, the issue is not the incompatibility between African mentalities and development; rather, the issue is whether the beliefs and values that are at the core of African identity can be validated in terms of policies of economic development […] (More importantly) autonomy should be seen in terms of dialectical interaction between the individuals and their community in search of better horizons, in particular, the horizon of development [.….] The substantive autonomy refers to the content of agency. According to Black and Mooney (2002:198), 'substantive autonomy emphasises the processes of formation of an agent's desires, beliefs and emotional attitudes, including attitudes and beliefs about herself'. (Ibid.)

The situation plays itself out within the so-called global markets, trade deals and pacts, which, in most cases, oppress and enslave local economies, making them "patients" of the unjust aggressor. Such situations manifest themselves as well in the politics of the allies. Such pacts, deals and relationships are not evil in themselves. However, a great deal of oppression and servitude lies within their aims and objectives. To proffer an alternative, specifically for Nigeria, is the objective of our research.

2.8.2.3 Rediscovery of the Agent's Autonomy Towards Accessing Development Vectors that Lead to Authentic Freedom and to Which They Have Rights

Access to development vectors within a given polity is a basic right of every *agent* on the one hand. On the other hand, the provision of the necessary institutions, in other words, creating opportunities that open access to socio-political and socio-economic vectors, providing institutions capable of absorbing the capabilities of the agents of development, become the constitutional responsibility of the policymakers, leadership and governance in general. The ability to access the interrelatedness and to invest in the network of varied development vectors form the core of Sen's understanding of development as freedom. In other words, "[...] development has to be assessed in terms of the expansion of capability or real freedoms, that is, in terms of how development enhances the freedom of people to lead the lives they value and have reason to value" (Ibid., p. 45).

Making a case for this rediscovery in the Nigerian context is to bemoan the failure of the policymakers, after over sixty years of independence, to harmonise, as a functional and structural unit, the three dimensions of politics, namely, the polity, politics and policy. Such a failure hits hard on the general human and socio-economic development, thereby giving rise to increasing rates of unemployment; disoriented empowerment policies in both human and economic resources; increase in crime and youth delinquencies due to joblessness and loss of perspective; abuse of drugs and narcotics in a bid to handle depression and frustration; increase in religious, faith and healing exploitations of the economically displaced citizens who often times turn to seek signs and wonders as a remedy as well as the massive exodus of indigenous manpower in search of economies that offer them the needed opportunities, etc. Autonomy or freedom to capability development and expansion, therefore, remains a prerequisite for any functional social, political and economic structure since development generally must be grounded on the principles of self-determination, under which the freedom of choice plays the key role as a basic right of the agents. In the Nigerian case, this is an urgent demand. The clarion call remains the need for development policies that recognise people's agency as the determinant factor that would eventually guarantee the concrete formation of people's own identities. Failure to do so leaves Nigeria hanging on a balance in issues of general development, particularly the empowerment of agents of development—*ndị Mmadụ*, the people.

These relevant approaches towards "rediscovery" discussed above represent the necessary measures towards the institution of policies and conventions that would work to maximise people's capabilities for the attainment of the relevant functionings that aid authentic development freedom. Denial of access to the means towards these rediscoveries amounts to abuse of the agent himself or herself and, consequently, stifles the capabilities within him or her. Such denial leads to failures in exploring and translating the heterogeneity and the relativity of these capabilities in the agents into social goods. It is exactly this denial that Sen describes as poverty (Sen 2010).

2.8.3 The Relevance of "Functionings" in the Nigerian Context

The goal of the capability approach is the attainment of the relevant functionings, which give the agents quality and dignified life, of course, only within the functional social structures that provide the relevant institutions. In other words, there is a greater percentage of the possibility of maximising functionings within a healthy and functional social structure than in a dysfunctional and defective one. Sen articulated the components that better define a functioning in relation to capability thus:

> A functioning is an achievement, whereas a capability is the ability to achieve. Functionings are, in a sense, more directly related to living conditions, since they are different aspects of living conditions. Capabilities, in contrast, are notions of freedom, in the positive sense: what real opportunities you have regarding the life you may lead. (Sen 2009)

To examine Functioning in the Nigerian context, one needs to analyse the living conditions of the agents as well as the extent to which they could autonomously access life opportunities that exert much influence on living conditions. One can appropriately infer that the better and the more people-oriented the living conditions of a given people are, the more the chances open themselves up for the achievement of the basic functionings, which improve the lives of the people and the society itself and ensure socio-political and socio-economic stability.

As the first step to this approach in the Nigerian context, one seeks to understand the concept of achievement. Achievement is a viable, relative and functional destination that continues to shape the life of the agent that has arrived at such destination. It is a state of "being" (Dasein) characterised by the agent's ability to access and bring the various vectors of the capabilities in and outside of him or her into practice. In achievement, the agent remains the pivot around which the interplay of development vectors revolves. In this constant interplay, what is achieved remains variable and not stagnant, in other words, what is achieved remains open to vestiges of possible actions for further development and sustainability. Again, the word interplay is well-chosen in this context since the agent's achievement cannot be ontologically separated from his or her community. In other words, the achieved functionings in the capability approach remain functionings with relevance to the broader vectors of development outside of the individual. Sen argued this interdependence thus:

> A person as an agent need not be guided only by her own well-being. [...] If a person aims at, say, the independence of her country, or the prosperity of her community, or some such general goal, her agency achievement would involve evaluation of states of affairs in the light of those objects, and not merely in the light of the extent to which those achievements would contribute to her own well-being. (Sen 1995)

This academic work identifies, in the Nigerian case, constant fracture in the interplay and interdependence with regard to the general well-being of the agents and, thus, presents the Nigerian system as structurally deficient in vectors of socio-political and economic success, achievement and development. In other words, the vectors leading to the well-being of the agents, *ndị Mmadụ* (the masses, people, individuals) in relation to the outer society, remain grossly wounded by factors that work against sustainable development. It is the failure to integrate the *internal capabilities* and the *combined capabilities* using the analyses given by Marta Nussbaum. For Nussbaum, therefore, "One job of a society that wants to promote the most important human capabilities is to support the development of internal capabilities—through education, resources to enhance physical and emotional health, support for family care and love, a system of education and much more" (Nussbaum 2013). The identifiable ideological and policy maladies in the Nigerian context have been examined and labelled structural sins, structural dysfunctions or structural defects, and they need both ideological and revolutionary cleansing, restructuring and remodelling.

The second step in this approach is to evaluate the state of affairs concerning the living conditions of the agents themselves. When, as at the time of our research, the socio-economic status of Nigeria in the international community is being projected to plunge even deeper into multiple crises, with a greater number of poor people living within her geo-political bounds in 2050, then the attestation would be affirmed, that the living conditions of her citizenry have no prospect. To talk about the living conditions of a people is to examine both the socio-political and socio-economic institutions, at both macro and micro levels, which serve as catalysts to the natural abilities of the autonomous agents—the masses—the political and socio-economic subjects and their diverse wealth of capabilities. To this effect, therefore, it should be taken as an axiom that those achievements, which are functionings in praxis, are necessarily predicated on the general living conditions of the agents, groups of agents or the society in general.

However, when the above-mentioned precondition, namely the presence of real functionings, is absent, a form of massive and, in most cases, an artificial blockade remains injected into the general system and becomes a deterrent to the general living conditions of the agents, since such blockade inhibit the social contingencies, which integrate variations and dynamics capable of translating income resources to march the kind of lives people can lead and have reasons as well to lead. Sen himself noted four of such important contingencies, namely: *Personal heterogeneities* with regard to those human characteristics relating to age, gender, disability and even susceptibility to illness; *Diversities in the physical environment*, with regard to the environmental conditions—climate, temperature, natural disasters—which are factors in the effort to convert personal resources to functionings; *Variations in social climate*, with regard to the specific nature and conditions available within the community relationship—healthcare, epidemiology, conditions of the public education system, etc.; and *Differences in rational perspectives*, with regard to patterns of behaviour which are taken to be conventional in a particular society or among society. The different standards of social variations substantially influence the translation of income and resources into functionings (Ibid., The Idea of Justice, pp. 255–256).

Perhaps one of the deadliest blows dealt to the general living conditions of the masses in the Nigerian context is the absence of relevant institutions for authentic human and economic development. Such deficiencies are blamed, to a greater extent, on the lack of the political will of the policymakers to deliver people-oriented development strategies. Such deficiencies retard the chances of the achievement of functionings, isolate the participatory rights of the agents and at the same time force them into becoming patients in a political economy within which they have natural rights, privileges and free access. The Power/Electricity institution where our fieldwork has been undertaken proves outstanding among other areas of recorded failures. Power/electricity generation failure has impacted so adversely on the general polity to the extent of reducing Nigeria to a mere consumer country due to discouragement and flight of investors, manufacturers and entrepreneurs. The collated data of the fieldwork shows a celebrated misplacement of values and man-made socio-economic inefficiency.

2.8.4 The Fieldwork: Proving the Nigeria's Fragile Structure

2.8.4.1 General Case Study

Inefficiency in Power Supply and the Proliferation of Private/Independent Power Generation – The China Connection. Proving the structural defect in the Nigerian power/energy policy system.

Ethics, in general, has both analytical and normative bindings and because our research was done in the area of Social Ethics in the Faculty of Catholic Theology and in collaboration with the Doctoral Programme on Environment and Society of the Rachel Carson Centre, LMU University, Munich, Germany—our research was present at the field to examine concrete facts and make conclusions based on available and analysable data. As a case study was the analyses of how and why the power/energy supply in Nigeria has failed or, in the main, has become chronically inefficient. This has given rise to the unfortunate proliferation of power generation for both domestic and commercial purposes by the masses themselves through generator sets, with all the challenges such practice poses to both health and the environment. This is done bearing in mind that the main purpose of fieldwork is to bring ideas and theories to bear on available facts, thereby purifying knowledge of just mere speculative assumptions. Therefore, our research has a specimen, an audience and a result.

2.8.4.2 Methodology

The research work made use of a Questionnaire, Data Collection/Collation, as well as Quantitative and Qualitative analyses. These four components of the research methodology used here describe the four steps that aided the completion of the fieldwork. The questionnaire was used as a research instrument, which contains a series of organised questions and/or other methods of enquiry used in gathering information from a group of targeted audience and respondents. The body of the set questions aided data collection in the questionnaire. A qualitative collation of the data later would present the data as gathered and measured within the targeted variables. Thus, they would aid us in attempting independent answers and evaluating the outcomes both quantitatively and qualitatively. Quantitative analysis, as employed here, helps in calculating and systematically measuring the collected data, often statistically, while indicating correlations and variables where they occur. Qualitative analysis also helped to read meaning into the collated data, and to define symbols and concepts associated with the research and data presented.

2.8.4.3 Location

This research was done in Nigeria, West Africa. The location was Anambra State, a State within the South-Eastern part of the country, particularly within two Local Government Areas: Nnewi (North) and Onitsha (North).

2.8.4.4 Cases: Targeted Audience

The targeted audience was made up of traders within the two biggest markets in the state—the Onitsha Main Market and Nnewi Main Market- as well as selected families, companies, organisations and some church communities.

2.8.4.5 Variables

To prove policy inefficiency in Nigeria as a structural dysfunction within the Nigerian polity, using the power/energy sector as a specimen, and to propose a possible, alternate and workable solution to constant power/energy inefficiency. Again, the call to develop and invest in Alternative and Clean-Energy Technologies.

2.8.4.6 Observations

Total number of questionnaire distributed: 300

Total number of collated data from respondents: 236 (representing varied opinions)

Total number of non-respondents: 64.

Table 2.3 (a) Case study 1: reasons for power/energy inefficiency

	Very great extent	Great extent	Average	Hardly	Not at all
The general problem of mangled megawatts in the power/energy sector	227	9	–	–	–
Poor and sub-standard power/energy infrastructure	203	27	6	–	–
Corruption	229	37	–	–	–
Vandalism	2	8	119	107	–
Embezzlement and misappropriation	198	34	4	–	–

2.8.4.7 Collated Data

Quantitative Analytical Presentation of Collated Data—The insightful research done by the *Deutsche Gesellschaft für Internationale Zusammenarbeit (GIZ)*,[5] a German Agency in connection with the European Union, would be substantially referenced as substantive and scientific authority.

The graphic design and sample of the questionnaire are presented in Appendix I, while the quantitative cum qualitative analysis of the collated information is given as follows (Table 2.3).

The case study (1), as collated, presents the opinion of many Nigerians that Nigeria's general power/energy output is a disgrace. That after sixty years of independence, Nigeria still battles to grapple with a meagre output capacity of about 6000 MW, leaves so many questions unanswered. We made this clear in another research work thus: "At the moment, incessant failure of household electric power supply has become chronic in Nigeria despite the federal government's announcement of huge investments in the power sector of the economy [...] Libya is about 3.5% and Morocco is about 18.8% of Nigeria's total population, yet we share the same capacity of household power output with them" (Ibid., Ethical Approach to Abuse of Common Good and Human Rights in Nigeria, The Polity, Oil, Violence and Niger Delta, pp. 74–75). Nigeria's population is the largest in Africa and the seventh in the world, estimated at over 200,000,000 (200 Million) people, yet countries like Egypt and Argentina, with a population of about 83,000,000 and 40,000,000, respectively, use, in each case, five times and four times the total MW capacity output in Nigeria. Secondly, the power/energy infrastructure in Nigeria is poorly maintained. If ever maintenance works would be carried out, the materials are mainly sub-standard due to the corrupt practices of those entrusted with such responsibilities. Of course, the identification of the presence of corruption, embezzlement and misappropriation of

[5] *Deutsche Gesellschaft für Internationale Zusammenarbeit (GIZ) GmbH is a German service provider with a focus on international relations and cooperation towards sustainable development. They work in partnership with the German government, EU institutions, United Nations and other private and governmental sectors.

funds within the institution shows a near unanimity among the respondents, as the collated data has shown. Vandalism is a reality and challenge, but it is not primarily a cancerous factor. A responsible administration can strategically checkmate the challenge (Table 2.4).

Case study (2) shows people's opinion of the generally observable effects of power/energy inefficiency. Investors' attraction in Nigeria has maintained a constant low ebb due to the fluctuating nature of the power/energy supply in Nigeria. Many Nigerian businesses have been reported to have taken flights to other safer and more supportive business havens in the neighbouring countries. The ones that decide to face the challenge squarely record low industrial outputs due to a lack of human resources, the manpower, whose employment opportunities must be substituted and exchanged with the companies' and organisations' struggles with the demands of powering the firm through power- /energy-generating sets which are cost-expensive. The consequence is predominantly an increase in the cost of production, of course, with its correlation seen in the high cost of commodities for ordinary citizens. On the social level, activities are often grounded to a halt both in private and public places and sectors due to power and energy failure. The collated data depict the general feeling of a nearly total breakdown of daily activities if private/individual power/energy generation were to become impracticable. Conversely, the ability to practise such has given rise to multi-faceted challenges to the Nigerian polity (Table 2.5).

For every action, there is an equal and opposite reaction. That is Newton's third law of motion. In a case study (3), the equal and opposite reaction to Nigeria's failed

Table 2.4 (b) Case study 2: effects of power/energy inefficiency

	Very great extent	Great extent	Average	Hardly	Not at all
Low investment attractions	207	29	–	–	–
Low industrial output (under-production)	77	156	3	–	–
Increase in cost of production	204	34	–	–	–
Techno-social and infrastructural malfunction	222	14	–	–	–

Table 2.5 (c) Case study 3: reasons for private power/energy generation

	Very great extent	Great extent	Average	Hardly	Not at all
Domestic purposes	231	5	–	–	–
Commercial purposes	2	17	221	–	–
Industrial purposes	–	98	138	–	–
Institutional purposes	216	20	–	–	–

energy system has become the culture of private power generation. The reasons, as identified, are domestic, commercial, industrial and institutional. Because the target of the questionnaire was mainly the common man on the street, a majority of respondents based their reasons on domestic and institutional purposes. The affordability of the Chinese Gen-Set technology, even for a majority of the citizens, has made the proliferation an existing social and economic problem in Nigeria. Individual families of between five to six persons sometimes experience the unfortunate situation whereby each person has his or her personal power-generating sets, each performing at the same time, emitting combusted fumes at the same time, and constituting noise pollution all at the same time. The Nigerian society has also recorded cases of death of individuals or entire family members due to intoxication through the combusted fumes while they were asleep. Varied pieces of evidence abound. Households are pressed deep into the need for private power/energy generation due to a cogent number of reasons, namely: keeping the refrigerators running for the preservation of food and other edible materials, combating the problem of heat and sweat due to the Nigerian climate with ventilators and air-conditioners which must be left to function almost 24-h so as to mellow the heat-wave, as well as helping in keeping lights on to checkmate, at least to an extent, the nocturnal acts of ill-motivated intruders, etc.

On the other hand, institutions, both governmental and non-governmental, are forced, in the Nigerian system, to resort to power-generating sets to combat power/energy failure, which has crippled the flow of activities in the majority of such institutions. The commercial and industrial areas tell their stories of threatening situations of the near extinction of their firms due to the excruciating cost of production occasioned by power failures and forced resort to private power/energy generation (Table 2.6).

In a case study (4), the collated data show unanimity to the fact that private power/energy generation in Nigeria has only one source, namely the generator set. The majority of 212 respondents also agree that solar panels hardly serve as available private power/energy generation sources. Under a workable policy for Nigeria, this should have been explored as an alternative, clean and supportive energy source, given the irradiation power and advantages Nigeria enjoys as a country located near the equator. However, the development of solar energy in Nigeria has been stalled due to a lack of vision of policymakers and an embedded culture of greed and inordinate amassment of wealth by all political and private persons who prefer to sacrifice

Table 2.6 (d) Case study 4: sources of private power/energy generation

	Very great extent	Great extent	Average	Hardly	Not at all
Generator sets	232	4	–	–	–
Solar panels	–	–	–	212	24
Gas plants	–	–	–	44	192
Wind energy	–	–	–	2	234
Bio-gas plants	–	–	–	–	236
Hydro power/energy	–	–	17	6	213

the growth and development of Nigeria at the altar of personal selfish gains and pursuance of the material. The disinterestedness and the low investment plans on solar panels keep the cost on the high side, even for middle-income earners. Other sources of energy named above are not projects for the individual or private persons, though they may not be exclusively impossible as private ventures. Recently, the federal government of Nigeria is set to commission a 10 MW Wind-Power energy in Katsina State—a project that began in 2010 but was halted along the line due to the abduction of one of the French workers on the site by kidnappers who demanded ransom. It was an unfortunate situation and, again, typical of the Nigerian society that is gradually drifting and degenerating into a lawless jungle. Nigerian society keeps watching and hoping that it does not turn to become the Nigerian *art of deception* again.

Furthermore, we made reference to a recent scientifically supportive and proven study and statistics from the International Energy Agency (IEA) published under the title, *Nigeria Energy Outlook*, to support our research opinions and propositions on the defective status of the Nigerian power and energy system and the need to restructure and develop its energy sources. The statistics show that Nigeria's energy sources and untapped potential feature a combination of coal, oil, back-up generators, gas, nuclear energy, hydropower, wind, solar PV, bioenergy and possible other renewables. An upward movement of the source of energy line under the term "*back-up generators*", confirms our findings and proves the persistence of the practice among the Nigerian populace due to inefficiency of the energy supply institution and the inertia to govern innovation and investment in the renewable potential. They show as well that possibilities and expectations, especially in the areas of solar- and expanded bioenergy sources, may only be in the initial phases, with the projection that they may be realised as projects come 2040. However, this would depend heavily on the political will to trigger the required policies within the Nigerian system (IEA 2019) (Table 2.7).

Case study (5) sought to analyse the socio-economic and market effects of private power/energy generation. The three economic terms above have been employed during the enquiry to determine the form of the *Consumer-Commodity* relationship within the general system of private power generation. Cost-efficiency deals with ways and strategies of expending money wisely while at the same time not losing the sense of saving. A cost-efficient commodity, therefore, ought to help the general savings and stability of the household economy. The statistics in this column above

Table 2.7 (e) Case study 5: cost-statistics of commercial and domestic uses of generator sets

	Very great extent	Great extent	Average	Hardly	Not at all
Cost-efficient	–	–	–	55	181
Cost-effective	–	–	6	57	173
Cost-inflating	182	41	13	–	–

show that the consumers' reactions point to the negative. For some, it is hardly cost-efficient, and for so many, cost-efficiency should never be associated with the practice. Many families are condemned to spend almost their entire monthly income on private power/energy generation, coupled with the special case with Nigeria, namely that a well-placed oil-producing country among other OPEC countries experiences incessant situations whereby the prices of such commodities like fuel, diesel and gas are both challenging to the average income earner and excruciating for the poor who are in the majority.

The imposed burden of maintaining and running a private power/energy generation through generator sets drags the cost-effectiveness to the mud. The statistics above show that many could hardly accept the fact that the system is cost-effective due to the financial burden it places both on household maintenance and commercial sustainability. Secondly, Nigeria's relationship with the Chinese market and technology has been that of economic enslavement and the conscious destabilisation of a developing economy. Most of such power-generating sets are sub-standardly and massively let loose into the Nigerian market. Their lifespan, therefore, is so minimal that the consumers are tied perpetually to a system of *purchase-discard-purchase*. Individual experiences show the cases of multiple purchases of the same power- / energy-generating commodity within the same year.

When a system or praxis is devoid of cost efficiency and effectiveness, a correlational factor would invariably be the inflation of costs. In the case under review—the private power/energy generation—this practice inflicts unprecedented cost inflation bearing on the services being received from such imposed alternative, which has kept so many individuals, families and institutions in bondage. The collated data above shows unanimity among the respondents in reference to the fact that this practice bequeaths structural deficiencies that defraud the entire polity, making daily life excruciating for the body politic (Table 2.8).

Economic Effects

The practice of private power/energy generation has a variety of effects, which impact the economy, environment and health. It is obvious from the feelings of the populace that, presently, Nigeria is becoming a slave to China's contemporary trend to conquer the lower economies and keep them perpetually tied to their well-designed economic subterfuge. Collated data show that Nigeria's presence in the Chinese market has reached the level of enslavement. Every little economic article is produced in China, including the focus of this fieldwork, namely, the power-/energy-generating set. Nigeria remains the widest market for the influx and trade of this technology. A German development agency, *Deutsche Gesellschaft für Internationale Zusammenarbeit*, GIZ reports: "The absence of on-grid energy supplies or unreliable power supply from the grid has led consumers from all categories to instal their own generation and even distribution equipment" (Deutsche Gesellschaft für Internationale Zusammenarbeit 2015b). Identifying this as a challenge, the above research institute bemoans the observable bankruptcy of reliable energy in Nigeria, despite all her natural potential. Thus, recognised as Africa's number one producer of crude oil

Table 2.8 (f) Case study 6: adverse effects of excessive private power/energy generation through generator sets

	Very great extent	Great extent	Average	Hardly	Not at all
Economic Effects					
(a) Enslavement to Chinese technology	179	57	–	–	–
(b) High cost of production	156	69	11	–	–
(c) Loss of revenue	40	172	24	–	–
Socio-environmental effects					
(a) Noise pollution	3	99	134	–	–
(b) Air pollution	66	148	22	–	–
(c) Environmental pollution	42	87	107	–	–
(d) Green-house effect	–	12	163	61	–
Health effects					
(a) Breathing impairment	–	227	9	–	–
(b) Lung infections	133	94	9	–	–

and with her vast potential for renewable energy, this independent research institute opined:

> […] the country's electricity supply remains unreliable. More than half of the population has no access to electricity, especially in rural areas where approximately 70% of dwellers do not have access to the national electricity grid. Those connected to the grid suffer from blackouts of several hours per day. Consequently, many households and businesses rely on expensive and ecologically harmful fuel generators. (Ibid., p. 1)

Furthermore, Lawal Gada, the Executive Director of Greenado International Limited and a former Renewable Energy Manager at Bank of Industry, examined the situation and, with collated data and experiences from the scene, Nigeria, lamented this conspicuous proliferation and remarked thus: "When you go to big organisations like the NNPC, CBN and BoI, you will see that they have up to four big generators, with a size of 500 KVA, operated on a 24-h basis, and they are burning fossil fuel that emits carbon dioxide" (Asu 2017).

The majority of respondents identify the high cost of production as one of the effects due to the extra demand that is pressured on the industries, which, in order to combat power failure, must resort to private efforts. Again, the situation is made more complex by the rising prices of fossil energy products in Nigeria, invariably giving rise to uncontrolled high costs of production. Thus, with the ever-increasing dependence on private power/energy generation, the Nigerian political economy shows, as at the period of this research, no laudable sign of seriousness towards diversification. The collated data identify the result to be constant and uncontrolled loss of revenue to other economies. GIZ reports after its research, "With several millions of privately installed diesel generators, Nigeria leads Africa as a generator importer and is one of the highest importers worldwide, with the total annual import figure being N17.9bn

($112 m)" (Ibid., The Nigerian Energy Sector, An Overview with a Special Emphasis on Renewable Energy, Energy Efficiency and Rural Electrification, p. 107). This is huge and discomforting to a developing economy.

Socio-Environmental Effects

The socio-environmental effects of private power/energy generation are obvious, ranging from noise to air and environmental pollution. The level of soundwaves being generated by the gen-sets and the manner in which they litter every nook and cranny of Nigerian streets, homes and premises constitute both noise and environmental pollution. This invariably affects people's health, psyche and behaviour. Lack of proper knowledge about the consequences keeps millions of Africans and, particularly, Nigerians exposed to varied potential health imbalances. A wider range of interrelated consequences has been studied and confirmed, such as hearing problems, nervous health issues, sleeping disorders, cardiovascular issues, troubled manner of communication as well as the effect on wildlife, acknowledging the opinion of Conserve Energy Future. The survey carried out met, in each case, the normal flow of life and daily activities among the people, yet within an environment deeply infested with all manner of generating sets. To the ears of so many, it has become normal.

A proliferated burning of fossil energy and the consequent emissions invariably and directly impact the condition and quality of air that living organisms breathe. Presenting a staggering report after in-depth scientific monitoring and research, Laura Tuck, Vice President for Sustainable Development at the World Bank, warns:

> Air pollution is a challenge that threatens basic human welfare, damages natural and physical capital, and constrains economic growth. We hope this study will translate the cost of premature deaths into an economic language that resonates with policymakers so that more resources will be devoted to improving air quality. By supporting healthier cities and investments in cleaner sources of energy, we can reduce dangerous emissions, slow climate change, and most importantly save lives," said Laura Tuck, Vice President for Sustainable Development at the World Bank. (World Bank Group 2016)

Furthermore, the inventory of polluted areas (most especially the metropolitans) of the world, according to the World Health Organisation, puts Nigeria at the fore as a country with densely polluted towns in Africa. While other factors contribute to this unfortunate development, excessive dependence on power- /energy-generating sets has been identified as deeply problematic. Respondents above write in the affirmative with experiences from the climatic realities and challenges that confront their varied immediate environments. Concerned about Nigerian air and health, the World Bank in "A Plea for Action against Pollution in Nigeria" reports that the Green Data Book issued in 2015 projected a 94% indication of the danger imposed on the population due to exposure to air pollution levels, which are measured in PM2.5. As indicated by World Bank analysis, the Nigerian situation exceeds the WHO guidelines and remains adversely cost-inflating to the Gross National Income (World Bank Group 2015b). Even though the scientific and technological approach to measuring and determining the level of the infested atmosphere is still in its infancy in Nigeria, the statistics above show the empirical consciousness of the respondents in this regard.

In other words, there is a wide range of ignorance about the "use and effect" of those power-/energy-generating commodities.

The greenhouse effect is a developing problem anchored on the inability of the post-industrial modern world to cut down emissions of greenhouse gases (mainly CO_2) or at least control it technologically. The individual societies and economies of today form a conglomerate of contributors, Nigeria inclusive. The Nigerian case of excessive dependence on generator sets (added to the problem of gas-flaring) risks her climatic sustainability and she remains a victim of the warning given by the Nobel laureate Kofi Annan, namely:

> Climate change demands that we rethink the relationship between energy and development. The carbon-intensive energy systems that drive our economies have set us on a collision course with our planetary boundaries. We can avoid that collision. As a global community, we have the technology, finance and ingenuity to make the transition to a low-carbon future, but so far, we lack the political leadership and practical policies needed to break the link between energy and emissions. (Annan 2015)

There is no doubt that the transition to a *lower carbon future* would go a long way to avert the impending socio-ecological pogrom, which Nigeria gradually paves its way through. However, although Nigeria does not lead the table of the index in the case of emissions, her position, however, starting from 2012, as the 35th country among 187 countries of the world with the volume of emissions calculated to the tune of 301,010.10 (kt of CO_2), remains worrisome.

Health Effects

The respondents remain unanimous on the reality of the health challenges posed by the excessive burning of hydrocarbon in Nigeria. As a country still developing scientifically and technologically, so many of its citizens remain trapped as victims of varied forms of hazards from these emissions due to ignorance and lack of access to relevant information. Breathing impairments and lung infections remain grossly associated with the cases of emission of carbon monoxide. In the extreme, such cases have led to suffocation and death. Many other countries of the world report cases of lung and breathing impairments, but mainly due to the habit of excessive smoking or other related factors. The Nigerian case, instead, while not denying the existence of smokers among the citizens, identifies such health problems as least related to the use of tobacco but predominantly as the aftermath of excessive burning of hydrocarbon. The data table in Appendix I presents documented pieces of evidence on the wastage of human resources in Nigeria due to the emission and inhalation of such poisonous gases, as reported across the nation.

The information in Appendix I represents only some of the cases among numerous tragic experiences that punctuate the densely populated Nigerian societies, all due to the negligence and greed that have stalled the needed policies to fix power failure in Nigeria. The known health hazard "asphyxiation" is let loose on a daily basis among the masses, whose quest to satisfy this basic social need of life, both private and commercial, is left only at the mercy of power-generating sets and in the manner with which it is proliferated in Nigeria. Obasi described the Nigerian scene thus:

> Positioning of these generators in these places and various social settings endangers lives through asphyxiation that results from the inhalation of the generated fumes. Families have woken up to the bizarre sight of deaths of some members or relations due to suffocation from inhaled fumes. Neighbours have been jolted to the horror of sight of their co-tenants choked to death by smoke from generators stationed in airtight enclosures. Churches have been alarmed at the death of members through generators placed dangerously within the precincts of their worshipping centres. The fumes from exhaust pipes of generators cluster in the lungs and block air passage leading to the death of victims. (Obasi 2018)

The huge economic burden and the health hazards heaped on the masses in Nigeria remain, therefore, one of the experimental proofs of multi-faceted inbuilt "structural sins" in a collective system of governance that is not *people-oriented*. When a polity remains bereft of the key drivers towards capacity building and sustainability, the possibility of the masses embracing all forms of mechanisms so as to counter organised and imposed socio-political and economic oppressions may sometimes reach its extremity. This is the naked truth to be learnt with the collations and expositions made on this fieldwork. The simple factual submission would be,

> [...] that uninterrupted power supply would reduce poverty, unemployment, (revive) massive foreign Direct Investments (FDIs) and improve standard of living and growth of indigenous human capital [...] The poor state of power and electricity in the country, and the suffering of common Nigerians are unimaginable. On a daily basis, most homes, businesses and institutions spend a huge part of their income on generator purchase, service and maintenance. Manufacturing industries end products' cost prices have increased and consumers are struggling to stay above the water. The story is also the same in the health sector, where many lives were lost due to the epileptic nature of electricity and power generation in the country [....] 2000 factories in the country were forced to shut down between the year 2000 and 2014 as a result of unsteady power supply and high cost of running generators. (Anyagafu 2014)

The reduction of poverty and unemployment, as well as the renewed attraction of foreign investments in the Nigerian context, prove rather socio-politically and socio-economically stagnated due to the intrigues of misguided policy implementation. While so many vision-driven countries embrace and invest in the increasing demands of power and energy, which drives growth and development through strong socio-economic institutions, Nigeria still marks time and refuses to rescue itself from a destructive lack of the political will to make positive change a reality. In twenty-first-century Nigeria, the power/energy minister still talks about Nigeria possessing a generating capacity of 7000 megawatts, yet with an embattled output capacity pegged at 5222 megawatts (Fashola 2018). Even the recent claims still trail behind the long-standing claim that Nigeria's power/energy sector reserves a capacity of 6000 megawatts. That is a political and socio-economic satire of the century.

2.8.4.8 Appraising the Data Presented

Because the case that has been studied has varied and multiple respondents in the fieldwork, which was also limited to selected groups of respondents, it is, therefore, necessary, in the final analysis, to apply the "mean calculation" so as to reduce exaggerated variations and make the work more verifiable. In the area of statistics,

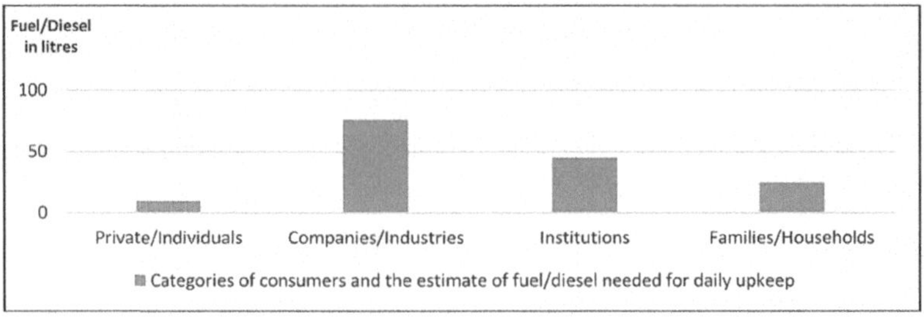

$$\bar{x} = \frac{1}{N}\sum_{i=1}^{N} x_i$$ using then the Mean formula: $$^{*}\ \frac{10 + 76 + 45 + 25}{4} = \frac{156}{4} = 39$$

****0 – 10; 0 – 76; 0 – 45; 0 – 25: representing the ranges of litres of fossil fuel expended daily on average within the categories of consumers above.***

Fig. 2.1 Mean calculation of daily expended fossil fuel by categories of consumers. *All illustrations and graphs are mine*

mean calculation helps especially "[...] when working with a large data set, [...] to represent the entire data set with a single value that describes the "middle" or "average" value of the entire set" (Rouse 2019). Again, it is needful to recall that the focus of the fieldwork has been to illustrate, with concrete facts on the political and socio-economic grounds in Nigeria, that the general system possesses some inbuilt man-made dysfunctions that have continued to hold the development and progress of this country hostage and, as a consequence, making life for the citizenry constantly excruciating and clueless. The *Mean-Formula*, as presented below, articulates our general appraisal of the situation in question (Fig. 2.1).

That means: Given the number of our respondents—236, distributed among the four categories of consumers, the mean calculation shows that in approximately every 236 persons, at least an average of 39 L of hydrocarbon and fossil fuel is burnt in varied intervals and circumstances. If, by extension, we apply the calculated average using the population estimate of Nigeria, which, as of 1 July 2020, stands at over 200 million, it all means that groups of 236 persons within a population of over 200 million people would be distributed approximately 847,000 (Eight Hundred and Forty-Seven Thousand) times. When this number is multiplied by the mean average of the burning of hydrocarbon in varied circumstances (847,000 × 39), it then means that an estimated 33,033.000 *lt* (Thirty-Three Million, Thirty-Three Thousand) litres of fossil fuel would be approximately combusted on a daily basis in Nigeria (in Germany, it is about 121 Millionen Litre Diesel daily, cf. Statistisches Bundesamt, 2017). The consumer duration and circumstances remain variant and distributed according to hours, days, weeks and months. This calculation excludes the number of litres burnt through locomotives—private vehicles, commercial vehicles and motorbikes. Nevertheless, the distributive factor—the affordability of and accessibility of consumers to power-generating sets—would invariably affect the consumer category. Consequently, there is a high probability that the estimated litre volume may fluctuate downwards due to the constraints of economic factors among the economically less privileged. This is taken cognisance of by our research.

The presentation above is not meant to represent a mere *mathematical syllogism* which gives a direct answer by mere addition and subtraction. On the contrary, it is meant to present verifiable facts which necessarily could vary, given its hypothetical nature and the general complexity of the Nigerian society. Again, it is a known fact that Nigeria is not alone in the use of fossil fuel energy for its varied civic, economic and commercial purposes. However, the emphasis is on the case of misplacement of the duty of the state in matters of energy supply which has enthroned arbitrary individual/private initiatives tied forcefully to the economic interests of both indigenous and external detractors. Added to such irresponsibility of the state is the ignorance of the poor and helpless masses who, on a daily basis, are the victims of the varied forms of toxicities associated with the very same domestic products for which they have little or no chance to avoid.

Furthermore, our research is not alone in its concern about Nigeria's defective structural and policy-based impairments. Thus, to substantiate the study done in the fieldwork, it becomes necessary to juxtapose the findings with acknowledged independent research of the Deutsche Gesellschaft für Internationale Zusammenarbeit, GIZ—an organisation with special focus on general sustainable development, especially in relation to developing economies and Africa in particular.

2.8.4.9 Qualitative Analytical Backup: Power/Energy Failure Amidst the Multiplicity of Energy Market Stakeholders as Structural Sin: Validating the Fieldwork with the Expert and Scientific Findings of the Deutsche Gesellschaft für Internationale Zusammenarbeit, GIZ

The Nigerian situation provides a viable specimen for both national and international research and study groups, and to reference a scientific study conducted independently on a similar background, would go a long way to validate our collated facts and figures. Thus, the Deutsche Gesellschaft für Internationale Zusammenarbeit, GIZ, represents one of such independent groups with a detailed analysis of the Nigerian situation presented under the caption: The Nigerian Energy Sector: An Overview with a Special Emphasis on Renewable Energy, Energy Efficiency and Rural Electrification, Second Edition, June 2015. Relevant to us, from the findings and report of the group, would be its analysis of the Nigerian energy sector, with special emphasis on the inadequacies associated with on-grip and off-grip energy supply, which has forced the majority of the masses to resort to independent/private power-energy generation.

Working within a population of over 200 million people, the group analysed the general energy consumption in Nigeria and presented the data and summarised its findings as follows:

> [...] Around 85% of Nigeria's consumed energy, 99.3 Mtoe annually, comes from biofuels and waste. Almost 90% of that energy is consumed for residential usage. This means that biofuels and waste cover about 98% of the energy demand in the residential sector. The lion's share of that most probably is for cooking purposes, as only thus can the predominant proportion of biofuels and waste be explained. The remaining energy in (the figure above) stems from conventional energy resources (~13%), with most of it being reimported oil products. The share of electricity in final energy consumption is almost marginal at less than 2%. (Ibid., The Nigerian Energy Sector, An Overview with a Special Emphasis on Renewable Energy, Energy Efficiency and Rural Electrification, p. 27)

The above scientific observation sets the stage right for a proper understanding of Nigeria's proliferated private power generation since almost 90% of energy consumption goes into residential matters. In other words, the greater part of her energy generation is needed mainly for domestic and household purposes. However, in a situation where the needed energy is not available or is grossly deficient in supply, people resort to the common practice of using generator sets with all known and proven dangers and side effects associated with it.

Despite the avowed reforms purportedly carried out in the energy/electricity sector between 2001 and 2015, the country's increasing energy consumption still hits misconstrued policies, which leaves the masses with no other option than to embrace any available alternative in a bid to satisfy this socio-economic need. GIZ reported about the situation on the ground thus:

> The existing fleet of power plants is a mix of plants built before the 1990's and plants built (or being built) since the mid-1990's. Since the older thermal power situations suffer considerably from poor maintenance, the available generating capacity was just under 6,200 MW in 2012 and has risen to 6,840 MW in 2015. However, unavailability of gas, breakdowns, water shortages and grid constraints severely limit the power plant performance, which means that despite an increase in the available installed capacity over the last years [...], only between 3000 MW to 4,500 MW are actually being generated (the highest peak generated ever in Nigeria was 4,517.6 MW on December 23, 2012) [....] The poor performance of the power plants has led to acute shortage of electricity across the country with power outages of several hours per day. The Presidential Task Force on Power regularly publishes the estimated peak demand and peak generation, whereby the former is with 12,800 MW regularly close to four times the latter. The only way the shortfall can be made up it by relying on off-grid electricity generation. (Ibid., pp. 33–34)

Off-grid electric generation, as referenced above and to which the masses are subdued, describes the general theme of private power/energy generation, which our research designates as socio-economic woe for Nigeria and her economic future. Unfortunately, the trend is still on the increase, gradually leading to a misguided social stratification whereby being able to purchase a private generator set invariably places one above his neighbours, who may not be buoyant enough to secure such a commodity. Consequently, such development has instituted a social and economic struggle which may seem to be satisfying the need for energy and power consumption but, at the same time, breeding economic and ecological hazards. GIZ noted further:

> According to a 2013 survey, approx. 80% of Nigerians use alternate sources of electricity supply such as generators or solar inverters. Estimates suggest that between 8 and 14 GW of decentralised diesel generator capacity is currently installed in the country. About 86%

> of the companies in Nigeria own or share a generator and 48% of their total electricity demand is covered by these private generators [GIZ; Mar 2015]. With several millions of private installed Diesel generators, Nigeria leads Africa as a generator importer and is one of the highest importers worldwide, with the total annual import figure being NGN 17.9 billion (US$ 112 million) [….] But there obviously are some major disadvantages. The use of decentralised Diesel generators is economically and environmentally questionable. (Ibid., p. 36)

The solar inverters mentioned above suffer gross neglect due to the visionless policies and the greed of all who are directly involved in keeping the nation's energy supply grossly dependent on their selfish economic gains. Consequently, although Nigeria reserves an enormous abundance of solar energy, installation of solar panels remains a mirage for an average income earner due to the cost of purchase, not to talk about the teeming section of the polity that wallows in abject poverty. In the main, the issues surrounding solar inverters should belong to the various contents of research and discussions on renewable energy programmes in Nigeria. Some notable past summits and deliberations happened under the themes Renewable Electricity Policy Guidelines (REPG), 2006; Renewable Electricity Action Programme (REAP), 2006; National Biofuel Policy and Incentives, 2007; Roadmap for Power Sector Reform, 2010 and 2013; Vision 20:2020, 2010, etc. (Ibid., pp. 69–72). However, unfortunately, many policy programmes in Nigeria are strategies to loot and deceive the uninformed citizenry, who are often lured into believing that something is being done. Many of these programmes suffer dormancy and abandonment at the end of the day. GIZ again gave credence to this fact while referring to the National Biofuel Policy and Incentives of 2007:

> Although the policy seems to be in force, no information is available as to the extent to which it has been implemented and whether adjustments have been made to it in light of experiences gained with the first steps and after feedback from players in the agricultural and petroleum subsectors. The major player, NNPC performed feasibility studies on ethanol production as well as cassava and palm-oil-based products but only mentions a few projects and initiatives on its websites. It can be concluded that the topic gained most interest during the peak years 2009-2012 of the global move towards biofuels but lost momentum after this source war largely criticised for its conflicting use of foodstuff and agricultural land for the cultivation thereof. (Ibid., p. 71)

This situation leaves one with the unanswered question as to whether criticism should lead to total abandonment or restructuring and reform. However, this remains the plight of the Nigerian system. Amidst all these inadequacies, the power/energy demand in Nigeria increases geometrically with Nigeria's fast-growing population.

Furthermore, GIZ presented a set of scientific monitoring and projections based on the calculations made by FMP, the Power Holding Comping of Nigeria and UN 2010 rural/urban population data. The observation shows a projection of grid and off-grid electricity demand for Nigeria taken the period between 2010 and 2035 into consideration. According to the analysis, while an increase in on-grid electricity demand within the given period of time, there is also a probable danger of an increase in off-grid electricity demand (referred to as Off-Grid D) should inefficiency and non-viable policymaking approaches become persistent within the sector (Ibid., p. 39).

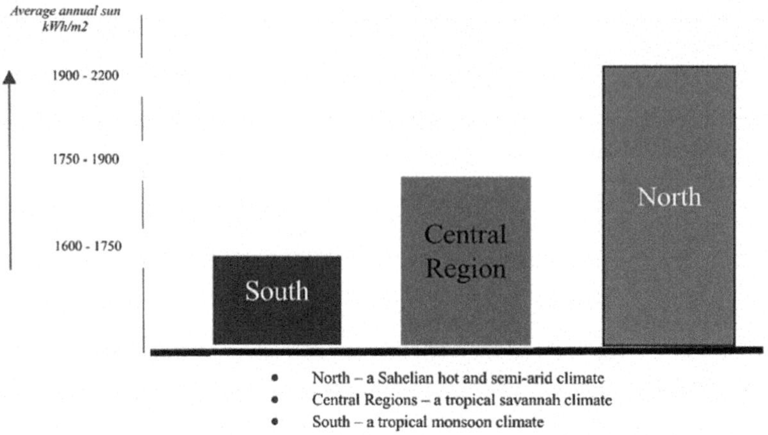

Fig. 2.2 Solar irradiation level capacities in Nigeria. *Illustration mine*

It all means that if nothing urgent is done to curb the off-grid system of power/energy generation, Nigeria would continue to immerse herself deep into economic, ecological and environmental woes associated with power-generating sets, added to the economic neo-imperialism, which China is introducing in the global market and economy, with strong dominance and influence in Africa.

Scientific-Based Possible Remediation for Nigeria's Power/Energy Quagmire: A Feasibility Study

With the varied feasibility studies that have been done in Nigeria, many proponents battle with the question of what constitutes the mainstream obstacle, which has kept Nigeria captive and comatose in taking the right steps towards meeting her energy demands. Feasibility studies are generally poised to determine and prove the viability of propositions and ideas based on their legal and economic values. Within the political and socio-economic settings of a geo-polity like Nigeria, feasibility studies have always shown viable alternatives that can remediate the present debilitating power/energy situation while offering a cleaner energy supply and boosting Nigerian internally generated revenue. However, the problem remains the Nigerian story within a highly corrupt system. GIZ and some other independent studies presented these feasible renewable remediations as (1) Bioenergy, (2) Geothermal energy, (3) Small hydropower, (4) Tidal energy, (5) Solar energy, and (6) Wind energy (cf. Ibid., p. 81). Our research joins these research groups to emphasise the great potential Nigeria has if its climatic conditions are optimally utilised with regard to solar energy. The table below presents statistically the scientific appraisal of solar irradiation level potent in Nigeria (cf. Ibid., p. 86) (Fig. 2.2)

While analysing the above feasibility study, GIZ stated emphatically,

> Nigeria has a potential for electricity production from solar PV technology in the range of 207,000 GWh per year if theoretically only 1% of the land area (e.g. 920 km^2 = 920*106 m^2) were covered with state-of-the-art polycrystalline PV modules, with an electricity yield of 1,500 Wh/Wp per year. This figure is tenfold the total electricity production of Nigeria in 2011 [....] (Such) Flat-plate photovoltaic devices utilise both diffuse and direct radiation [....]. (Ibid., pp. 85–86)

The generating capacity of the northern part of the country in terms of solar energy, as seen in the diagram above, is yet to be harnessed. GIZ added: "This high solar irradiation (in northern Nigeria) is comparable to very high-yield sites in southern Spain, northern Africa, Australia and Latin America" (Ibid., p. 87).

The opportunity to spend some time in Europe or elsewhere in the so-called developed economies leaves one with the question of "*why countries that have a lesser amount of annual sunshine or solar irradiation keep investing in clean solar energy*" while their African counterparts trail very much behind, or in many cases, show no interest and effort to optimise their natural solar endowments. In the Nigerian case, the resolve to get such moves concretised and realised as projects hits so many challenges on the pathways, ranging from lack of the political will to invest in the demands for large quantities of studies to the corrupt transactions associated with access grants, licence submissions and approvals. The bureaucratic and, in varied cases, illegal and unnecessary structures embedded in the Nigerian system and public services are often packs of corrupt practices. With particular reference to the power and energy sector, GIZ observes further:

> To date, a large number of such licenses have been granted, for both on- and off-grid independent power producers (IPPs). However, the number of the projects that have since progressed to financial close is buttressed by PPAs is limited (in the case of renewable energy there is none), suggesting that licenses were obtained as negotiating chips for future sale rather than with the firm commitment to build and operate a power plant. (Ibid., p. 102)

Another complex challenge within the system is the multiplicity of partners, stakeholders, providers and energy vendors within the sector, alongside the never-ending cases of inefficiency, power/energy epilepsy, sub-standard output and service provision, etc. The network of multiple policy documents and stakeholders in the Nigerian system could easily deceive the uninformed to think that a lot is being invested so as to redeem the sector. On the contrary, piles of experiences have shown that within a corrupt system of governance (like in Nigeria), multiplied governmental parastatal serve as deceptive avenues for misappropriation and embezzlement. At the time of writing this work, the energy market in Nigeria is flooded with numerous governmental and non-governmental players on the scene. This is exhaustively illustrated in Appendix IV as attached.

The network of institutional interplay and relationship within the sector may not have shown ideological indolence, but the problem in the Nigerian case revolves around quality, substance, authentic intent and purpose. To multiply institutions would never have a bearing on the transformation of the real sectors, especially when the policies and agendas of such institutions are far from being purposeful and coercive. Granted that the Nigerian population is intimidating and diverse, it would be a culpable omission if we do not bemoan the persistent lack of inclusive policies capable of responding to the basic needs of the common men and women on the streets, who, at the same time, form the body politic at the grassroots. In line with this statement, GIZ, in a concluding remark after its scientific research work, wrote:

> While Nigeria has overtaken South Africa as the biggest economy on the continent, the electricity sector as the backbone of the economy significantly lags behind. Power outages are a daily routine, and so are the cues in front of the petrol stations [....]The rural electrification sector still awaits the approval of the Rural Electrification Strategy and Plan. Operationalisation of the Rural Electrification Fund is as important as a coherent planning and regulatory framework in order to promote investments and streamline the actions of federal, state and private actors towards a mass roll-out of electrification projects across the country. Energy efficiency still needs to be embraced by stakeholders and the population at large as a cost-effective tool to save electricity and to improve the reliability of the power system. Awareness creation measures are as important as mandatory standards and voluntary pilot projects in order to demonstrate the viability of this approach. (Ibid., p. 129)

In the final analysis, our purpose of embarking on the fieldwork is justified, namely, that there exist inbuilt structural defects within the Nigerian general system which politically and socio-economically weaken all plans and efforts in getting at the right policies that would answer the basic questions on the daily lives of the citizenry and make development people-oriented. Our research holds the truth that such structural defects are man-made, consciously strategised and imposed to control the system. Feasibility studies have exposed alternatives, but the "how" and the form an approach would take remain the lingering questions. We would later revisit these questions in order to proffer some practical sustainable inroads. However, there would be a need at this point to give a concise appraisal and assessment of the apparent *step in the right direction* associated with the proposed project "Energising Education Programme", which Nigeria has allegedly approved and ratified. This is to support our claim that the mainstream problem in Nigeria is not a lack of policy. It is, rather, the failure of leadership seen in the lack of the political will to apply those policies and make their principles and ethics coercive.

The Green Bond: Energising Education Programme for Green Energy and Afforestation Initiatives

In December 2017, the Federal Government of Nigeria took an apparent step in the right direction and issued a sovereign green bond with the caption: Energising Education Programme. This is not a direct academic programme for the educational institutions but an indirect educational and institutional support programme in terms of the energy transition, energy revolution and infrastructural electrification. The very recent R20 World Summit held in Austria in 2019 gave a description of the programme in its Special Report:

> In December 2017, Nigeria became the first (and so far only) African nation to issue a sovereign green bond, and in doing so became only the fourth nation in the world to issue one after Poland, France and Fiji. The US$ 30 million tranche (in local currency) with a 5-year-tenor was the "pilot sovereign" for the country's anticipated several hundred-million-dollar green bond program. The proceeds of the bond are used to provide green electricity to

> rural communities that were previously in darkness and support a government afforestation initiative. One of the projects, the Energizing Education Programme, aims to provide reliable, off-grid power supplies to 37 universities and 7 university hospitals across the country. These will utilise renewable energy in the process, particularly solar and hydropower [....] The green bond issuance marks a breakthrough not only for the oil-rich Nigeria, contributing towards the commitments it made under the Paris Agreement, but also for other developing and emerging nations. With international support the Government of Nigeria has put in place a comprehensive governance structure and framework that is aligned with the country's newly developed green bond guidelines and international best practices. This created a benchmark for subsequent issuances of green bonds by state governments as well as corporates.[6]

As laudable and plausible as this programme may be, many Nigerians continue to watch with scepticism to see what would again become of the programme given the Nigerian habit of negligence and nonchalance. As it stands now, many would want to remind the government and indeed the whole world that this same country alleged that a former President, Olusegun Obasanjo, spent over 5.4 trillion Naira (about US$16 billion) between 1999 and 2007, all on the same story of trying to fix the energy and power sector of the economy. Yet, despite the above claim and the expended huge sum, the energy and power sector of Nigeria keeps retreating to instability and inefficiency. However, the newly ratified Green Bond, in view of power generation, is estimated to generate an off-grid energy supply to the tune of 89.6 megawatts for the earmarked institutions. The advantage would be two-fold, namely: (1) an obvious reduction in the load of connections, which the on-grid sector is presently compelled to carry; (2) a long-awaited positive step towards development and investment in clean and green energy.

Critiques of the ratification of the Green Bond, our research for one, stand strong with the opinion that such a project has both revolutionary and revaluation powers to apply the policy and transform the targeted institutions positively and sustainably. What Nigeria needs is the right implementation strategy and the sincere personnel who can hit the field of work and achieve results. Nevertheless, if the politics of greed and corruption should further have a grip on the project, it will again go down the Nigerian history as unrealised, partly realised, completed or uncompleted, or again as a project, the resources of which have again been massively embezzled. Both the local and international communities are watching.

---

[6] Federal Chancellery, Republic of Austria. R20 Austrian World Summit Special Report 2019: Financing of Climate Projects. Available via https://www.catalyticfinance.org/resources. *Green Bonds are fixed income, liquid financial instruments used to raise funds dedicated to environment-friendly projects. Historically, most green bonds have been used to finance climate-change mitigation measures. Over the last two years, the financing of climate-change adaptation and resilience projects has also gained increased attention. Issuers of bonds can be private companies, supranational institutions (such as multilateral banks) and public entities (municipal, state or federal). Development banks (such as the European Investment Bank and the World Bank) and corporates have been the largest issuers of labelled green bonds to date. Green Bonds can help catalyse the market and unlock investment for private sector projects. A lack of appropriate institutional arrangements for green bond management, the issue of minimum size, and the high transaction costs associated with green bond issuance are the key barriers to the development of green bonds in developing countries. According to the Climate Bonds Initiative, emerging market (EM) green bonds account for about a quarter of total issuance to date.

Evaluation: The Nigerian State of In-built Structural Sins (Defects) and Structural Corporate *Irresponsibility* as the Known Causes of Failed and Non-Sustainable State: A Socio-Ethical Analysis

We have tried to approach the internal dysfunction within the Nigerian socio-political and economic structures with the development principles of the capability approach. The approach has helped to expose the failures and the reasons for instability as well as the non-sustainability of the polity or state. First, it is the issue of the generally conceptual but defective ideology that is inbuilt within the system. Secondly, it is the corporate irresponsibility of the political clue, who have for long chosen to drown the development of the Nigerian State in the deep ocean of their inordinate and selfish political ambitions and recklessness.

Social corporate responsibility is a socio-economic language employed to emphasise the necessity of the inclusion policy if any political economy would want to associate itself with the demands and principles of sustainable development. It is taken as a "[...] philosophy of conduct and a concept of doing business applied by the business community, companies and individual businessmen for sustainable development and preservation of resources for future generations [....]" ((Moscow 2006)). The concept is strategised to ensure quality products and services; to create decent jobs alongside production and human resources development; to ensure optimal compliance with laws as regards taxes and other ecological issues affecting businesses and production; to maintain integrity and reciprocity with stakeholders; to hold on to the dictates of efficiency and economic values, while introducing fairness and healthy competitiveness, the aim of which should be the benefit of shareholders and society; and to partner business ethics and social development programmes which would ensure sustainable evolution and development of the civil society (cf. Ibid.). All of these are geared towards the well-being of the moral agent himself or herself—the *Mmadụ*, and towards sustainable growth and development.

Although the concept of Corporate Social Responsibility is predominantly employed in business ethics, its normative approach and principles give objectivity to our dealings with sustainable development, relationships, interactions and interdependences, which are core objective concepts in the issue of general human and societal development. In evaluating, therefore, the situation of the principal agent of development, *the Mmadụ*, within the Nigerian socio-political and economic structures, our research observes and delineates rather *structural corporate irresponsibility* in policymaking, which negates the dictates of responsible leadership in the Nigerian system.

One thing that is obvious in the definition given above is that the activity of cooperating is predicated on the backdrop of planned and organised actions, which are geared towards achieving sustainability in the development and well-being of the agents in relation to the entire society. When the reverse is the case, as with structural corporate irresponsibility, the whole activity becomes a planned but manipulated structural deficit, which debilitates the entire structure, keeping it misguided and misdirected. The consequence is often the enslavement of the masses alongside the political and economic triumph of a corporate few, who are the perpetrators and drivers. Breaking such a debilitating corporate circle has become a persistent challenge to Nigeria's democracy and the same situation keeps Nigeria stagnant amidst the global agenda for sustainable nations, economies and policies. There is no doubt that the modern trend and clarion call on the nations of the world to embrace and be open to more people-oriented policies also has prospects and potential in Nigeria, given her vast ocean of resources—human, natural and mineral. However, our research, having exposed a network of structural deficiencies, presents in the next chapter the current ideological trends associated with the concept, *Sustainable Development*. Our reflection would be based on the questions surrounding the meaning, principles, ethics and modes of its application. A relevant research question would be: Why has it eluded the Nigerian system, and what prospects could be remedial to the Nigerian case?

References

Achebe C (2013) Chinua Achebe on corruption and hope in Nigeria (Interview). In: The Christian Science Monitor, 22 Mar 2013. Available via https://www.csmonitor.com/World/Africa/2013/0322/Chinua-Achebe-on-corruption-and-hope-in-Nigeria Accessed 18 Aug 2019

Adefolaju T (2014) Repositioning health insurance in Nigeria: prospects and challenges. Int J Health Sci 2(2):152, 151–162

Adeoye DO (2016) Challenges of urban housing quality: insights and experiences of Akure, Nigeria. In: Urban planning and architecture design for sustainable development, UPADSD. Procedia Soc Behav Sci 216:261, 260–268. Available via https://doi.org/10.1016/j.sbspro.2015.12.036. Accessed 3 Feb 2019

Ake C (1995) Socio-political approaches and policies for sustainable development in Africa. In: Paper presentation, annual meeting symposium of the African development bank, Abuja, 25 May 1995. Available via https://ouci.dntb.gov.ua. Accessed 3 Feb 2020

Akpor-Robaro M (2012) The impact of socio-cultural environment on entrepreneurial emergence: a theoretical analysis of Nigerian society. Eur J Bus Manage 4(16):173, 172–182. Available via https://www.iiste.org/Journals/index.php/EJBM/article/view/3016. Accessed 17 Sept 2020

Annan K (2015) Forward. In: Global goals, African realities: building a sustainable future for all. Available via https://m.polity.org.za/article/global-goals-african-realities---building-a-sustainable-future-for-all-october-2015-2015-10-01. Accessed 5 Mar 2020

Anyagafu SV (2014) Nigeria records over 10,000 deaths through 'generator fumes'. Quoting Festus Mbisiogu, President, Good Governance Initiative, GGI. In: Vanguard Nigeria, Nigerian Daily, 8 Aug 2014. Available via https://www.vanguardngr.com/2014/08/nigeria-records-10000-deaths-generator-fumes/. Accessed 3 Feb 2019

Asu F (2017) The heavy use of generators puts Nigeria's climate plans in jeopardy. In: Punch Nigeria, Nigerian Daily, 28 Dec 2017. Available via https://punchng.com/heavy-use-of-generators-puts-nigerias-climate-plans-in-jeopardy/. Accessed 3 Feb 2019

Breen JM (2008) John Paul II, the structures of sin and the limits of law. Saint Louis Univ Law J 52(2):318–371. Available via https://scholarship.law.slu.edu/lj/vol52/iss2/3. Accessed 23 Oct 2019

Brookings Institution Report (2018) Re: growth spikes (5.01 per cent), but people are hungry? In: ThisDay, Nigerian Daily. Available via https://www.thisdaylive.com. Accessed 10 Feb 2020

Catechism of the Catholic Church (1999) The fall. Part One, Section Two, Chapter One, Parag. 7. Theological Publications, India, p 76

Christman J (2018) Autonomy in moral and political philosophy. In: Zalta EN (ed) The Standford encyclopaedia of philosophy (Spring 2018 Edition). Available via https://plato.standford.edu/archives/spr2018/entries/autonomy-moral/. Accessed 23 Aug 2019

Crossman A (2020) The concept of social structure in sociology. In: ThoughtCo, 27 Aug 2020. Available via thoughtco.com/social-structure-defined-3026594. Accessed 3 Feb 2019

De Muro P, Tridico P (2008) The role of institutions for human development. Università deli Studi Roma Tre, vol 4, pp 1–15. Available via https://www.researchgate.net. Accessed 3 Aug 2020

Deutsche Gesellschaft für Internationale Zusammenarbeit, hg. v. (d.) (2015a). Bonn. The Nigerian energy sector, an overview with a special emphasis on renewable energy, energy efficiency and rural electrification, 2nd edn. Nigerian Energy Support Programme (NESP). Available via www.giz.de. Accessed 10 Feb 2020

Deutsche Gesellschaft für Internationale Zusammenarbeit, hg. v. (d.) (2015b) Bonn. The nigerian energy sector, an overview with a special emphasis on renewable energy, energy efficiency and rural electrification. Nigerian Energy Support Programme (NESP), Second edn, June 2015. Available via https://www.giz.de/en/downloads/giz2015-en-nigerian-energy-sector.pdf. Accessed 10 Feb 2020

Drèze J, Sen A (2002) India: development and participation. Available via https://academic.oup.com/book/6815. Accessed 4 Sept 2019

Eric P (2016) The amalgamation of Nigeria: revisiting 1914 and the centenary celebrations. CSCanada, Can Soc Sci 12(12): 66, 66–68. Available via http://www.cscanada.net/index.php/css/issue/archive?issuesPage=3. Accessed 12 Nov 2021

Fashola R (2017) Why poor electricity is still a problem in Nigeria. Comment of the Nigerian Minister of Power, Works and Housing, 28 May 2017. https://dailypost.ng/2017/05/28/poor-electricity-still-problem-nigeria-power-minister-fashola/. Accessed 12 Apr 2019

Fashola R (2018) Power generation rises to 7000 megawatts. In: Vanguard Nigeria, Nigerian Daily, 3 Aug 2018. Available via https://www.vanguardngr.com/2018/08/power-generation-rises-to-7000-mega-watts-fashola/. Accessed 14 Mar 2019

Furze B, Savy P, Webb RD, James SL, Petray T, Lie J, Brym RJ (eds) (2015) Sociology in today's world. Cengage Learning, Australia, p 7

Gandhi A (2013) The seven deadly social sins. In: The collected works of Mahatma Gandhi (electronic ed), vol 33, pp 133–134. Available via http://www.gandhiinstitute.org. Accessed 3 Aug 2020

Glatzer W (1969) Sozialstruktur. In: Endruweit G, Trommsdorf G (eds) Wörterbuch der Soziologie (2. Auflage). UTB, Stuttgart, pp 534–538

Ibe RU (2016) What's really wrong with the economy? The Guardian, Nigerian Daily, 5 Oct 2016. Available via https://guardian.ng/opinion/whats-really-wrong-with-the-economy/ Accessed 3 Mar 2020

IEA (2019) Nigeria energy outlook, IEA 2019, Paris. Available via https://www.iea.org/articles/nigeria-energy-outlook. Accessed 8 Aug 2022

Ilo UJ (2012) Crisis of rule of law and culture of self-help in Nigeria. In: Premium Times, Nigerian Daily, 20 Oct 2012. Available via https://www.premiumtimesng.com/opinion/104256-crisis-of-rule-of-law-and-culture-of-self-help-in-nigeria-by-udo-jude-ilo.html. Accessed 3 Aug 2020

Katzer M (2010) The basis of universal liberal principles in Nussbaum's political philosophy. Pub Reas 2(2):60–75. Available via https://www.publicreason.ro/pdfa/34. Accessed 23 Aug 2019

Knight K (2017) Question 82, original sin, as to its essence. The Summa Theologiae of St. Thomas Aquinas, Second and Revised Edition. Literally translated by the fathers of the English Dominican Province. Available via https://www.newadvent.org. Accessed 16 Aug 2019

Leimgruber Y (2016) Polity, policy, politics. PB-Tools. Available via http://politischebildung.ch/fuer-lehrpersonen/grundlagen-politische-bildung/polity-policy-politics. Accessed 4 Feb 2021.

Madueke D (2015) Why Nigeria's refineries are in terrible shape. Opinion of Diezani Alison-Madueke, Former Nigerian Minister of Petroleum Resources. Premium Times, Nigerian Daily, 22 Feb 2015. Available via https://www.premiumtimesng.com/business/175498-nigerias-refineries-terrible-shape-petroleum-minister.html. Accessed 3 Feb 2019

Mbiti JS (1969) African religions and philosophy. Heinemann, London, p 1
Monash University (2019) D—substructure and superstructure. Available via https://www.monash.edu/__data/assets/pdf_file/0011/1807436/MDCS-SD-2019-Final.pdf. Accessed 17 Aug 2019
Morishima M (2016) Why has Japan succeeded? Western technology and Japanese Ethos. In: Ntibagririrwa S (ed) Local cultural values and projects of economic development: an interpretation in the light of the capability approach. Globethics, Geneva, p 38
Moscow (2006) Memorandum on principles of corporate social responsibility. Russian Managers Association, p 4. Available via https://www.yumpu.com/en/document/view/44269638/on-principles-of-corporate-social-responsibility-csr-weltweit. Accessed 11 Feb 2019
Nkrumah K (1974) Neo-colonialism: the last stage of imperialism. Panaf, London, p 3
Ntibagiriwa S (2014) Philosophical premises for African economic development: Sen's capability approach. Globethics, Geneva, pp 271–272
Nussbaum MC (2001) Women and human development. In: The capabilities approach, Revised edn. Cambridge University Press, United Kingdom, p 70
Nussbaum M (2013) Creating capabilities. The human development approach. Harvard University Press, p 21
Nussbaum M, Clover J (eds) (1996) Women, culture and development: a study in human capabilities. Oxford University Press, USA, p 185 (emphasis on bold, mine)
Nwosu CF (2016a) Ethical approach to abuse of common good and human rights in Nigeria, the polity, oil, violence and Niger delta. AV Akademikerverlag, Belin, p 67
Nwosu CF (2016b) Ethical approach to abuse of common good and human rights in Nigeria, the polity, oil, violence and Niger Delta. AV Akademikerverlag, Belin, p 75
Obasi EB (2018) Deaths and electricity generator fumes. In: The Guardian, Nigerian Daily, 25 Sept 2018. Available via https://guardian.ng/opinion/deaths-and-electricity-generator-fumes/ Accessed 3 Feb 2019
Oforka VC (2015) The bleeding continent: how Africa became impoverished and why it remains poor. Xlibris, United States, p 123
Ojukwu CO (1989) Because i am involved. Spectrum Books, Ibadan
Olaseni M, Alade W (2012) Vision 20:2020 and the challenges of infrastructural development in Nigeria. J Sustain Dev 5(2):69, 63–76
Omolulu F (2007) Corruption, governance and political instability in Nigeria: a dysfunctional conundrum (research paper). Afr J Polit Sci Int Relat 1(2):28, 28–37
Omotayo Y (2015) Five reasons Nigerian students perform excellently abroad but fail to do so in Nigerian universities. In: Legit.ng, Opinion, Education News. Available via https://www.naij.com/educationnews/. Accessed 3 Aug 2020
Osasona T (2017) Generators: providing powers and taking lives. In: The Guardian, Nigerian Daily, 27 July 2017. Available via https://guardian.ng/opinion/generators-providing-power-and-taking-lives/. Accessed 3 Feb 2019
Oyelaran-Oyeyinka B (2011) Nigerian universities and global trends in higher education. In: Paper presentation, 26th annual conference of the association of vice chancellors of Nigerian Universities (AVCNU), Covenant University, Ota, Nigeria, 27–30 June 2011. In: Olaseni M et al (eds) Vision 20:2020 and the challenges of infrastructural development in Nigeria. J Sustain Dev 5(2):69
Paul II J (1984) Reconciliatio et Penitentiae. On Reconciliation and Penance in the Mission of the Church Today, Rome, 2 Dec 1984, no 14
Paul II J (1987) Encyclical Letter Solicitudo rei Socialis. On the Twentieth Anniversary of Populorum Progressio, Rome, 30 Dec 1987, no 7
Pettinger T (2019) Structural adjustment definition. In: Economics—helping to simplify economics, 3 Mar 2019. Available via https://www.economicshelp.org/blog/2139/economics/structural-adjustment/. Accessed 6 Aug 2020
Pius XII (1946) Discorsi e Radiomessaggi VIII, Radio Message to the U.S. National Catechetical Congress in Boston. In: Reconciliatio et Penitentiae, no 18. Rome, 26 Oct 1946

Robeyns I (2019) The capability approach: an interdisciplinary introduction. Available via https://scholar.google.com/citations?user=OD5XAAsAAAAJ&hl=en. Accessed 3 Feb 2019

Robeyns I, Byskov M (2021) The capability approach. In: Zalta EN (ed) The Stanford encyclopaedia of philosophy (Winter 2021 Edition). Available via URL https://plato.stanford.edu/archives/win2021/entries/capability-approach/. Accessed 23 Aug 2019

Rouse M (2019) Statistical mean, media, mode and range. In: Course hero, compliance and governance, programming. Available via https://www.coursehero.com/file/51207882/What-are-Mean-Median-Mode-and-Range-pdf/. Accessed 12 Sept 2020

Saidu T (2015) Leadership challenges and economic development in Nigeria. J Polit Gov 4(1–4):25–35. Available via https://www.semanticscholar.org/paper/Leadership-Challenges-and-Economic-Development-inSaidu/3e3d9e8e68aa3da459212a702a736c2d6326f9bf

Seligson MA, Passe-Smith JT (eds) (2013) Development and underdevelopment: the political economy of global inequality. Lynne Rienner Publishers, United States, p 80

Sen A (1993) Capability and well-being. In: Nussbaum M, Sen A (eds) The quality of life, pp 30–53. Available via https://doi.org/10.1093/0198287976.003.0003. Accessed 23 Aug 2019

Sen A (1995) Inequality re-examined. Harvard University Press, MA, p 56

Sen A (1997) Resources, values and development. Harvard University Press, MA, p 494

Sen A (2009) The standard of living: lecture II, lives and capabilities. In: Tenner lectures in human. The Standard of Living, vol 36, pp 20–38. Available via https://www.cambridge.org/core/books/tanner-lectures-in-human-values/standard-of-living-lecture-ii-lives-and-capabilities/CFF69D3D06E75FD8F942857280BD3FDA. Accessed 23 Aug 2019

Sen A (2010) The idea of justice. Penguin Books, London, p 254

Soyinka W (1976) The man died, the prison notes of Wole Soyinka. Vintage, United States

Statista (2022) Unemployment rate in Nigeria 2021, Published by Aaron O'Neill, 19 Oct 2022. https://www.statista.com/statistics/382366/unemployment-rate-in-nigeria/. Accessed 29 Dec 2022

Strong AH (2017) Systematic theology, volume II: the doctrine of man revised and enlarged. Devoted Pub., Ontario, p 165

Teferra D (2000) Revisiting the doctrine of human capital mobility in the information age. In: Paper presentation, regional conference on brain drain and capacity building in Africa, 22–24 Jan 2000. In: Oforka VC (ed) The bleeding continent: how Africa became impoverished and why it remains poor. Xlibris 2015, United States, pp 84–85

The Federal Republic of Nigeria (2009) Economic transformation blueprint. Nigeria Vision 20:2020. National Planning Commission, pp 19–20. Available via https://nairametrics.com/wp-content/uploads/2013/06/nigeria-vision-20_2020_draftetb.pdf. Accessed 5 Ma r2020

The Federal Republic of Nigeria (2014) The national health act. Available via https://nigeriahealthwatch.com/wp-content/uploads/bsk-pdf-manager/2018/07/01-Official-Gazette-of-the-National-Health-Act-FGN.pdf. Accessed 5 Mar 2020

United Nations (1987) Our common future towards sustainable development. Report of the World Commission on Environment and Development. The Brundtland Commission. Available via https://sustainabledevelopment.un.org/content/documents/5987our-common-future.pdf. Accessed 3 Feb 2020

United Nations Development Programme (UNDP) (1990) Human development report 1990. Available via https://hdr.undp.org/content/human-development-report-1990. Accessed 16 June 2019

United Nations Development Programme (UNDP) (2018) Human development indices and indicators. 2018 Statistical Update. Available via https://hdr.undp.org/content/statistical-update-2018. Accessed 5 Oct 2019

United States Agency for International Development, USAID, Nigeria (2021) Abuja, 3 June 2021. Available via https://www.usaid.gov/nigeria. Accessed 5 Nov 2021

United States Catholic Bishops (2009) Economic justice for all: pastoral letter on Catholic social teaching and the U.S. economy. United States Conference of Catholic Bishops, Washington

DC. Available via https://www.usccb.org/upload/economic_justice_for_all.pdf. Accessed 3 Feb 2020, no 185

Vogt M (2013) Prinzip Nachhaltigkeit, Ein Entwurf aus theologisch-ethischer Perspektive, 3. Auflage. Oekom Verlag, München, p 110

Wiemeyer J (2015) Keine Freiheit ohne Gerechtigkeit, Christliche Sozialethik angesichts globaler Herausforderungen. Verlag Herder, Freiburg im Breisgau, p 100

Wikipedia (2019) Greenhouse gas. Available via https://en.wikipedia.org/wiki/Greenhouse_gas. Accessed 6 Aug 2019

World Bank Group (1994) Nigeria—structural adjustment programme—policies, implementation and impact. Washington DC. Available via http://documents.worldbank.org/nigeria-structural-adjustment-programme. Accessed 5 Nov 2021

World Bank Group (2000) Federal republic of Nigeria: water supply and sanitation interim strategy note, No 19, pp 5–6. Available via http://docplayer.net/20806603-Federal-republic-of-nigeria-water-supply-sanitation-interim-strategy-note-november-2000.html. Accessed 6 Aug 2019

World Bank Group (2015a) Nigeria economic report, No 3. Available via https://www.worldbank.org›Nigeria/. Accessed 3 Feb 2020

World Bank Group (2015b) A plea for action against pollution in Nigeria. Feature Story, 16 June 2015. Available via https://www.worldbank.org/en/news/feature/2015/06/16/in-lagos-nigeria-a-plea-for-action-against-pollution. Accessed 12 Nov 2019

World Bank Group (2016) Air pollution deaths cost global economy US$225 Billion. Washington DC, 8 Sept 2016. Available via https://www.worldbank.org›press-release›2016/09/08. Accessed 12 Nov 2019

World Bank Group (2017) Governance and the law. In: World development report. Available via https://www.worldbank.org›wdr2.../. Accessed 3 March 2020

Chapter 3
Sustainable Development and the Nigerian State

Abstract Development concepts have seen, in recent decades, broader and more scientific approaches. One of these approaches, which dominates and informs the approaches of the United Nations, is to interpret and apply development discourses with intergenerational character. From the post-war development approach to the innovative minds that made up the Brundtland's Commission, the focus has been to integrate and interpret the three dimensions of sustainable development within available signs of the time—poverty and destitution, ever-widening economic gaps between the global North and South, the challenges of terrorism, climate change, etc. Within these general and shared concerns, the journey so far shows that there is still the need to shift to new paradigms that would be more inclusive and post-colonial. To do this, normative sustainability ethics should lead the way to investing in core values and developing action plans.

Keywords Development concepts · Brundtland commission · Intergenerational · Post-colonial

3.1 The General Concept of Development

The human world has always seen events, progress and facts about building and re-building, as well as events of destruction. Within the observable system of the cosmos lies the interrelated process of generation, regeneration, degeneration and, again, a constant repeat of the whole process. The whole system goes into what is described as development in the general sense of the word. While some of these concrete facts could be described as events with positive impacts, others are seen in their negative forms and, of course, with negative consequences in the world of events. The theme associated with development has undergone a long process of study and systematic applications. At the practical level, the concept is abstracted from many socio-phenomenal areas that provide empirical facts about whether development has occurred or not. As human history and experiences show:

C. F. Nwosu, *Systemic Remediation*, Environmental Humanities: Transformation, Governance, Ethics, Law 4, https://doi.org/10.1007/978-3-031-61909-0_3

> Development has come a long way in the past six decades. As both an enterprise and a scholarly discipline, development became significant in the period immediately following World War II. The western world confronted the challenges of rebuilding countries—and in Europe, a continent—that had been shattered by war. The institutions that would help manage this process, such as the International Bank for Reconstruction and Development (which soon came to be known as the World Bank), were created for the task. Alongside them arose a tradition of theorising about the special challenges facing backward regions and countries, and the means by which these challenges could be met in such a way as to put these areas on sustainable paths to industrialisation. (Rapley 2007)

The above-mentioned analysis describes the very point at which renewed concerns for development took centre stage practically all over the world, with each section developing the approach that best suits the challenges of their time and worldview. However, the truth about levels of independence as well as dependability in the general concept of development remains a heated argument. More important is the timeline of events that saw the world through to its present stage. The very beginnings present development as synonymous with industrialisation, first in the Western economies and, much later, in the developing economies that started gaining independence from the colonial empires. For Rapley, therefore, a new boost was given to development with experiences within the general push to industrialise, namely,

> [...] (through) the coming of independence to the former colonial empires of Europe, a process that picked up speed in the wake of the war. By and large, Asian and African countries came to independence poor, and were eager for two reasons to speed up their development. One was the obvious fact that they sought to provide better lives for their citizens. The second was the obvious need to consolidate their independence, to convert newly won nominal political equality with the rich countries into an economic equality that would earn them the respect and sense of self-dignity they felt had been denied them under colonialism. And the lessons of the early postcolonial age, particularly those recently learned in Latin America (where independence had come in the previous century), crystallised around a common set of assumptions. The scholarly literature of the time only reinforced this push: development was about using the state to spearhead the process of modernising the society and raising its incomes. (Ibid., pp. 1–2)

These very beginnings in the approach to development, especially in the so-called first world economies, concentrated much on the maximisation of profits and with a focus on prosperity measured in dollar figures (Ibid., p. 1). Such fundaments in the general concept of development still influence the studies and ways of application by varied developers, institutions, governments, ideologies and policies. But central to the general concept, regardless of institutions of application, is the constant employment and unification of efforts towards set and targeted goals in the process of modernising and stabilising the society. Such resolve for unified efforts spurred varied ideologies into action, creating camps of opinions that, at one time, might be complementary, at other times conflicting. Consequently, people and groups began the pursuit of development, either as the Lefts, Rights, Liberals, Communists, Conservatives, Democrats, Blue or Red States, Coalitions and Allies, etc. All these represent groups of ideologies and policies believed and designed to approach development, with varied policies that either emphasise the primacy of the state over the market or vice versa. Thus, "In the twentieth century, the left—which included not only socialists and communists but also modern liberals—generally, if not always, favoured using the state as an agent of social transformation" (Ibid., p. 2). The right remains on the side-line and could only begin their influence with the decline of post-war world economic boom which also saw a drastic decline in the "third world" (now global South) economies considered as "... growing more slowly than required to continue improving the standard of living of the world's poorest citizen" (Ibid., p. 3). The approach to development, therefore, would have to change for the developing economies—the third-world economies. At this time, the language of the approach became: Less state, more market. Thus,

> [...] this was the essential thrust of the strategy known as structural adjustment, which was seen applied in much of the third world. The idea seemed sound, but as time would tell, structural adjustment contained its own problems. Its shortcomings, which grew more evident with the passage of time, shed a new and damaging light on neoclassical theory. Structural adjustment yielded some positive gains in some of the more advanced third-world countries. However, in the poorer countries, those most in need of rapid change, it was less effective, and in some places actually did more harm than good. (Ibid., pp. 3–4)

The foregoing antecedents are facts that help in an integrated understanding of what development should generally mean and the journey that has been made in unfolding the concept up to the present.

Development in our research, therefore, would be taken in its global form, with hints on the indelible timeline of events that impacted human history globally and necessitated the pulling of political and socio-economic institutions together to chart a lasting course for peace, progress and understanding, if ever the human society would be able to avoid extinct. Any authentic approach to the concept of development must have the political and socio-economic dimensions as fundamental. However, it would be a great danger if humanity ever insists on summarising development to mean only these two aspects. As an exhaustive approach to the general definition of development, the different development theories offer us, first, an integrated and holistic platform upon which the facets of development could be justifiably explored. Such known classical theories are the concepts behind modernisation theory, neoclassical theory, structuralism, basic needs theory, dependency theory, structural adjustment theory, human development theory, post-development theory, as well as sustainable development theory. However, the main focus of our research would be to explore the current tenets of the sustainable development theory and apply them to the structural situation of Nigeria as a geopolitical and economic entity.

As all recent and modern approaches show, there has been the need for a shift of emphasis from a market-based, state-based or money-based approach to development to a rediscovery of the true agent of development—the "Mmadụ", the human person, who is and remains the bearer and the functional catalyst for development. Thus, the most recent approaches to development lay more emphasis on the issues of sustainability and how sustainable development impacts positively on the concrete lives of the citizens. For a more systematised approach to our research, we appeal to some of the key factors of Rapley's presentation of development theories in the post-war period, highlighting the relevant antecedents that set the stage for the ever-dynamic approaches to development.

3.1.1 The Post-War Period and the Development Theory

This period saw the final efforts of the allies and groups of allies during World War II, who, with victory in view, started developing concerns about what would be left of the devastated world and human society if no concrete structure for the world's economy was deliberated on. They were concerned about a world that was about to recover from the great depression but which was again brought down, crumbling on its knees, by World War II. This show of concern gave rise to the meeting, which took place in a hotel in Bretton Woods, New Hampshire (Ibid., p. 13). With the intellectual key figure to the meeting, John Maynard Keynes, these Allied leaders approached development with an ideological blueprint that would later give rise to a post-war capitalist economy. The focus was "[...] the desire to create a favourable international trading environment" (Ibid)—an approach to forestall all adverse conditions that inflicted the *Great Depression*.[1]

As a step in the right direction towards development, the Bretton Woods conference, as it is still known to date, strived to concretise its ideologies and plans by creating two socio-economic institutions, namely: the International Monetary Fund (IMF) and the International Bank for Reconstruction and Development, the latter of which became known as the World Bank. The aim was, on the one hand, to help create an international trading environment that would be relatively stable amidst the efforts to rebuild the devastated economies due to conflicts and war. On the other hand, it was to help stabilise the market through the creation of economic interdependence, whereby the stronger economies could mutually absolve the weaker ones. Thus, through the policies of both institutions, but especially the IMF, governments would be able to obtain loans and empowerment benefits, which were channelled to the reconstruction of the many European states ravaged by war. The conference and its focus on development, in the end, gave rise to the General Agreement on Tariffs and Trade (GATT). The implementation and the challenges that faced the agreement in the face of varied institutions and governments were both strong and impacting. A deeper analysis would be beyond the scope of our research. However, of key importance is to recognise the intellectual and ideological efforts that the post-war economists undertook to make development happen and as well work for its sustenance.

The general ideology of the post-war development pursuit was chiefly concerned with an economic development strategy that would ensure the free market. Thus, individual economic liberty was not emphasised within the general political economy. This was the exact point at which Keynes' influence became evident. While emphasising the State and market economy in development, the role and influence of the state must be clearly and minimally defined, and following Smith's opinion, such a role could be summarised thus:

---

[1] Great Depression, worldwide economic downturn that began in 1929 and lasted until about 1939. It was the longest and most severe depression ever experienced by the industrialised Western world, sparking fundamental changes in economic institutions, macroeconomic policy, and economic theory. Available via https://www.britannica.com/event/Great-Depression.

> [...] defence of national sovereignty, protection of citizens' rights against violation by one another, and provision of public and collective goods. (Furthermore), [...] Individuals unfettered by state interference, would use their ingenuity to the greatest extent. Division and specialisation of labour would allow resources to be used in the most efficient and productive manner possible. If individuals pursued their narrow self-interests, all of society would benefit inadvertently. (Ibid., *Understanding Development: Theory and Practice in the Third World*, p. 15)

It means that a smoothly running capitalist economy tends to place a much greater role for the state than had been tolerated in classical and neoclassical models of development, which focussed more on the development of the free market. However, following Adam Smith, it would be expedient to emphasise both the role of the free market and individual liberty in economic success (Ibid).

The foregoing discourse would be seen as the foundational antecedent to the modern concept of an inclusive economy, which, when realised, would purify capitalism of its excesses, give the role of the state its appropriate place and boundaries and place the means of production in use for the human agents and not the human agents for the means of production, in other words, a form of economic tyranny. This serves as the fundament for the post-war development in the industrial world, and following their opinion,

> [...] capitalism was to be not only more efficient, but indeed more humane. It was a recipe for social peace like none seen before: investors would grow richer—Keynes himself had grown rich on the stock market—but so too would workers, and poverty would become a thing of the past. Scholars proclaimed that correct economic management would prevent there ever being another worldwide depression, and that the high growth rates that followed in the 1950's were a permanent feature. All of this was possible because the ingredient missing from earlier capitalism—an appropriate interventionist role by the state—was not in place. (Ibid., p. 18)

A whole lot of other struggles, challenges, radically opposing and opposed ideologies, rejections and adoptions of further treaties and agreements that followed the post-war development ideology remain essential parts of the broad development dynamics that impacted the Western world and shaped its path to development. As the industrial world kept its struggles between Capitalism and Soviet Communism, the capitalist West moved on with a form of statist liberalism. It was, therefore, within this conglomerate of economic and socio-political approaches that the concept of "developing countries or economies", or the so-called "global south economies", came to the limelight (Ibid). However, before this work focuses on the global South, it would be important to examine the Church's line of thought within the general development process, given her position as a moral voice in the society.

3.1.2 The Catholic Church's Social Reflections on the General Development Processes

While not denying the varied forms and degrees of negligence and moral failures on the part of the catholic church, which, as a consequence, continues to question her authority and competence with the social teachings and ethics, her contributions to the general development discourse remain critical and impactful. With the publication of the encyclical, Populorum Progression, On the Development of Peoples, Paul VI made it a vocational project of the Church to deeply identify with all "[...] peoples who are trying to escape the ravages of hunger, poverty, endemic disease and ignorance; of those who are seeking a larger share in the benefits of civilisation and a more active improvement of their human qualities; of those who are consciously striving for fuller growth" (Paul 1967). In the foregoing observations, the church articulated and highlighted some of the core elements that drive development struggles among peoples and within societies. As a fundamental action and approach, the encyclical emphasises the need for concerted action since social questions, based on experiences, demand collaboration, which must tie all men together—regardless of race, colour and language (Ibid., no. 3).

Remarkably, Paul VI made an optimistic appraisal of the concrete known efforts, which have been taken up as concerns for the *development of the peoples*. To this regard and with particular reference to development efforts in the global South, the encyclical, while affirming some concrete failures, exploitations and imbalances associated with the historical social, cultural and economic exploits between the *global North* and the *global South*—colonialism (not exempting as well some Christian missionary failures and mistakes, cf. PP, no. 12), reserved some words of appreciation and acknowledgement for the traces of developmental efforts made as authentic concerns for progress and growth. The encyclical teaches:

> On the other hand, we must also reserve a word of praise for those colonisers whose skills and technical know-how brought benefits to many untamed lands, and whose work survives to this day. The structural machinery they introduced was not fully developed or perfected, but it did help to reduce ignorance and disease, to promote communication, and to improve living conditions. (Ibid., no. 7)

Central to the general theme of development for the encyclical is the need to unite the efforts of all men and women in the approach to the social, economic, cultural and spiritual development of peoples. Thus, authentic development cannot be an isolated event of human life and history. It can never really qualify to be real development when its impacts remain predominantly exclusive. It must, as a prerequisite, be as inclusive as it is to be authentic. The encyclical used the concepts: *to be well rounded, for each man, for the whole man*—there could be no economics outside human realities and no development outside the civilisation that must integrate the humans themselves and their world (Ibid., no. 14). "*Exclusive*" development is a one-sided project that keeps the human world balkanised and codified into extremely rich and abjectly poor, the rich global North and the poor or struggling global South, the privileged and the underprivileged, etc. However, to bridge this gap, there must be the need to marge development approaches with ethics and morals to avoid the latent dangers of avarice, selfish interests, soul-stifling materialism, dissension and disunity—whether by individuals themselves or among the various nations (Ibid., no. 19). Authentic development of the people requires the "gleaning", not "grabbing", of our common world and the natural capitals (Ibid., no. 22). Therefore, a holistic approach to social and corporate relations in the world order would avert so many banes to authentic development. Hence, with an appeal to concrete experiences, especially about developing economies, the encyclical warns: "Makeshift agrarian reforms may fall short of their goal. Hasty industrialisation can undermine vital institutions and produce social evils, causing a setback to true human value" (Ibid., no. 29).

As a plan and programme of action towards authentic development, the encyclical proposes true leadership built on a deep sense of subsidiarity, since

> It is for the public authorities to establish and lay down the desired goals, the plans to be followed, and the method to be used in fulfilling them; and it is also their task to stimulate the efforts of those involved in this common activity. But they must also see to it that private initiative and intermediary organisations are involved in this work. In this way they will avoid total collectivisation and the dangers of a planned economy which might threaten human liberty and obstruct the exercise of man's basic human rights. (Ibid., no. 33)

The cooperation implicit in the above statement requires basic education and enlightenment as the first objective of any nation aspiring towards authentic development. According to the encyclical, "[…] the illiterate is a starved spirit" (Ibid., no. 35).

The optimism reflected in Paul VI and his call for the united effort of all men towards authentic development of the people was followed up by John Paul II's worries about the persistence and broadening of global poverty. Thus, with the publication of the encyclical, Solicitudo rei Socialis, for the twentieth anniversary of Populorum Progression, he reminded the world about the appeal of Paul VI for a united effort in approach to development. However, he bemoaned the various man-made modern obstacles to authentic development, which he carefully articulated thus:

> […] how can one justify the fact that huge sums of money, which could and should be used for increasing the development of peoples, are instead utilised for the enrichment of individuals or groups, or assigned to the increase of stockpiles of weapons, both in developed countries and in the developing ones, thereby upsetting the real priorities? This is even more serious given the difficulties which often hinder the direct transfer of capital set aside for helping needy countries. If 'development is the new name for peace,' war and military preparations are the major enemy of the integral development of peoples. (John Paul 1987)

The calling for action against the misplacement of development values came as a strong critique from the encyclical at a time when the boom in technological know-how had begun to signal danger within the scientific world. Consequently, atomic technological power took centre stage as the modern identity of the so-called "strong nation", leading to illegal atomic tests, environmental hazards, heavy financial investments in lethal technologies, arms deals, as well as the politically and economically driven propaganda of supporting warring nations and peoples with lethal and atomic technologies. For the modern world, therefore, there is, to this extent, a breakthrough in human knowledge and capabilities within the development discourse. Yet, in the real sense of the word, authentic development keeps eluding epochs, nations and leaderships. The major dichotomies seen in institutions, nations, individuals, groups and peoples keep on widening, and the result has been the persistent social inequalities that have been the bane of authentic development. The whole situation gives rise to an organised socio-political and socio-economic course towards man-made underdevelopment—a man-made course towards self-inflicted social, political and economic enslavement. This does not end here. Thus, the Pope observed further:

> However, the picture just given would be incomplete if one failed to add to the "economic and social indices" of underdevelopment other indices which are equally negative and indeed even more disturbing, beginning with the cultural level. These are illiteracy, the difficulty or impossibility of obtaining higher education, the inability to share in the building of one's own nation, the various forms of exploitation and of economic, social, political and even religious oppression of the individual and his or her rights, discrimination of every type, especially the exceptional odious form based on difference of race. (Ibid., *Populorum Progression*, no. 15)

In the main, the total worries of John Paul II culminated in the definition of what authentic development should, in reality, mean. Thus, to be authentic (sustainable), "[...] either all the nations of the world participate, or it will not be true development" (Ibid., no. 17). Unfortunately, the human world always chooses to emphasise divisions, impose inequalities, practise exclusions, and develop themes and concepts that constantly impose exploitative blocs and imaginary identities. These often culminate in a dichotomy of love and justice, which remains central to the teachings of Benedict XVI on integral human development.

Amidst the dastardly imposed dichotomy of love and justice in our world, the encyclical Caritas in Veritate took up the theme "love" as the integral driving force behind the authentic development of both human persons and, in a broader sense, humanity itself (Benedict 2009). Thus, to set the world's development process right, love and justice must be made to unite and influence not only the microrelationships but the macro-relationships as well. The encyclical teaches:

> [...] (love) is the principle not only of micro-relationships (with friends, with family members or within small groups) but also of macro-relationships (social, economic and political ones). I am aware of the ways in which charity has been and continues to be misconstrued and emptied of meaning, with the consequent risk of being misinterpreted, detached from ethical living and, in any event, undervalued. In the social, juridical, cultural, political and economic fields—the contexts, in other words, that are most exposed to this danger—it is easily dismissed as irrelevant for interpreting and giving direction to moral responsibility. (Ibid., no. 2)

The denial of the driving force of love, which must be translated into concrete forms like responsibility towards charity, respect for life and nature in general, the sense of common good, and above all, passion towards the solidarity of all men, leads to forms of exclusivity in development processes, which create extremities among economies, political and social ideas, peoples and nations. With such extremities, the development processes of our world have constantly, in the past two decades, been developing ideas and connotations based on racial, supremacist and imperialist institutions believed to be guarantors of our world's right to development. However, the experiences of our world have proved such exclusive and imposed institutions as failures. The intensified clamour and emphasis on solidarity towards the sense of sustainability in the development process of our world would prove the encyclical right that those institutions can never be drivers of authentic development since "[…] integral human development is primarily a vocation, and therefore it involves a free assumption of responsibility in solidarity on the part of everyone" (Ibid., no. 11). There could be no authentic development if our world chooses to live with an impoverished global South alongside an extremely rich global North; if the language and concept of globalisation remain selective and exclusive and not inclusive and distributive; and if the modern development ideologies do not remove impositions, political, social and economic "*inferiorisations*", exploitative capitalism, extreme consumerism, etc. It all points to the fact that solidarity needs *communality* (commonality).

The need to pay attention to and resolve the imposed dichotomy, therefore, would receive contemporary attention with Pope Francis' call for a change of the development narratives of our modern world towards the "care for our common home"—nature, the mother earth. Thus, the concept of a *common home* defines the sense of dependability and union existing between humanity and nature itself. Thus, authentic development can never be a reality outside this union. For "[…] all creatures are connected, (therefore) each must be cherished with love and respect, […] all of us as living creatures are dependent on one another" (Francis 2015). Alongside the central call on humanity to respect nature in general and begin practical efforts to save our world from environmental deterioration, the encyclical, Laudato Si, called attention to the re-examination of the so-called growth (development strategy) of our past two centuries, which, in multi-faceted realms, has proved insufficient for the authentic development of our world. The Pope teaches:

> The social dimensions of global change include the effects of technological innovations on employment, social exclusion, an inequitable distribution and consumption of energy and other services, social breakdown, increased violence and a rise in new forms of social aggression, drug trafficking, growing drug use by young people, and the loss of identity. These are signs that the growth of the past two centuries has not always led to an integral development and an improvement in the quality of life. Some of these signs are also symptomatic of real social decline, the silent rupture of the bonds of integration and social cohesion. (Ibid., no. 46)

With the above observation, the encyclical joins the contemporary world's political agenda for concrete development programmes in the post-colonial era. Its language would remain integral to the Sustainable Development Goals, the Agenda 2030 and the African development programme, the Agenda 2063. Furthermore, the publication of the encyclical in the year 2015 showed an observable effort of our modern world to revisit its development agendas and apply principles that would ensure sustainable results. In this regard, the year 2015 goes down the annals of history as the year of consolidating development discourses since the same year also witnessed, alongside the encyclical, the issuance of the Sustainable Development Goal of the United Nations and the ratification of the Paris Agreement on climate change (Sachs 2022). In the main, therefore, the church has maintained its pace, at least vocally and in solidarity with the development trends of our world as they affect humanity and nature itself. However, our research expects more and dedicates its contribution to the unreserved effort of the church to translate discourses into concrete and planned actions for more definitive results.

3.1.3 Development Indices Within the Global South

The very beginnings of development studies saw industrialisation as the hallmark of development. Thus, the industrialised world with its advanced capitalist ideology was designated as the first world; the anti-capitalist and communist economies were ranked in the second world, while other forms of economies and states, which later came to experience colonisation and Balkanisation, are all current members of the developing economies or the economies of the global South. Rapley made an exhaustive analysis of some characteristic features of the so-called third-world countries:

> First, by comparison with the advanced capitalist economies of Western Europe and North America, their per capita incomes are low. This poverty translates into shorter life expectancies, high rates of infant mortality, and lower levels of educational attainment. Typically, a high proportion of the population is engaged in agriculture. The secondary or manufacturing sector occupies a relatively less important place in the economy than it does in the first world, and exports come mainly from the primary sector (the cultivation or extraction of natural resources, as in farming and mining). Such a characterisation, of course, fails to capture the great variety within the world [....] Yet as a rule, there is a correlation between national income and a country's ability to improve the social indicators of its citizenry. With the exception of the few countries endowed with an abundance of natural resources, there is also a correlation between industrialisation and growing national income. There are factors other than economic that are common to third world countries, including a tendency towards high population growth rates. However, perhaps the world country began its modern history as a colony of one of the former imperial powers of Europe or Asia (Britain, France, Belgium, Germany, Spain, Portugal, the Netherlands, and the Ottoman Empire). (Ibid., pp. 18–19)

The above-examined realities present development as being confronted with deeper challenges in developing economies. As the colonies started gaining independence from the colonial powers, the very first breeds of politicians, who were invariably products of the colonial regimes, became energised and kept focussing on efforts to launch the different newly independent states into the modern world of industrialisation and technological know-how. But the market capitalist dominance and the persistent imperial powers proved (and perhaps, still prove) that the actual independence of the third-world economies is still far from being realisable. One of the foremost thinkers borne out of this first generation of leaders, Kwame Nkrumah, articulated this about Africa thus:

> Africa is having to pay a huge price once more for the historical accident that this vast and compact continent brought fabulous profits to western capitalism, first out of the trade in its people and then out of imperialist exploitation. This enrichment of one side of the world out of the exploitation of the other has left the African economy without the means to industrialise. (Nkrumah 1974)

Development within the third-world economies, therefore, took off within the persistent tenets of neo-colonial capitalist grips, which continues to resist indigenisation and practicable independent resource control for the indigenous people. Of course, the post-independence nationalists remained unanimous in their view that authentic independence would continue to elude the global South economies until colonial economic structures become overthrown alongside colonialism and its agents. Rapley elaborated further on this neo-colonial grip on the third-world economies in the following expression:

> Third-world nationalist argued that by using the colonies as sources of raw materials and markets for finished goods, and by establishing intra-imperial free trade blocs that prevented colonial administrations from using protective barriers to nurture industrial development, the imperial countries had actually impoverished the third world in order to enrich the first. Where shoots of industrialisation had begun to sprout, as in precolonial India, the imperialists rolled them back by swamping the colonial markets with the cheap manufactures of their factories. (Ibid., *Understanding Development: Theory and Practice in the Third World*, p. 20)

The whole system and struggle became shrouded in the fight against the one-sided dominance of the market and economic means. The ideological and empirical efforts to break the jinx of such imperialism have become a persistent challenge, which the economics of the global South, as well as its general development process, are still to surmount. For Nkrumah, this could only happen when the developing economies in question make it a duty to rediscover and exert the needed *economic will* that would be able to ignite and force an economic adjustment that would be able to combat and break all neo-colonial patterns and dominance (Ibid., *Neo-Colonialism: The Last Stage of Imperialism*, p. 9).

The forgoing antecedents and thought patterns with regard to development within the developing economies, one must say, have undergone reforms, restructuring, redesigning and re-definitions. However, whether there has been a total liberation of the global South or the developing economies from the in-built principles of neo-colonial and imperialist structures remains widely debated today. Thus, in presenting development within Africa and with predication to Nigeria, we propose that the general development indices show disorientation, inauthenticity, imperialist undertones, corruption and deception, especially in connection with external players, exploitation and alienation, etc. Nigeria, which is the main working specimen of our research, is fully cut in the web, and she would remain trapped unless there is a true struggle against "[…] the external forces which have a vested interest in keeping it underdeveloped" (Ibid., p. xx). Such external forces, it must be said, do not work alone. Their nefarious actions have enemies and defective structures within which serve them as catalysts and partners in crime. To understand vividly the situation on the ground within our specimen, Nigeria, one of our foremost political analysts and the Catholic Bishop of Sokoto Diocese, Matthew Hassan Kukah, described the Nigeria met by our research as a geo-polity *flying without a black box*. In his opinion:

> There has been a nagging concern as to how and why our country has found itself in this state of immobility and decay. This is not the place to reel out the statistics concerning our rut. Everyone, young and old, in the city or in the remotest part of the country knows and feels the pain of what is wrong with our country. How do we explain the fact that after over 50

> years, we are unable to generate and distribute electricity, supply water to our people, reverse the ugly and avoidably high infant mortality, set up and run an effective educational system, agree on rules of engagement for getting into power, reverse the circle of violence that attends our elections, contain corruption, instil national discipline and create a more humane and caring society? [...] (Failure to provide answers to these questions) laid a foundation for a nation that I refer to as flying without a black box. Thus, today, we have nothing to draw from, no inspiration about the past to engender sacrifice and patriotism. (Kukah 2012)

All the colonial, global South or developing economies had to or are still facing the challenge of identity in the global economy since the whole experience of colonialism "[...] left behind immature capitalist classes" (Ibid., *Understanding Development: Theory and Practice in the Third World*, p. 29). To stay aloof and only watch the trend of events remains as challenging as every effort to gain a position in the global market, which is controlled by industrialised economies. Thus, designing a development strategy for the developing economies met a stand-off with three available options, namely: (1) Disengagement with the world economy while building indigenous processes for development; (2) Partnership with foreign agents and economic actors to buildup the industrial sector, and (3) Total dependence on the state through taxation, borrowing or resource control to mobilise capital (Ibid).

For many development socialists and economists, the first option, which is known as *autarky*[2] was, from the very beginnings, problematic and very challenging since "By the time countries in Africa became independent, the costs of establishing a new industrial venture were estimated, in relative terms, to be 250 times what they had been for an entrepreneur in the early days of the Industrial Revolution" (Ibid., *Understanding Development: Theory and Practice in the Third World*, p. 29). The Cambodian experience was regarded as a tragedy that left the option weak and less attractive. In our modern world, autarky would have to face deeper challenges as well, especially within a world economy that has become globalised and digitalised, although its tenets and principles could never be seen as totally inapplicable. The other two options are seen as operational and policy-wise embedded in the socio-economic dynamics of states and governments in the developing economies, but they keep these economies heavily dependent. In most cases, the general struggle to find identity remains stifled by the greater development and economic powers that are optimally protective and conscious of their leading status and benefits. In all, therefore, our vantage point in the approach to the concept of development has been to establish that development in Africa still faces many challenges, which are a combination of many factors, both internal and external. The structures and factors discussed above have affected and deflected indigenised development in most African countries and their economies. This is exactly the development situation met by our research.

---

[2] Autarky is the quality of being self-sufficient; the term is usually applied to political states or their economic systems. Autarky exists whenever an entity can survive or continue its activities without external assistance or international trade. Available via https://en.wikipedia.org/wiki/Autarky.

3.2 Sustainability as the Modern Correlation to the Concept of Development

Development theories saw a major shift and more radical transformation beginning with the world and ideas of the twentieth century. Thinkers, and indeed the realities in our modern world, discovered the need and timely as well to make a paradigm shift in the general approach to development and to give the concept an added qualification. Consequently, the modern concept of development receives the word *sustainability* as a correlating terminology. This has happened, first, in the bid to present development as a *continuum* rather than a process that begins, ends and dies off; second, to sensitise the modern world about the danger involved when one tries to reduce functional development only to fiscal socio-economic and socio-political matters while neglecting other essential areas of the bigger cooperate social responsibility like ecology, environment and the human person. The new terminology, Sustainability, therefore, becomes "[…] a value statement, a vision of people and society and a very strong moral and ideological choice" (De Janvry and Sadoulet 2016).

This new direction brought to the fore innovative developers of the concept, and very influential among them was Amartya Sen. However, the early foundations for the entrance of this concept into the dynamics of development theories gained stronger recognition with the Brundtland Commission—a World Commission on Environment and Development. For the commission, Sustainability is "[…] concerned with intergenerational equity: that the well-being of future generations should not be inferior to that of the current generation as a consequence of the current generation's behaviour towards the use of natural resources and the environment" (Ibid). Sen's approach, on the other hand, acknowledged the Brundtland impact but moved further to give the concept a more inclusive meaning. Thus, in his 2013 journal article on the Ends and Means of Sustainability, Sen opined:

> The idea of 'sustainability' received serious attention in the so-called Brundtland Commission Report that has many attractive features. In particular, it highlighted the importance of intergenerational justice while maintaining a concern for the poor of each generation and shifted the focus away from resources to human beings. I argue that this way of understanding sustainability, while a great improvement, is still incomplete. There are important grounds for favouring a freedom-oriented view, focusing on crucial freedoms that people have reason to value. Human freedoms include the fulfilment of needs, but also the liberty to define and pursue our own goals, objectives and commitments, no matter how they link with our own particular needs […] A fuller concept of sustainability has to aim at sustaining human freedoms, rather than only at our ability to fulfil our felt needs. (Sen 2013)

The intergenerational dynamics associated with sustainability place responsibility on our present mode of life, which must be lived in a manner that should safeguard the future. In doing this, therefore, any sustainable approach has to embrace both distributive and inclusive justice, whereby efforts to sustain and respond to the needs of the present generation should not jeopardise the right of the future generation to come into being and meet conditions necessary and conducive for existence. Put in another form; it means that the resources of our mother earth are placed at the disposal of our present world, and caretakers must ensure that they are used in such a way that would be able to offer an inhabitable future for the world after us. Furthermore, the inclusivity of the concept should focus not just on the fulfilment of societal and human needs but also on the ability to emphasise human persons' freedom of choice. In other words, the focus should include each individual's capability to make *value choices* for which he or she has the freedom and every reason to choose. This form of approach to the concept of sustainability has come to take centre stage in our twenty-first-century approach to development and has helped to shape varied socio-political and socio-economic concepts of our world today. Thus, the paradigm shift is the fact that human societies should not only be concerned about achieving development but rather should seek the path of development that remains sustainable. This would, therefore, be able to guarantee inclusivity and guard against the enemies of a progressive society in the forms of poverty, inequality and inequity, the varied forms of economic slavery, exploitation and imperialism, etc. The tenets behind this form of approach, which have given rise to the Global Goals of the United Nations and are shaping the ideologies of modern governance, are discussed in this chapter. The purpose is to discuss sustainable development in its dual components, namely: (a) the temporal component, which emphasises the need for the preservation of enough natural capitals such that the future generations and their living conditions, including the climatic ones, would not be endangered; (b) the substantive component, after which the natural cycles and regenerative components of the natural capitals are resourcefully preserved (Siep 2016). We would then examine the development prospects proposed in each of the goals within our research specimen, Nigeria.

3.2.1 Sustainable Development Strategies: Historical and Literary Appraisal

Before the current emphasis on sustaining development came to take centre stage in our contemporary world, a series of international relationships and conferences, declarations, pacts, as well as treaties helped to draw the attention of humanity in general to the essentials of development theories and application. The following literary appraisal shows the efforts of various epochs and policy-making institutions in trying to respond to the demands of seeing development theories translate to phenomena that impact the human world and lives positively. Generally, and with regard to development, it has always been the process of policy formation and definition, negotiations and renegotiations, conflict management and resolutions, and in extreme cases, wars and other forms of humanitarian crises. The whole process and events have come a long way and have helped and continue to help shape the development history of our world. Our approach focuses deeply on the efforts of the United Nations within the development history. However, the work finds the Malthusian Essay on the Principle of Population an antecedent worth mentioning in discussing the prospects and challenges of development.

3.2.1.1 Malthusianism: A Pre-united Nations' Model

The Malthusian theory, as an antecedent of sustainable development theory, drew attention to the essential role, which the population can play in the dynamics of development. This theory was named after its proponent, Reverend Thomas Robert Malthus, who opined that within a social structure, population growth tends to be *potentially exponential* alongside the growth of the food supply, which remains *linear*. Linked to the dynamics of development, Malthus sought to confront the challenges posed by poverty in society on the one hand. On the other hand, he was convinced that the population growth must be made to balance with the growth of food supply if ever the gap created by poverty could be breached or at least be brought to the barest minimum. Thus, writing in *An Essay on the Principle of Population*, Malthus recognised two types of "checks" that help the exponential behaviour of population growth, namely: "'preventive checks', such as moral restraints (abstinence, delayed marriage until finances become balanced), and restricting marriage against persons suffering poverty or perceived as defective, and 'positive checks', which lead to premature death such as disease, starvation and war, resulting in what is called a Malthusian catastrophe. The catastrophe would return the population to a lower, more "sustainable", level" (Malthus et al. 2018). Malthusianism, therefore, has a major impact on issues and concerns about population growth. In his tenets concerning the preventive checks, Malthus, being a devout Christian, is invariably against contraception, thereby proposing abstinence and self-control, as against the neo-Malthusians' favour of the same through artificial birth control (Ibid).

Later, a modern off-shoot of Malthusianism—neo-Malthusianism—shifted the paradigm from a mere focus on poverty and population growth to embracing the need for concern towards "[...] environmental degradation and catastrophic famine" (Ibid). In general, the different tenets and concepts seen running through Malthusianism and its off-shoots would later help in specifying, structuring and restructuring the basic tenets and ideology behind sustainable development theories as applied to our human world. Malthusianism would later face divergent criticism of the theory from the perspective of other social sciences and approaches. However, our focus here is on extracting the very antecedents that have contributed to the shaping of modern sustainable development theories.

3.2.1.2 The 1972 Conference on the Human Environment: Stockholm, Sweden 1972

This 1972 conference first applauded the invaluable evolution of the human race and its ability to transform the world and the environment through scientific and technological breakthroughs. Excerpts from the opening reflections read: "Man is both creature and moulder of his environment...In the long and tortuous evolution of the human race on this planet, a stage has been reached when, through the rapid acceleration of science and technology, man has acquired the power to transform his environment in countless ways and on an unprecedented scale" (United Nations

1972). But moving further, the conference immediately set the tone that would guide the actual focus of the conference by drawing attention to the fact that the rapid acceleration of science and technology by the human race could, by application, be a double-edged sword. In other words, science and technology as tools in the hands of humans could be wisely applied to reap benefits, or they could as well be wrongly used to the enslavement of the user and the phenomenal world outside of him or her. In the words of the conference,

> In our time, man's capability to transform his surroundings, if used wisely, can bring to all peoples the benefits of development and the opportunity to enhance the quality of life. Wrongly or heedlessly applied, the same power can do incalculable harm to human beings and the human environment. We see around us growing evidence of man-made harm in many regions of the earth: dangerous levels of pollution in water, air, earth and living beings; major and undesirable disturbances to the ecological balance of the biosphere; destruction and depletion of irreplaceable resources; and gross deficiencies, harmful to the physical, mental and social health of man, in the man-made environment, particularly in the living and working environment. (Ibid)

Having identified the empirical consequences of human negative impact on our human world, the conference also drew attention to the specific situations in the developing world, where environmental problems are made more complex due to underdevelopment situations. Within such developing countries and economies, "[...] Millions continue to live far below the minimum levels required for a decent human existence, deprived of adequate food and clothing, shelter and education, health and sanitation" (Ibid). Evoking, therefore, the call for collective action, the conference made it clear that the humans, the people, must be placed at the centre of discussion about progress and development since "it is the people that propel social progress, create social wealth, develop science and technology and, through their hard work, continuously transform the human environment" (Ibid). However, the human capacity to do this must not be left unchecked to avoid excessiveness and abuse. In clearer terms, the conference raises alarm and teaches that,

> Through ignorance or indifference, we can do massive and irreversible harm to the earthly environment on which our life and well-being depend. Conversely, through fuller knowledge and wiser action, we can achieve for ourselves and our posterity a better life in an environment more in keeping with human needs and hopes. (Ibid., Excerpt from Preamble, Paragraph 6)

The probably endangered conditions—life and well-being—are not only those of the present generation but also those of future generations, and the proper preservation of such conditions has become both imperative and prerequisite. To achieve results, therefore, the conference proposed twenty-six practical and applicable working principles that should guide the concrete actions of nations and peoples towards protecting and safeguarding the human environment (Ibid., *Report of the United Nations Conference on the Human Environment,* pp. 4–5).

Furthermore, the principles covered issues of the fundamental right to freedom, right to quality and humane living conditions, respectful and wise dealings with natural resources, judicious use of non-renewable resources and thinking *renewably*, care for wildlife and nature conservation, pertinent need to protect the environment and ecosystem from toxic substances and other forms of pollution, practical agenda for economic and social development that impacts the quality of life, solidarity of states against disaster-prone areas and less privileged geographical areas, subsidiarity consciousness among states with regard to developing states (Principle 11), integrative and coordinated approach to development policies for optimal and integral results, rational planning, remedial demographic policies, empowerment and commissioning of appropriate national institutions for proper realisation of development goals, concern for environmental-friendly science and technology policies, enlightenment programmes on environment and sustainability, the autonomy of States in the exploitation of natural resources, with high sense of respect for other States' autonomy and boundaries, defined policies for compensation of victims of pollutions and other environmental damages, bilateral and multilateral cooperation of States in matters of improvement and protection of the environment, as well as the united efforts of all States against the adverse effects of nuclear weapons and other means of mass destruction.

3.2.1.3 The World Commission on Environment and Development: "Our Common Future"—The Brundtland Commission Report, October 1987

The Brundtland Commission was constituted by the United Nations Organisation and given the task of designing *a global agenda for change*—change in human attitude and approach in dealing with the environment, which would positively translate to sustainable development; change that would suggest and "recommend ways concern for the environment may be translated into greater cooperation among developing countries and between countries of different stages of economic and social development" (United Nations 1981), leading to support for common objectives; changes and recommendations that would be capable of uniting the efforts of the international community in dealing effectively with environmental concerns; as well as changes that would ensure "[...] long term agenda for action during the coming decades and aspirational goals for the world community" (Ibid). The set agenda for this commission guided its work and produced the famous report with the caption: *Our Common Future: From One Earth to One World*. Another popular connotation associated with this report is *The Brundtland Report*, in recognition of the impact made by the commission under the leadership role of its chairperson, Gro Harlem Brundtland.

A wider and more inclusive definition of the term sustainable development has come to be one of the invaluable contributions of this commission. Thus, going by the commission's definition, the development dynamics that are worth being sustainable are those capable of stepping in to stem the drifting tides associated with our present time and age, through which policies, politics and economics are made "[...] capable of diverting the paths that evolution has followed for millions of years [...] (thereby, and forcibly too) creating a planet our ancestors would not recognise" (Ibid). The commission, therefore, avoided the grave mistake of narrowing its approach only to core exclusive environmental issues. The members chose a rather broadened path and approach based on the understanding that "The environment does not exist as a sphere separate from human actions, ambitions, and needs [...], (and any attempt) to defend it (the environment) in isolation from human concerns have given the very word "environment" a connotation of naivety in some political circles" (Ibid). The need for growth and development and, of course, for sustainable paths towards progress need not be ideologically exclusive but rather inclusive. Such inclusive approaches do not need to be forceful and practicable. They should, as a matter of urgency in our modern world, offer sustenance and legitimacy to all indices of human and social development.

In the final analysis, therefore, the commission views sustainable development as the "development that meets the needs of the present without compromising the ability of future generations to meet their own needs" (Ibid). The call for concrete actions becomes obvious from the commission's approach, with emphasis on the fact that the path would be tough, painful and challenging. Come what may, it still lies in the hands of humans to embrace such intergenerational demand in order to create a more habitable earth for both the present and the future.

3.2.1.4 Rio Declaration on Environment and Development: The Earth Summit and *Agenda 21*—Rio de Janeiro, Brazil, 1992

After 20 years of the insightful conference on the human environment in 1972 in Stockholm, the United Nations deemed it fit not to relent in the pursuance of its agenda on development. Thus, the year 1992 saw yet another conference, the deliberations of which gave rise to what is today known as the *Rio Declaration on Environment and Development.* The conference is also known as "the Earth Summit and Agenda 21", following its 40-Chapter blueprint that provided the commission with its areas of concern and deliberation. The focus of the commission is stated clearly in its *Principle 1* among the other 27 Principles that make up the declaration, thus: "Human beings are at the centre of concerns for sustainable development. They are entitled to a healthy and productive life in harmony with nature" (United Nations 1992).

The expression "life in harmony with nature" strikes immediately the chord of reaffirmation, especially in reference to the Brundtland Commission, which sought to place the human agents at the centre of development dynamics but in strict relationship to the natural environment that provides the same humanity with shelter and nourishment. The principles and commitment of the commission sought to form strong coalitions among States as well as unite the relevant political will, which would translate the raised concerns into practices that would ensure sustainable development. Thus, Agenda 21, a 700-page long blueprint with four sections detailing areas of concern and projecting costs of implementation, became a *Plan-for-Action* document placed in the hands of the delegates of 172 sovereign nations that took part, and of course, in the hands of our world leaders, who are prepared to make the difference in the development indices. An exhaustive summary of the document can be given thus:

A. **Section One—Social and Economic Dimensions**: Containing the preamble, this section, with its eight (Ibid) chapters, views Agenda 21 as the very beginning of a new global partnership. Thus, it calls for dynamic programmes towards Accelerating Sustainable Development, Combating Poverty, Changing Consumption Patterns, integrating Population and Sustainability, actions for Protecting and Promoting Human Health, caring for Sustainable Human Settlements, and Making Decisions for Sustainable Development.
B. **Section Two—Conservation and Management of Resources**: This section contains chapters 9 through 22. It brought to the fore the need for humanity to be more responsible by rediscovering the attitude of conserving, which leads to proper management. The main focus is on divergent but related areas: Protecting the Atmosphere; Managing Land Sustainability; Combating Deforestation; Combating Desertification and Drought; Sustainable Mountain Development; Sustainable Agriculture and Rural Development; Conservation of Biological Diversity; Management of Biotechnology; Protecting and Managing the Oceans; Protecting and Managing Fresh Water; Safer Use of Toxic Chemicals; Managing Hazardous Wastes; Managing Solid Wastes and Sew, and Managing Radioactive Wastes.
C. **Section Three—Strengthening the Role of Major Groups**: The concern in this section was to highlight the need for a more integrative approach in the general dynamics, which would ensure the active participation of the various social groups and stratifications. The different areas of focus are as follows: Women in Sustainable Development; Children and Youth in Sustainable Development; Strengthening the Role of Indigenous People; Partnership with Non-governmental Groups (Civic Groups); Local Authorities; Workers and Trade Unions; Business and Industry; Scientist and Technologists; and Strengthening the Role of Farmers.
D. **Section Four—Means of Implementation**: The commission in this section encouraged the nations not to be scared about the cost of implementing sustainable development since "…the cost of inaction could outweigh the financial costs of implementing Agenda 21". A coalition of efforts is therefore needed, especially towards proper integration of the developing nations and economies. This section, therefore, dealt with: Financing Sustainable Development; Technology Transfer; Science for Sustainable Development; Education, Training and Public Awareness; Creating Capacity for Sustainable Development; Organising for Sustainable Development; International Law; and Information for Decision-Making (Ibid).[3]

From the specifications above, this commission made a thorough networking of ideas to reiterate the urgent demand placed on our world of today not to "[…] think of environment and economic and social development as isolated fields" (Ibid). Therefore, more integration of human actions with nature itself, as well as the broadening

[3] This form of summary and adaptation is based on the article entitled *The Earth Summit's Agenda for Change* by Michael Keating, September 1992. Published by the Centre for Our Common Future, Geneva, Switzerland. Available via https://library.lincoln.ac.uk/items/57959.

of partnership among nations, peoples and economies, were made a prerequisite towards clinching a better developed and more sustainable human society.

3.2.1.5 The International Conference on Population and Development—Cairo, Egypt, 1993

The conference in 1993 brought back the theme of the population to the limelight. On the one hand, the United Nations, through the conference in Cairo, sought to remind the world that, while some tenets of Malthusianism may not directly apply in concrete terms, its proposition about the impact of population growth on the general dynamics of development cannot be swept so negligently under the carpet. In the preamble to the document of the conference, the focus of the assignment was made very clear:

> The 1994 International Conference on Population and Development occurs at a defining moment in the history of international cooperation. With the growing recognition of global population, development and environmental interdependence, the opportunity to adopt suitable macro- and socio-economic policies to promote sustained economic growth in the context of sustainable development in all countries and to mobilize human and financial resources for global problem-solving has never been greater. Never before has the world community had so many resources, so much knowledge and such powerful technologies at its disposal which, if suitably redirected, could foster sustained economic growth and sustainable development. Nonetheless, the effective use of resources, knowledge and technologies is conditioned by political and economic obstacles at the national and international levels [....] The world population is currently estimated at 5.6 billion. While the rate of growth is on the decline, absolute increments have been increasing, currently exceeding 86 million persons per annum. (United Nations 1994)

The recognition of the interdependence of different economic and social variables, which inevitably bring macro and socio-economic policies over the edge and, in many cases, at conflict with the surging population of the world, remained the focus of the conference. The conference, therefore, sought to unite principles that would help the nations of the world direct and advance policies that would be able to integrate population growth with workable as well as sustained economic growth strategies in the context of sustainable development. The failure to strike such a balance would have the consequence of pushing the poverty level to the extreme, and the most devastating effects would be on developing economies and societies. It called for partnership and dedicated interest of all and sundry for the achievement of results. But fully aware of the challenges that are bound to accompany varied forms of implementation, the conference issued a strong statement about the fact that this may be,

> [...] the last opportunity in the twentieth century for the international community to collectively address the critical challenges and interrelationships between population and development. The Programme of Action will require the establishment of common ground, with full respect for the various religious and ethical values and cultural backgrounds. The impact of this Conference will be measured by the strength of the specific commitments made here and the consequent actions to fulfil them, as part of a new global partnership among all the

> world's countries and peoples, based on a sense of shared but differentiated responsibility for each other and for our planetary home. (Ibid., Chap. 1, Preamble 1.15)

The adaptation below summarises the developed principles that the commission deliberated upon and set out as modalities towards concrete results in all efforts to integrate socio-political and economic development in the face of all existing challenges of population growth (Table 3.1).

Table 3.1 15 Principles of implementation: the international conference on population and development (Ibid., *Report of the International Conference on Population and Development*, pp. 11–14)

	15 Principles for proper implementation of the report of the International Conference on Population and Development, Cairo, 5–13 September, 1994
1	All human beings are born free and equal and they have equal right to life, liberty and security of person
2	At the centre of sustainable development are the human beings; they have every right to adequate standard of living
3	Right to development must be made to meet the population, development and environmental needs of our time, but must not jeopardise the needs of the future generation
4	Gender equality and equity must be advanced, especially with regard to the empowerment of women and protecting them against acts of violence
5	To better improve the quality of people's lives, practicable population policies must play an important role
6	For sustainable development and progress to happen, States must guard against all forms of unsustainable production and consumption
7	Special attention and effort must be given to the eradication of poverty and the widening of economic gaps; more advanced economies should unite to help the developing ones
8	Access to both health care and the right of couples to the decision on the number and spacing of their children is paramount
9	The Family remains the nucleus of the society and must be protected; Family life must be the result of a truly and freely consenting couples
10	Education is a priority and the parents have the first responsibility; All, both male- and female-child have equal access to it
11	Appropriate legislative, administrative and educational measures must all be employed as relevant instruments towards the protection of every child
12	Migrants should be treated as part of the Universal Declaration of Human Rights, and therefore, should have access to protection and security of their lives
13	Asylum seeking for safety and protection purposes is permitted; States should unite in the implementation of the Geneva 1967 Protocol
14	Indigenous people's identity, culture and interest should be integrated by the State, in all concerns about population and development
15	Sustainable economic growth that would benefit all requires the cooperation of both advanced and developing economies; efforts must be made to narrow imbalances

Illustration mine

With the principles laid out, the conference drew the world's attention to the fact that no further time should be wasted in understanding the interconnectedness among patterns of our society, which must work in unison towards development or disunited towards socio-political, economic and ecological crises. Therefore, States' partnership with one another in a united effort would reflect the very beginning of "[...] the growing awareness that population, poverty, patterns of production and consumption and the environment are so closely interconnected that none of them can be considered in isolation" (Ibid., Chap. 1, Preamble 1.5).

3.2.1.6 World Summit for Social Development, Copenhagen, Denmark 1995

The 1995 summit on social development was designed to bring together world leaders to deliberate on the need for authentic social development and the advancement of human well-being as the world then prepared to enter the twenty-first century. As a guiding agenda for the summit, three social problems demanding the urgent attention of world leaders were put forward, namely: poverty, unemployment and social exclusion. The human person stays as a common denominator at the centre of these social ills, and, according to the report, the situation places an urgent demand on the world leaders "[...] to address both their underlying and structural causes and their distressing consequences in order to reduce uncertainty and insecurity in the life of people" (United Nations 1995a).

The focus of the summit was on a designed plan of action for results that would be unshakable and sustainable for a better future, stronger nations, more inclusive economies, as well as a human society that means a real home for the human person. Apart from the identification of the social problems, the summit still acknowledges the dynamics associated with globalisation as positive tools in the hands of the world's political economy but which could as well be misused as political ideologies. In its words,

> We are witnessing in countries throughout the world the expansion of prosperity for some, unfortunately accompanied by an expansion of unspeakable poverty for others. This glaring contradiction is unacceptable and needs to be corrected through urgent actions. 14. Globalization, which is a consequence of increased human mobility, enhanced Communications, greatly increased trade and capital flows, and technological developments, opens new opportunities for sustained economic growth and development of the world economy, particularly in developing countries. Globalization also permits countries to share experiences and to learn from one another's achievements and difficulties, and promotes a cross-fertilization of ideas, cultural values and aspirations. At the same time, the rapid processes of change and adjustment have been accompanied by intensified poverty, unemployment and social disintegration. (Ibid., no. 14)

The demand for a united effort in combating the mentioned social ills also came with the summit's recognition of the positive and concrete efforts made in the areas of global wealth growth, life expectancy, literacy and education, as well as the democratic institutions that have gained entrance into many countries and societies. But amidst such accolades, the summit sees the persistent lack that exists when our world today fails to recognise the fact that "...far too many people, particularly women and children, are vulnerable to stress and deprivation. Poverty, unemployment and social disintegration too often result in isolation, marginalisation and violence. The insecurity that many people, in particular, the more vulnerable face about the future about their own future and that of their children is intensifying" (Ibid., no. 16). In the final analysis, the united world leaders see the only chance for a sustainable and stable society embedded only in unreserved "[...] committed to a political, economic, ethical and spiritual vision for social development that is based on human dignity, human rights, equality, respect, peace, democracy, mutual responsibility and cooperation, and full respect for the various religious and ethical values and cultural backgrounds of people" (Ibid., no. 25).

3.2.1.7 The Beijing Platform for Action, Beijing, China, 1995

It was exactly by the fiftieth anniversary of the founding of the United Nations that Beijing, China, provided the platform on which this Fourth World Conference on Women was held in 1995. The focus was to draw attention to the fact that development, in general, can never be sustainable if our world continues to experience inequality and neglect in regard to women and children. Alongside varied recognitions of some effective impacts that have been made by various governments and people in fostering the status of women and children in various countries and societies, the conference pledged its unwavering commitment to addressing persisting constraints and obstacles to the empowerment of women and protecting the children all over the world, an assignment which requires "urgent action in the spirit of determination, hope, cooperation and solidarity[....]" (United Nations 1995b). The spirit behind such unanimous commitment was articulated thus:

> The equal rights and inherent human dignity of women and men and other purposes and principles enshrined in the Charter of the United Nations, to the Universal Declaration of Human Rights and other international human rights instruments, in particular the Convention on the Elimination of All Forms of Discrimination against Women and the Convention on the Rights of the Child, as well as the Declaration on the Elimination of Violence against Women and the Declaration on the Right to Development; [...] (these should lead to ensuring) [...] the full implementation of the human rights of women and of the girl child as an inalienable, integral and indivisible part of all human rights and fundamental freedoms; Build on consensus and progress made at previous United Nations conferences and summits [...]; and, achieve the full and effective implementation [...] for the advancement of women[....]. (Ibid., nos. 8–11)

Programmes for the empowerment of women, which, as the conference proposed, should unite the collective efforts of governments, institutions and international organisations at all levels, must be designed to combat all obstacles that impede the proper development of women and children, with special emphasis on girl-child development. Thus, such united efforts should work to ensure "[...] women's active participation in all spheres of public and private life through a full and equal share in economic, social, cultural and political decision-making" (Ibid., *Mission Statement*). A fundamental principle of the conference is to be seen in its use of the concept, *Transformed Partnership*, meaning a partnership based on equality and not seclusion and which invariably paves the way for people-centred sustainable development (Ibid).

3.2.1.8 Kyoto Protocol to the United Nations Framework Convention on Climate Change: Kyoto, Japan 1997

The Kyoto Protocol was the document of the United Nations signed in accordance with the United Nation's yearly meeting of the parties involved. The yearly convention of the parties is referred to as the Conference of the Parties and it is designed to keep a constant review of the issues concerning climate change and other related matters. At its convention on climate change in Kyoto, Japan, in 1997, the focus of the convention was on the developed world and economies and the need to highlight the imposed danger of greenhouse gas accumulation, which is grossly impacting our world due to excessive industrial and economic activities going on in those developed (or developing) spaces. At the centre of the convention's set goal is to keep the issues and problems surrounding climate change evergreen in the minds of policymakers and the world in general. Consequently, the core philosophy of the Kyoto Protocol seeks ways,

> [...] to reduce environmental pollutants while at the same time altering the freedoms of some citizens [...]; it seeks to impose regulations that will reduce production of pollutants towards the environment; In one side it imposes bigger regulations towards companies and reducing their profits as they need to fulfil such regulations with, which are oftentimes more expensive, alternatives for production. On the other hand, it seeks to reduce the emissions that potentially cause the rapid environmental change called climate change. The conditions of the Kyoto Protocol consist of mandatory targets on greenhouse gas emissions for the world's leading economies. As provided by the United Nations Framework Convention on Climate Change, "These targets range from −8 to +10% of the countries' individual 1990 emissions levels with a view to reducing their overall emissions of such gases by at least 5% below existing 1990 levels in the commitment period 2008–2012. (Wikipedia 2019a)

From the above citation, one sees the convention's attempt to apply the principles of social cooperate responsibility by exploring strategies to unite and place responsibility on government and non-governmental organisations, especially in matters of ecology and the environment. Consequently, in Article 10 of the protocol, the parties called each other to a deeper cooperation,

> [...] in the promotion of effective modalities for the development, application and diffusion of, and take all practicable steps to promote, facilitate and finance, as appropriate, the transfer of, or access to, environmentally sound technologies, know-how, practices and processes pertinent to climate change, in particular to developing countries, including the formulation of policies and programmes for the effective transfer of environmentally sound technologies that are publicly owned or in the public domain and the creation of an enabling environment for the private sector, to promote and enhance the transfer of, and access to, environmentally sound technologies[....]. (United Nations 1997)

The emphasis on sound technologies by the convention brings the developed and rapidly developing economies to the centre of discussion, and in this way, the parties brought to the fore the basic fact that a more sustainable world would be one that uses technology as the abstraction of nature and not the destroyer. In other words, economic and technological activities for the benefit of man and society must, at the same time, avoid the excessive exploitation and depletion of the resource itself. *Nature* itself needs the corresponding *nurture* and this must be fundamental to the sustainable development concept.

3.2.1.9 The World Summit on Sustainable Development: Johannesburg, South Africa 2002

This summit focussed on the actual situation of people in the world. Three forms of outcomes emerged from this summit, namely: (a) Johannesburg Declaration on Sustainable Development; (b) Johannesburg Plan of Implementation; (c) Type II Commitments by Governments and other stakeholders. The summit was fundamentally a forum that united Heads of State, governments and interest groups to speak with one voice and deliberate on the issues of poverty; the need for global stability; globalisation; global disparity; support for stronger regional groupings; private sectors' commitment; call for effective implementation of Agenda 21; option for more and effective democratic and multilateral institutions; united efforts in monitoring progress in our world; basic sanitation and environmental protection; concern for biodiversity; as well as responsible and sustainable consumption and production (United Nations 2002).

Varied opinions view the set agenda for this summit as a concrete response to the vital speech of the then Secretary General of the United Nations, Mr. Kofi Annan, a few months before the convocation of this summit, in which he ardently identified five key areas where the world's attention is needed for a sustainable future, namely: Water and Sanitation, Energy, Health, Agricultural Production, Biodiversity and Ecosystem Management (Sulabh International 2002). Thus, the world leaders came together to speak with one voice and apportion themselves the committed responsibilities as people's representatives, whose efforts must be formidable to promote progress, stabilise societies and give the world and human life, both present and future, more sustenance.

In setting the agenda for this summit, the world leaders placed before them a world that has been brought to the crossroads by human activities as well as human neglect. The anticipated resolve was to liberalise a cause of action that would be able to introduce new dynamics of action, which would impact positively on the world's socio-economic and socio-political spaces. But to do this, the inevitability of cooperative efforts must be held to be paramount. The committed leaders, therefore, recognised that "[...] poverty eradication, changing consumption and production patterns, protecting and managing the natural resource base for economic and social development, were overarching objectives of and essential requirements for sustainable development" (Ibid). Pledging then their commitment to action, the leaders called on both government and private sectors to make the sustainability of economic, social and political dynamics the guiding principle against all forms of societal "deep fault line", which constantly deepens the widening gaps between the rich and poor and plunges our world further into instability.

3.2.1.10 The Fourth United Nations Conference on the Least-Developed Countries: Istanbul, Turkey 2011

To unite the world's focus also on the *least-developed countries* is a deeper step in making the sustainable development concept functionally more inclusive. Authentic development of the world's peoples and societies cannot be properly achieved if the varied and multi-faceted situations of peoples are not taken into consideration, and this is within the nuances and realities of social, political, economic and cultural vectors of human existence. Thus, the rich and poor dichotomy, the developed and underdeveloped forms of classification, the civilised race against the uncivilised, the educated few versus the illiterate majority, etc., these disparities do not and should not fit into the concept of a global sustainable world.

A conference of the United Nations, therefore, with a focus on the least-developed countries of the world, must have the intent to remediate the root causes of a world in social, political and economic turmoil due to man-made oppressive inequalities existing in both national and international relations. The conference must have the intent to declare such disparities incongruent with the precepts of sustainable development. Thus, the uniting voices in Istanbul set on course a programme of action for combat against such inequality in a programme slated for a decade—2011 to 2020. First, they recognised that under the term "least-developed countries" are 48 countries, inhabiting around 880 million population of the world. The main characteristic features of such least-developed countries are seen in such constraints as "[…] low per capita income, low level of human development, and economic and structural handicaps to growth that limit resilience to vulnerabilities" (United Nations 2011) (Fig. 3.1).

The conference, therefore, took its objectives as the garnishing of efforts "…to overcome the structural challenges faced by the least-developed countries in order to eradicate poverty, achieve internationally agreed development goals and enable graduation from the least-developed country category" (Ibid., *Fourth United Nations Conference on the Least Developed Countries*, no. 27). This set objective, it must be emphasised, builds into the plan of the United Nations the importance of autonomy and respect for the indigenous people of every nation, thereby bringing closer the concept of sustainable development to bear on the dynamics of people's values, cultures and worldviews. To make the objectives clearer, the conference gave a clear vision of the key priority areas for action (Fig. 3.2).

To be noted above, these priority areas for actions contain the basic elements that go into both human and social development dynamics. Thus, in approaching the challenges posed by and within the least-developed countries and economies, the conference sees the need for proper integration of these elements while at the same

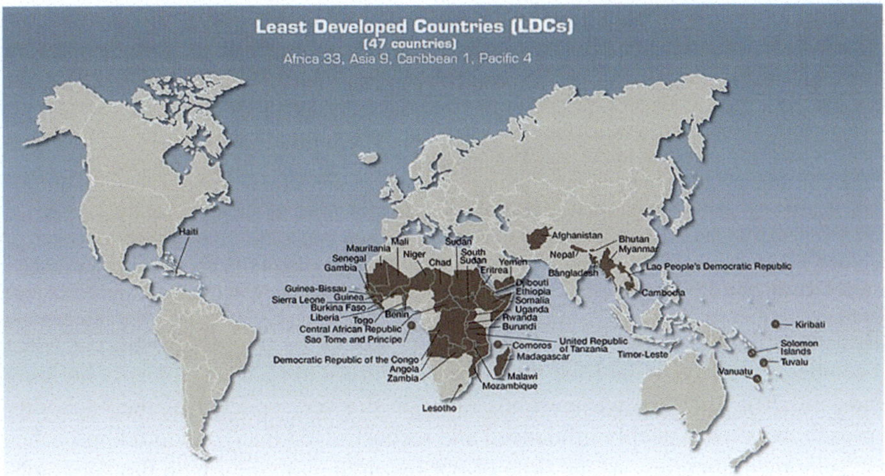

Fig. 3.1 47 Least-developed countries (LDCs) of the world. *Source* United Nations Conference on Trade and Development (United Nations 1964)

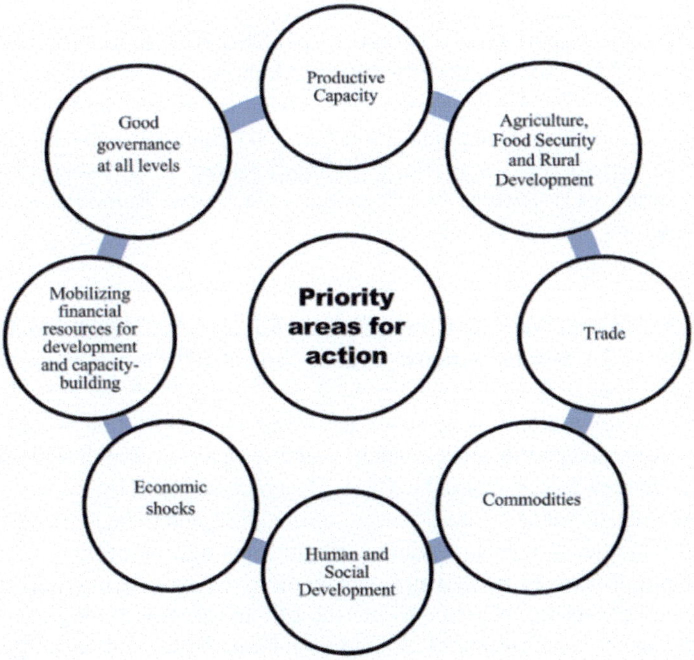

Fig. 3.2 Priority areas during the fourth United Nations conference on the least-developed countries (Ibid). *Source* United Nations Conference on the Least-Developed Countries; *Illustration Mine

time working for more participative, inclusive and equitable international relations, which would be more open and tolerant to the developing and underdeveloped. In the vision of the conference, this should be referred to as:

> An integrated approach in which the development process in least developed countries should be viewed in a comprehensive and holistic manner. The promotion of policy coherence and consistency of the international economic, financial and trading systems with the aim of increasing the quantity, quality and effectiveness of least developed country-focused international support measures and mechanisms is of key importance. The implementation of the Programme of Action should be integrated into all relevant international processes [….]. (Ibid., no. 29 (b))

All sustainable impacts to be made, therefore, would toe the part of success when it becomes a priority, in the least-developed countries, to focus on capacity building (Ibid., no. 70). In this way, one finds intrinsic the very concept of the capability approach, which has deeply influenced and reconstructed many modern approaches to development theories. At the time of our research and going with the geographical map shown above, the identified LDCs include (1) Afghanistan; (2) Angola; (3) Bangladesh; (4) Benin; (5) Bhutan; (6) Burkina Faso; (7) Burundi; (8) Cambodia; (9) Central African Republic; (10) Chad; (11) Comoros; (12) The Democratic Republic of the Congo; (13) Djibouti; (14) Eritrea; (15) Ethiopia; (16) Gambia; (17) Guinea; (18) Guinea-Bissau; (19) Haiti; (20) Kiribati; (21) Lao People's Democratic Republic; (22) Lesotho; (23) Liberia; (24) Madagascar; (25) Malawi; (26) Mali; (27) Mauritania; (28) Mozambique; (29) Myanmar; (30) Nepal; (31) Niger; (32) Rwanda; (33) Sao Tome and Principe; (34) Senegal; (35) Sierra Leone; (36) Solomon Islands; (37) Somalia; (38) South Sudan; (39) Sudan; (40) Timor-Leste; (41) Togo; (42) Tuvalu; (43) Uganda; (44) United Republic of Tanzania; (45) Vanuatu; (46) Yemen; (47) Zambia.

3.2.1.11 The United Nations Conference on Sustainable Development: Rio + 20, Rio De Janeiro, Brazil, June 2012

This conference drew attention to seven key areas of priority that would help our world to achieve sustainable development, namely: Energy, which should be both affordable and renewable; Sustainable Cities, designed and built for improved quality of life; Food Security and Sustainable Agriculture, designed to be more justifiable to all and reduce detractive barriers; Water, with the view of promoting more universal access to it and improving wastewater management; Oceans, with emphasis on the sense of sustainability in reference to marine life and marine biological diversity; Disaster Readiness, with emphasis on helping nations build resilience against both man-made and natural disasters; and Decent Jobs, with emphasis on cooperative efforts in the provision of good conditions of work and respect for workers' rights (IEEE EARTHZINE 2012). These priority areas, as can be observed, integrate the three basic areas of application in sustainable development goals, namely social, economic and political.

The vision of the conference can be interpreted summarily thus: Sustainable energy prospects are invariably the driving forces behind humane economic growth on the one hand and a healthy and regenerative environment on the other hand; there remain no sustainable cities when human socio-economic and political activities remain grossly arbitrary and therefore impact negatively on the agents of development—the "Mmadụ", human person; food security is none other than the development of subsistent agro-economic methods that can respond to the dynamics of population growth and resource sustainability; water and oceans hold varied basic elements which serve the basic human needs, therefore, life in the waters must be regenerated and not terminated; experiences in our world show the inevitability and unavoidable nature of some natural disasters and occurrences and the human race is known to possess the scientific capabilities to contain as well as combat such occurrences. Sustainable development, therefore, cannot be extricated from the constant readiness to deal with such disasters; all strategies for sustainable socio-economic, political and ecological development revolve around the issue of sustainable human ecology. Therefore, the economic empowerment of the agent of development—the human person, remains an essential factor of sustainable development.

With the recognition of the centrality of the human person within the processes of sustainability, the conference summarised its focus with an emphasis on green economic practices towards ensuring poverty eradication. In other words, every sustainable effort should focus on making economic activities more humane than oppressive by employing means that better respond to both the safety and empowerment of the human agent—the "Mmadụ". In the words of the conference:

> We affirm that there are different approaches, visions, models and tools available to each country, in accordance with its national circumstances and priorities, to achieve sustainable development in its three dimensions which is our overarching goal. In this regard, we consider green economy in the context of sustainable development and poverty eradication as one of the important tools available for achieving sustainable development and that it could provide options for policymaking but should not be a rigid set of rules. We emphasize that it should contribute to eradicating poverty as well as sustained economic growth, enhancing social inclusion, improving human welfare and creating opportunities for employment and decent work for all, while maintaining the healthy functioning of the Earth's ecosystems. (United Nations 2012)

The approach, as recognised by the conference, remains broadly diversified. However, the international community must remain united in demanding cooperation and consistency of all with the international law, respect for autonomy and indigenous people and the advancement of international relations (Ibid). Proper observation of these rules for a more sustainable world remains heavily dependent on good governance and visionary leaders whose interests are action- and result-oriented. In this way, therefore, they take as their aim unreserved regard for every relevant issue aiming at promoting sustainable development (Ibid., IV. 76(b)).

3.2.1.12 The Third International Conference on Small Island Developing States: Apia, Samoa, 2014

In 2014, a Private Sector Partnership Forum (PSPF) was convened in Apia, the capital city of the South Pacific Island nation of Samoa. It was a conference that aimed at creating awareness of the fact that no part of our world would be left out in any approach to sustainable development. It was a cooperative work of the Government of the Independent State of Samoa, the United Nations Office of the High Representative for the Least-Developed Countries, Landlocked Developing Countries and Small Island Developing States (UN-OHRLLS) and the Samoa Chamber of Commerce and Industry. The keynote address of the then Secretary General of the United Nations, Mr. Ban Ki-moon, clearly defined the focus of the conference. For him,

> [...] it was a collective responsibility to protect the small islands given the numerous challenges they face, in particular the negative impact of climate change. He remarked that SIDS Conference is timely given that the Climate Change Summit will convene in New York in September. He noted that Member States are progressing toward consolidating the future development agenda, which should reflect the unique challenges and aspirations of the SIDS. He reiterated the outcome of the Rio +20 Conference on Sustainable Development, where it was recognised that collective action was impossible without partnerships. He highlighted that partnerships formed the backbone of collective action. He remarked that the Third UN Conference on SIDS was an opportunity to move the planet toward a sustainable path. (United Nations 2014a)

The forum made it clear that, while these Islands inhabit so many natural treasures of our world, their existence is constantly threatened by varied physical constraints, which, through partnership and cooperative efforts, could be controlled and augmented so as to preserve their existence. Therefore, as a sustainable approach, the forum focussed on six areas that demand protection and attention within the Small Island Developing States, namely: Sustainable Agriculture, Fisheries and Marine Resources, Connectivity through Transport and ICTs, Renewable Energy, Sustainable Tourism, and Disaster Risk Reduction. The aim was to partner with the private sector, which would be able to work with the administrations of governments of the affected Islands to ensure a more inclusive administration and preservation of the treasures of such Islands. The set objectives remained: (1) Provide a platform through which genuine and durable partnerships to support the advancement of sustainable development of SIDS can be announced; (2) Facilitate the sharing of information and best practices within SIDS regarding policies, programmes and practices geared towards enabling business and industry in the pursuit of sustainable development, strengthening policy frameworks for business facilitation; (3) Showcase tools available to the microsized, small- and medium-sized enterprises (MSMEs) in SIDS for promoting more resource-efficient processes and practices with particular emphasis on energy efficiency; (4) Launch discussions on possible follow-up mechanisms, including the establishment of a "SIDS Global Business Network" that would contribute towards enabling strengthened business contacts, sharing of best practices including market access, enhance private sector partnerships and increase inter-regional SIDS private sector cooperation and collaboration, and promote increased partnerships with the global business community (Ibid., *Concept Note*).

3.2.1.13 Second United Nations Conference on Landlocked Developing Countries, Vienna, Austria, 2014

This conference, which was held in Vienna, was the second of its kind. It focussed on how to make development sustainable within countries described as "landlocked" due to their location and position within the global community. Thirty-six countries with such features are identified within the continents of Africa, Asia, Europe and South America and the affected areas are habitats to over 440 million people. The objective of the conference was to deliberate and discover ways to harness international partnerships dealing with ways to help augment the deficiencies within those areas, which could pose hindrances to sustainable development prospects. Presenting the challenges involved, the conference recognised that,

> [...] face special challenges that are associated with their lack of direct territorial access to the sea, and their remoteness and isolation from world markets. Their international trade depends on transit through other countries. Additional border crossings and long distances from major markets, coupled with cumbersome transit procedures and inadequate infrastructure, substantially increase the total expenses for transport and other transaction costs, which erodes the competitive edge of landlocked developing countries, reduces economic growth and subsequently negatively affects their capacity to promote sustained economic development, human and social progress and environmental sustainability. Landlockedness is a major contributor to the relatively high incidence of extreme poverty and structural constraints in landlocked developing countries. Landlocked developing countries, as a group, are among the poorest of developing countries, and many of them are also least developed countries, with limited capacities and dependence on a very limited number of commodities for their export earnings. (United Nations 2014b)

The conference, therefore, reiterated the need to evaluate the united and recent past efforts of the international community with regard to the challenges and needs of the landlocked countries. Thus, a revaluation of the Amnesty Programme of Action (2003), the World Summit (2005) and the United Nations Conference on Sustainable Development (2012) was considered needful. At the centre of developing viable strategies that would be able to offer workable systems and solutions was the conference's call for a functional partnership between the affected countries and the possible transit countries capable of offering outlets and access. In the words of the conference:

> Partnership between landlocked developing countries and transit countries is mutually beneficial for the improvement and constant maintenance of their infrastructure connectivity and of technical and administrative arrangements in their transport, customs and logistic systems. Efficient transit transport systems, strong collaborative efforts in multi-modal transport infrastructure development and interlinkage, the promotion of an enabling legal environment and institutional arrangements and strong national leadership on cooperative arrangements between landlocked developing countries and transit countries are also crucial for achieving structural transformation and sustainable economic growth and development. (Ibid., III (17))

In the end, six priorities for action became the focus of the conference, namely: (1) Fundamental transit policy issues; (2) Infrastructure development and maintenance, namely—Transport infrastructure, Energy and information and communications technology infrastructure; (3) International trade and trade facilitation: (a) International Trade; (b) Trade facilitation; (4) Regional integration and cooperation; (5) Structural economic transformation; (6) Means of implementation (Ibid., V).

3.2.1.14 The Third United Nations World Conference on Disaster Risk Reduction: WCDRR, Sendai City, Japan 2015

This conference was convoked in 2015 in Sendai City, Japan, in solidarity with the country and its people, who were gradually recovering from the Great East Japan Earthquake of 2011. It was a renewed effort to recognise the growing impacts of physical and natural disasters on our planet, the earth, and to reawaken the consciousness of the international community that varied efforts could be garnered to both predict occurrence as well as explore scientific and technological methods to reduce risks, especially in the cases of natural occurrences. Four priorities were then set forth to guide the deliberations of the conference, namely: (1) Understanding disaster risk, (2) Strengthening disaster risk governance to manage disaster risk; (3) Investing in disaster risk reduction for resilience; and (4) Enhancing disaster preparedness for effective response and to "Build Back Better" in recovery, rehabilitation and reconstruction (United Nations 2015a).

The conference was unanimous in talking about guiding principles in matters of handling disasters and reducing the risks they pose to our world. It becomes, therefore, the primary responsibility of each State to develop principles and practices that would be able to give timely and corresponding responses in matters of disaster and the risks that follow. In the words of the conference:

> Each State has the primary responsibility to prevent and reduce disaster risk, including through international, regional, subregional, transboundary and bilateral cooperation. The reduction of disaster risk is a common concern for all States and the extent to which developing countries are able to effectively enhance and implement national disaster risk reduction policies and measures in the context of their respective circumstances and capabilities can be further enhanced through the provision of sustainable international cooperation [....]. (Ibid., III. On Guiding Principles (a))

Being able to apply methods and principles to handle disaster risks, according to the conference, remains the only viable instrument to protect property, health, livelihoods and productive assets while at the same time promoting and protecting all human rights, including the right to development (Ibid., III (c)). However, while the role of leaders and policymakers is being emphasised, the conference places importance on the need to educate the grassroots of every society on the reality and possibility of disaster risks, as well as the need to keep ready for the rainy day. In its words: "While the enabling, guiding and coordinating role of national and federal State Governments remain essential, it is necessary to empower local authorities and local communities to reduce disaster risk including through resources, incentives and decision-making responsibilities, as appropriate (Ibid., III (f)). Furthermore, such combined efforts would only be sustainable when adequate provisions are made to enhance inclusive and integrative policies at all levels, such that the commitment to both prevention and reduction of disaster risks could be optimally realisable "[…] through the implementation of integrated and inclusive economic, structural, legal, social, health, cultural, educational, environmental, technological, political and institutional measures that prevent and reduce hazard exposure and vulnerability to disaster, increase preparedness for response and recovery, and thus strengthen resilience (Ibid., II (17)).

3.2.1.15 The Paris Agreement: Paris, France 2015

> "A great tide has turned. Finally, the world stands united against the central environmental challenge of our time, committed to cutting the carbon pollution that's driving climate change. This agreement sets us on a course of verifiable gains we can build on over time. It provides real protection for people on the front lines of climate chaos. It speeds the global shift away from dirty fossil fuels and toward cleaner, smarter energy options to power our future without imperilling our world. And it sends a clear message to our children: we will not abandon you to pay the price for reckless habits that wreak havoc and ruin on our planet and lives. A crisis that took centuries to get here won't go away overnight. But climate change has met its match in the collective will of a united world. Our challenge now, in our country and all others, is to make good on the promise of Paris, by turning the action we've pledged into the progress we need". (Suh 2015)
>
> —*Rhea Suh, President, Natural Resources Defence Council*

The above citation was the very well-articulated words of the president of the Natural Resources Defence Council (NRDC), Rhea Suh, as soon as the final text of the Paris Agreement was assented to by the member parties. This agreement, under the United Nations Framework Convention on Climate Change (UNFCCC), was a landmark in the effort to sustain development at all levels of human society, but with particular reference to the urgency of recognising climate change as a reality that carries detrimental consequences for our world with it, both present and future. The general focus was to bring countries and parties together to agree on a common course towards saving our planet Earth from impending devastation due to climate change. Thus, the envisaged practical aim is "[…] to keep the increase in global average temperature to well below 2 °C above pre-industrial levels; and to limit the increase to 1.5 °C, since this would substantially reduce the risks and effects of climate change" (Wikipedia 2019b).

The set objectives, therefore, were clearly defined in the blueprint for the adoption of the agreement thus:

> […] This Agreement, in enhancing the implementation of the Convention, including its objective, aims to strengthen the global response to the threat of climate change, in the context of sustainable development and efforts to eradicate poverty, including by: (a) Holding the increase in the global average temperature to well below 2 °C above pre-industrial

> levels and pursuing efforts to limit the temperature increase to 1.5 °C above pre-industrial levels, recognizing that this would significantly reduce the risks and impacts of climate change; (b) Increasing the ability to adapt to the adverse impacts of climate change and foster climate resilience and low greenhouse gas emissions development, in a manner that does not threaten food production; and (c) Making finance flows consistent with a pathway towards low greenhouse gas emissions and climate-resilient development. 2. This Agreement will be implemented to reflect equity and the principle of common but differentiated responsibilities and respective capabilities, in the light of different national circumstances. (United Nations 2015b)

It is to be noted that the most debated issue in the concept of sustainable development has been the environment, within which the effects of human activities, with all the positive and negative consequences, are felt. Thus, in recent times, the issue of climate change, seen from the perspective of our world's scientific and technological impact on the atmosphere, has awoken the attention of many groups and activists, who feel it is time to act fast to guard against all probable impending doom. Thus, the Paris Agreement remains a landmark in the search for sustainable developed or developing worlds and economies, which would invariably affect positively both the environment and human life in general. In doing this, emphasis was laid on the need for capacity building, in other words, the creation of awareness, which eventually leads to empowering the society to keep conscious in mind of people the reality of climate change and its effects on both the environment and human subjects. Under Article 11, therefore, the Agreement explained:

> Capacity-building under this Agreement should enhance the capacity and ability of developing country Parties, in particular countries with the least capacity, such as the least developed countries, and those that are particularly vulnerable to the adverse effects of climate change, such as small island developing States, to take effective climate change action, including, inter alia, to implement adaptation and mitigation actions, and should facilitate technology development, dissemination and deployment, access to climate finance, relevant aspects of education, training and public awareness, and the transparent, timely and accurate communication of information. (Ibid., Article 11(1))

To achieve the above, the agreement calls for cooperative and participatory efforts among parties and member states to institute the united efforts of both the strong and the weak countries and economies in a manner that would be mutual, innovative and problem-solving.

3.2.1.16 Moving Forward the 2030 Agenda for Sustainable Development: The UNESCO Approach, 2017

In 2017, the United Nations Educational, Scientific and Cultural Organization (UNESCO) took a bold step to institutionalise strategies meant to integrate the Sustainable Development Goals (SDGs) in its approach to the circumstances of our time and world. As a body of the United Nations, the purpose of which "[...] is to contribute to peace and security by promoting international collaboration through educational, scientific, and cultural reforms in order to increase universal respect for justice, the rule of law, and human rights along with fundamental freedom proclaimed in the United Nations Charter" (Wikipedia 2019c), it sought in 2017 to embed the principles of the 2030 agenda very deep into its plans for action. The defined focus was on Sustainable Development—shifting the world on to a sustainable and resilient path; Universality and Interdependence—reflection on a changing world and a new appreciation for the growing complexity and interconnectedness of development challenges reaching beyond borders; Inclusivity—"leaving no-one behind", including ending poverty in all its forms and reducing poverty within and among countries; Rights-based—the realisation of human rights and the promotion of gender equality; Inequalities—cooperation for inclusive education and access to digitalisation; National Ownership—incorporation of sustainable development concepts into national planning processes, policies and strategies; Integration and Interlinkages—integral response to the complex challenges of our time; Peace, Intercultural Understanding, Global Citizenship—towards the ethics of global citizenship, integrated and shared cooperation and responsibilities; Crisis, Conflicts and Disasters—solidarity with nations with crises and conflict situations; Knowledge, Science, Technology and Innovation—building skills and human capacities for information and communication technologies; and Multi-Stakeholder Partnerships—partnerships for integrated approaches and cooperation among States and stakeholders (United Nations 2019) (Ibid).

Furthermore, UNESCO gave special attention to nine of the 17 Goals, namely Goal 16: On Peace, Justice and Strong Institutions, Goal 4: On Quality Education, Goal 5: On Gender Equality, Goal 15: On Life on Land, Goal 6: On Clean Water and Sanitation, Goal 14: On Life Below Water, Goal 13: On Climate Action, Goal 11: On Sustainable Cities and Communities, and Goal 9: On Industry, Innovation and Infrastructure. It reflected on the way forward as represented below (Fig. 3.3).

In approach to the above-defined principles, UNESCO places emphasis on the recognition, respect and promotion of people's cultural values as the most authentic driver to the achievement of sustainable development in our world. Its supportive

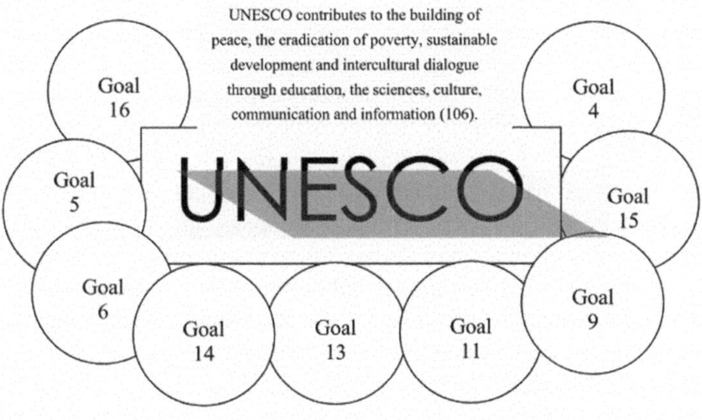

Fig. 3.3 UNESCO direct contribution to nine SDGs. *Illustration mine*

argument is that culture as peoples' way of life and understanding of the world around them remains "[...] a driver and enabler of sustainable development and essential for achieving the 2030 Agenda as it contributes to human and socio-economic development, quality education, social inclusion, sustainable cities, environmental sustainability, and peaceful societies" (Ibid). Furthermore, the embrace of culture as a viable approach and ground of application would, therefore, provide the safest approach to human and socio-economic development, quality education, sustainable cities, environmental sustainability and peaceful societies and inclusion (Ibid). The Agenda 2030, with the indigenised model as constituted by the African Union (AU), the Agenda 2063, would serve as vade mecum in its contribution to the right course for sustainable development within the African nations and, in particular, Nigeria. A sub-theme would be developed within the context of the African Union's Agenda 2063.

Evaluation

Embarking on a literary appraisal of the issue of sustainable development provides us with a functional network of ideas surrounding it as an ideology in our twenty-first-century world. It shows a history of the concerted efforts that have been made and are still being made to keep the socio-political, socio-economic and socio-ecological human world stable and functional. Sustainable policies, therefore, as shown in the various conventions and conferences, are ways to garner efforts and pull together partners and policymakers to help reconceptualise the "human versus nature" relationship of our present world, the impact of which has been deemed parasitic and destructive. The focus on sustainability serves both as a socio-political and socio-economic approach to stem the tide of human excessiveness on the one hand and, on the other hand, to institute a consistent mastery of nature and its internal workings so as to be better placed to respond accordingly to its sporadic challenges.

Applied to this present work, this academic work finds it important to draw from precedence and antecedents that are actual and trending within our global world, especially with special emphasis on the issue of sustainable development. This is in an effort to prove the structural dysfunction present in the socio-political, socio-economic and socio-ecological Nigeria. Although our work does not intend to prove a total absence of development efforts in the daily experiences of the country Nigeria, however, it maintains unequivocally that the expected pace and manner of approach have been grossly impaired due to in-built and man-made structural and administrative indiscipline, which has eaten deep into the system and is almost proving to be a way of life. Consequently, the most populous country in Africa, Nigeria, battles a persistent culture of negligence and varied policy ambivalences, which are the root causes of her internal weaknesses and ethical failures. Further application of the term sustainability in our research would now look at its three dimensions, some probable inherent challenges, the sustainable development goals, as well as the evaluative theses given by Markus Vogt.

3.3 The Three Dimensions of Sustainable Development

The issue of sustainable development centres on the economic, social and environmental dimensions of our world. These three dimensions bring together the totality of the human-nature relationship, which builds up our world and everyday life. To keep these dimensions sustainable is to create a balanced and integrated relationship among them. A sustainable economy is an integral part of a functional social system and no social system can positively thrive without balanced and inclusive economic empowerment. Furthermore, human socio-economic activities can never be sustainable without a defined policy of integrating the environment in a manner that respects the principles of corporate social responsibility (CSR). Such a call for integration has become a major focus of the United Nations in recent times and has been described as a prerequisite shift in policy due to the persistent widening of the gap between the rich and poor in our world, the observable recklessness in the dealing with the resources of nature, as well as the actual and impending consequences of environmental devastation and climate change. The United Nations Economic and Social Commission for Asia and the Pacific observed:

> Where public policy either implicitly or explicitly prioritizes investments in economic activity without regard for the impact on the stocks of natural, human or social capital, negative externalities result, such as pollution, emissions, waste and social clashes. But policy responses to these phenomena that do not address the root causes of the policy conflicts and trade-offs ultimately will not support achievement of sustainable development. (United Nations 2015c)

The corporate efforts seen within institutions and policies to help integrate the uniqueness of these three dimensions give rise to three possible classifications of sustainable development dynamics, namely: Economic Sustainability, Social Sustainability and Environmental Sustainability. It otherwise means that the sustainable development concept is a discourse on the relationship existing within the economy, society and environment in a symbiotic and integral manner. These three elements are described as the three pillars of sustainability (Fig. 3.4).

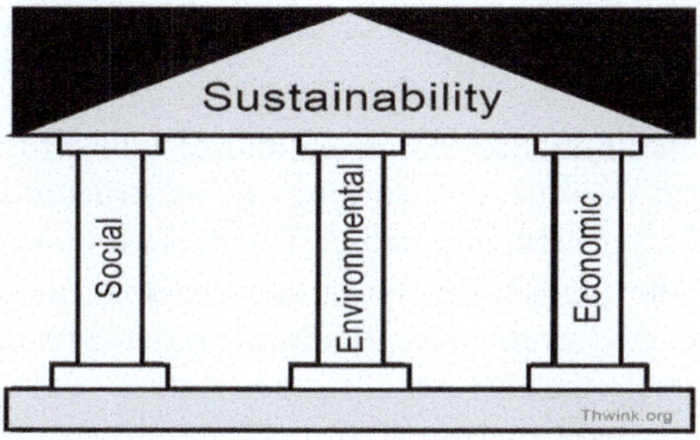

Fig. 3.4 The three pillars of sustainability. *Source* thwink.org

3.3.1 Economic Sustainability

Every nation, people and society in our world engages in one economic activity or the other. Some nations are naturally privileged with resources that drive their economic activities, while others may be dependent on and participative in the economic activities of the naturally privileged economies. These disparities create observable imbalances in the global markets and, in many circumstances, give rise to widening cases of poverty; impeded access to the basic means of livelihood (as in the cases of the so-called developing and underdeveloped economies); monopoly of the global market; excessive capitalist influence; economic wars and woes; inflicted and imposed austerity measures; total loss of local economies; forced migration and the consequent enslavement; etc. At the global level, such problems have led to a shift in both political and economic paradigms and have necessitated the modern discussion on globalisation. At the national level, they have led to the demand for more inclusive and people-oriented economic policies.

Economic sustainability, therefore, should mean a revolutionary trend towards getting the world economies—at the national, international and intra-national levels—to properly respond to the principles of political inclusion, which should guarantee rights to political participation and representation, and the principles of economic inclusions, which invariably ensures more distributive access to resources and the means of production, leading to integral transition towards sustainable market economies (The European Bank for Reconstruction and Development 2019). The European Bank for Reconstruction and Development opined:

> An inclusive market economy ensures that anyone regardless of their gender, place of birth, family background, age or other circumstances, over which they have no control, has full and fair access to labour markets, finance and entrepreneurship and, more generally, economic opportunity. Inclusion is thus an intrinsic element of a sustainable market economy [....] Promoting an inclusive market-based system is about efficient (human) resource allocation rather than being a social policy choice [....] But there is also a political dimension to inclusion. Beyond its contribution to efficient markets, fair and equitable access to economic opportunity is essential in fostering broad support for market reforms and, ultimately, sustainable market economies. Economic inclusion is therefore an important transition quality. If people are given a chance to succeed, they are more likely to participate in the workforce, pursue education, or engage in other activities that lead to economic growth. This, in turn, strengthens the transition process. (Ibid)

From the foregoing, the inference that economic sustainability serves to integrate society as a whole and stabilise the workings of its structural and administrative patterns can stand as a plausible argument. The general aim is towards *economic efficiency*. However, the plausibility must be verifiable through the application of economic strategies and evaluative principles. Balaman suggests three of such guiding principles, namely the cost-effectiveness analysis (CEA), cost–benefit analysis (CBA), and multi-criteria analysis (MCA) (The European Bank for Reconstruction and Development 2019). A deeper discussion on these principles would be outside the scope of our research. Nevertheless, discussing economic sustainability in our approach is to justify the focus of our research, namely the identification and proposing of all practical inroads towards achieving workable, sustainable development strategies within the Nigerian state. It is a necessary attempt to emphasise the need to form "[...] consciences on the moral dimensions of economic decision-making and to articulate a moral perspective in the general societal and political debate that surrounds these questions" (Balaman 2019).

3.3.2 *Social Sustainability*

Social sustainability is a discourse on the need for social security and stability. An unstable and insecure social system remains bereft of the needed institutions that would help create sustainable development. Social sustainability, therefore, is the situation whereby there exists the possibility of opportunities for people, humane quality of life and sustainable human development. Such development should liberate individual capacities and fulfil human needs, thus ending poverty and improving individuals' quality of life. Such a social situation offers a secure life with full rights and liberties in the long term and could guarantee social cohesion (United States Catholic Bishops 2009). Social sustainability deals with policies that ensure a better quality of life in society, inclusive access to education, progressive affinity to community development, and provision of equal and distributive opportunities for all, including gender equality and respect for the rule of law and ethics. Sen gave the dimensions of social sustainability as *Equity*, the equal provision of opportunities for every member of a society; *Diversity*, the acknowledgement of unity in diversity; *Interconnected/ Social Cohesion*, the promotion of processes, systems and structures that support social inclusions; *Quality of life*, to be made available for all through the community; *Democracy and Governance*, which should be fundamental to the leadership of the community, and *Maturity*, which should influence individual persons' lives to the level of being a virtue (Globalisation and Livelihood Options of People Living in Poverty, GLOPP 2019).

Varied forms of social unrest, in extreme cases both political and economic wars, are known threats to social sustainability. Bad governance, as well as dictatorial regimes, have never and can never achieve social sustainability. Thus, social sustainability is necessarily predicated on the political ideology and situation of a given society. Only through the reality of the fusion of these two—the social and political aspects—would one think about the realisation of the sustainable dynamics of authentic development. This social dimension does not only concern itself with the affairs and welfare of the present generation. It rather makes the dynamics and policies of development from today bear responsibility for the future generations' rights and privileges as well. Thus, giving an exhaustive background to this concept, Schönborn G. et al (with reference to Sachs Spangenberg and Omann) opined:

> The overall social development, which meets the needs of both present and future generations, is generally known to the public as "sustainable" and is acknowledged as a necessity in all strata of society. Spangenberg and Omann propose a system dynamics concept of social sustainability, writing that it is a qualitative criterion that requires multi-criteria assessment

> to be fairly captured in economy, society and population (in a European, and particularly a German, policy context). Their work suggested the following criteria for social sustainability: self-determined life and mixed work; satisfaction of basic needs; reliable and sufficient security system; equal opportunities, participation, democracy; social innovations; intergenerational equity; objective and subjective basic needs; objective and subjective social resources; objective and subjective equal opportunities; subjective participation; objective sustaining oneself; cultural diversity; and solidarity and tolerance. As evidenced by the crosscutting nature of these criteria, all societal strata are touched upon. (Wikipedia 2019d)

Therefore, social sustainability encompasses all that goes into the creation of the social space that would be able to respond correspondingly to societal needs, be it human, political, economic, as well as environmental demands. It "[...] combines design of the physical realm with design of the social world—infrastructure to support social and cultural life, social amenities, systems for citizen engagement and space for people and places to evolve" (Schönborn et al. 2019).

3.3.3 Environmental Sustainability

Issues concerning the environment and the impact of human activity on it have recently taken centre stage and have been widely discussed in all facets of human knowledge and academic disciplines. This particular dimension of sustainable development theories and concepts has seen the rise of so many activists and interest groups, who unite their voices to alert our present world about the danger of environmental as well as ecological recklessness. The impact of human scientific and economic activities on the environment has reached a turning point in our twenty-first century such that any protracted inaction would be placing the planet Earth on top of a *Timebomb* that awaits explosion in due time. Environmental sustainability describes, therefore, all concerted efforts towards a humane and safe approach to nature in general as well as the individual elements that constitute the natural habitat—earth. The sustainability concept with reference to the environment demands planned efforts towards a form of maintenance culture capable of safeguarding nature and its natural capitals. Thus, the focus is on social, economic, development- and resource-based actions that could be continued indefinitely without grave consequences to both human and natural habitats. In other words, it is all about what can be sustained on the one hand and, on the other hand, others that place demands on our society to take action against unsustainable policies and practices.

The need to sustain the environment places emphasis on the elements of nature, upon which humanity, as an integrative and interactive part, is dependent and which deeply places unconditional responsibility on humanity to protect, maintain and help both in resuscitation and replenishment. It is, therefore, a normative ethical approach to the nature-man relationship, which otherwise makes the principle of relation complementary instead of exploitative. The normative (traditional) ethical theories, namely teleologism, utilitarianism and deontologism, therefore, seek to establish the direct connection that exists between man and nature, thereby placing an intrinsic bond of responsibility, which must flow from the human agent to the natural (environmental) capitals, with the two being united symbiotically. Such dependency and mutual interaction have been identified as the safest approach to safeguard the future of humanity and make the environment and ecology, both for the present and future generations, sustainable. Writing in "Environmental Ethics: An African Understanding", Ojomo brought this demand for environmental sustainability to bear on Africa as the continent with a special situation, thus:

> It is a known fact today that the environment crisis is one of the most pressing and timely concern of our planet in the turn of the twenty-first century. As a global phenomenon, no society is totally immune against the threats and dangers, which the environmental crisis poses to humanity and our collective planet, the earth. But with respect to the African experience, a vast area of land rich in natural resources of all categories, flora and fauna of immense diversities, the dimension of the global environmental crisis in the continent has a peculiar character. The causes of environmental pollution and degradation, environmental injustice, poverty of effective coping and management strategies in challenging the environmental crisis, and lack of a viable environmental ethics that takes cognizance of the peculiar dynamics of the environmental crisis in Africa are issues worth courting with philosophically. (Ibid)

The philosophical approach, therefore, is to be understood as the need to use the necessary ethical tools, first, to help checkmate human recklessness and excessiveness, second, to explore the most effective technological and scientific methods available, which would help to achieve results as well as save and protect the natural capitals.

3.3.4 Two Basic Inherent Challenges of Sustainability Discourses

As plausible as all approaches to sustainability or sustainable development are in reality, the concept itself pulls along some inherent challenges, which have always set opinions against one another. For some, the whole concept is like a hoax, an organised socio-political campaign meant to limit both human and economic freedom, or a socio-political *mantra* of the United Nations, which has been unnecessarily overemphasised. For others, the world has, for long, neglected such a relevant topic and, as a consequence, has turned itself into an enemy of the natural capitals, to which it is supposed to be a steward.

3.3.4.1 Goal-Ideal Approach

The goal-ideal approach leads our studies to the need not to play down the inconsistencies that are associated with the concept and application of sustainability. In other words, the general view on sustainability, with particular reference to the Brundtland Commission's definition, presents the concept, at the same time, as a goal and an ideal. Goals, in the sense of plans and strategies or potential achievements, which can be fulfilled. Ideals, on the other hand, are expectations and "*oughts*", which present the ideological prototypes that may lead to actual achievements. Applied, therefore, to the issue of Sustainability, the whole concept presents an inherent tension that leads to doubt as to whether there could ever be authentic and realisable sustainable development. This is because, according to some critical opinions, "[…] it (maybe) impossible in practice to meet the needs of the present without compromising the ability of future generations to meet their own needs—no matter whether these needs deal with nourishment, housing, transportation, drinking water, energy, and so on" (Ojomo 2011).

However, while agreeing to some extent with the foregoing sceptical opinion, especially in reference to the non-renewable resource capitals of nature, our research would, on the other hand, see this inherent challenge as integrative and never exclusive. The goals of sustainable development are rather achievable or achieved abstractions from the ideals that provide the guiding principles for the achievement of such goals. As Pruzan puts it, "Ideals are aims of perfection that provide guidance for our decisions but are essentially unachievable (e.g. peace on earth, perfect justice, equity, tolerance—and sustainability)" (Pruzan et al. 2014). The sustainability principle, therefore, does not aim at perfection but rather at achievable, relative and empirical goals like means towards modulation, conservation, preservation, regeneration, reproduction, etc. In this way, the goal-ideal dichotomy remains inherent but at the same time capable of producing some concrete positive results.

3.3.4.2 "Housekeeping" Versus "Nature-Keeping" Approach

The two concepts above—*housekeeping and nature-keeping*—are two ways to place economics and ecology side by side. The issue surrounding economics versus ecology builds up another form among the inherent challenges associated with sustainability. Presenting economics and ecology as dichotomised, of course, with regard to the principle of sustainability, therefore, is to acknowledge the need and the challenge one encounters in "[…] understanding not only what we are seeking to "sustain" (the diversity and functioning of biological systems that exist on earth), but also how and in what way man's activities and needs are implicated in and can threaten the future of these biological systems" (Mukherjee and Stückelberger 2014a). In other words, it is the issue of acknowledging that man's economic activities have negative impacts on the ecology, especially where such activities are not modulated and ethically structured and oriented.

While economists focus on efficiency in production, "[…] distribution and consumption of goods and services, focussing (also) on the market, its prices and competition" (Ibid., p. 13), ecology contrasts this with a focus on "[…] the interrelationship between organisms and their surroundings, drawing heavily on the biosciences and empirical studies of nature and the environment" (Ibid., p. 304). The simple inherent tension is: To what extent should *housekeeping* be continued that it does not exploit the *nature-keeping*? The ecologist would seek and demand the quantification and regulation of the economist's maximisation principle, while the economist demands the obliteration of the ecologist's strict conservation principles, which he views as constraints. In other words, while ecology sees a threat (through economic activities) to the ability of these natural capitals to regenerate themselves, the economy feels threatened by all extreme conservative approaches that limit its increase of consumption possibilities (Ibid). In this way, such inherent conflicts result mainly in conflicts of ideas, policies, approaches, interests, opinions, etc. The issues surrounding sustainability as both economic and ecological principles put the concept in question at the frontline of these conflicts and challenges. Nevertheless, the relevance of the concept in our present epoch outweighs all forms of conflict and challenges it must have ignited. In further discussions, we would present a critique of the sustainability principle as a relevant approach to the social situations of our modern world.

3.4 The Principle of Sustainability: Towards a New Social Principle on Development

Facing a society with numerous and multi-faceted life situations and opportunities, which themselves influence the human agents of the society, Vogt calls for a renewed focus on sustainability principles as a new and relevant social principle. Such a renewal becomes a demand given the social and life situations our world has to grapple with, especially in the wake of the twenty-first century. In concrete terms, our twenty-first century experiences a powerful grip on globalisation, which meets different forms of human societies within their various socio-economic and socio-political situations. This same reality carries with it obvious and overt conflicts that lead to divisions, feelings of marginalisation and exploitation, injustice and similar factors (Ibid).

Given the above challenges, Vogt found the time already ripe to discover a principle that would be able to integrate society, faith, the Church and concrete actions to address the challenges of globalisation. Such a principle would be oriented towards finding sustainable approaches to the adversities and threats plaguing our modern world, which externalise themselves in various social dichotomies such as the immense adversities of a section of humanity in the face of excessive prosperity of the other, usually the minority; the unending search for unity and peace; persistent distortion, disruption and destruction of societies and lives, etc. These uncertainties place doubts on the authentic future and existence of both our human world and the generality of other earthly natural capitals. For Vogt, therefore, since fascinating opportunities and dangers that extend far into the future, immeasurable misery and immeasurable prosperity, unity and conflict, hope and despair lie closely together, the general evaluation of these processes remains subject to considerable differences (Vogt 2013).

The principle of sustainability, therefore, must be made the capable tool to halt the drifting society of our time, which witnesses much socio-political and socio-economic progress alongside a staggering and increasing number of the world's poorest people and societies. It should serve as a protective policy against all forms of degradation and destruction involving the ecology, environment and biodiversity. It should serve as a functional uniting principle for a productive integration of economic welfare, social justice and ecological viability. On the contrary, with a deficit of the principle of relationship among the already named phenomena, the three realms run in constant conflict with one another. In the opinion of Vogt, economic prosperity, social justice and ecological sustainability are today so interdependent due to rising global interdependencies that they cannot be secured individually or even against each other. Without a systematic connection and worldwide perspective expansion, the concepts of economic, social and environmental ethics would remain only short-lived (Ibid). Therefore, as a principle and approach, and in line with the united voices of the world leaders during the World Conference on Environment and Development, Rio de Janeiro 1992, only sustainable action theories of development would be able to guarantee social, economic and ecological efficiency (Ibid, p. 19). Our research considers a healthy interaction among these constituent elements, namely social justice, economic welfare and ecological viability as the Integrative Force of Sustainability Principle.

The demand for more efforts in the establishment of this principle in the mainstream socio-political and socio-economic policies of our world remains a clarion call. Concrete situations and experiences often show strong rejection, snubbing and subordination of this principle by various societal institutions in the areas of the economy, management and administration, energy and power, construction and engineering, mobility behaviour or general development politics (Ibid). These mainstream institutions in any political and geographical entity would find sustainability principles conflicting, especially when the focus and objective degenerate to mere gain and maximisation of profit. In contrast, the sense of sustainability should introduce sanity in such a network of institutions to ensure a mutual interrelation and interaction, which defines a healthy and sustainable social structure and introduces the fundamentals of ethical orientations for the best approaches and behaviours within the economic, social and ecological realms.

3.4.1 Globalisation and Inherent Challenges in Understanding Sustainability

The sustainability principle, when probably applied, would give rise to sustainable economies, sustainable development politics and policies, sustainable cities, sustainable natural capitals, sustainable and more distributive finances, as well as sustainable markets. However, this last economic institution—the global market—has received aspersions as the bane of globalisation. Our research, through the opinion of Markus

Vogt, sees inherent ethical challenges to globalisation when one tries to evaluate the general concept of sustainability within the already prepared and, in some divergent opinions and voices, dominated and manipulated global markets and economies. The inherent ethical challenges give rise to huge disparities between and within the top and the bottom—the prosperous Western and European economies and the indigent developing and underdeveloped economies. Analysing the situation with the involvement of the United States and other advanced economies, Stiglitz opined, "On the face of it, this is a remarkable claim. The United States and other advanced countries wrote the rules of globalisation, and they run the international organisations that govern it. The complaint of those in the developing world was that the advanced countries had written the rules and managed these international organisations in ways that disadvantaged them" (Stiglitz 2017). Thus, as an insider in the global economic mainstream, Stiglitz observed:

> In the beginning of this century, I wrote...to explain the unhappiness with globalisation on display in so many countries in the developing world that I had been able to observe closely from my perch as chief economist of the World Bank. This is the part of the word with 85% of the world's population but with 39% of the world's income. The unhappiness was greatest in Sub-Sahara Africa, often rightly called a forgotten region, with a burgeoning population expected to reach 2.1 billion by 2050—close to seven times that of the United States today; its rich human and natural resources have been stolen from it for centuries, leaving it today with a per capita income 2.5% of that of the United States. (Ibid., pp. xi–xii)

On the other hand, Vogt argued that the current system of globalisation lacks the ethical guidance that would be able to give a normative standard for the proper handling of the general structure. Therefore, there is an urgent need for a principle that could address such inherent challenges and help to make, clarify, systematise and structuralise more transparently global relations (Ibid., *Prinzip Nachhaltigkeit: Ein Entwurf aus Theologisch-Ethischer Perspektive,* p. 21). In search for this normative standard, there is a need for a shift of paradigm about the classical social principles of personality, solidarity and subsidiarity. This is because, while these principles maintain their normative relevance and validity, they are heavily interpersonal. A new entrance in response to the ethical challenges of globalisation, therefore, would not be the abandonment of the classical social principles but rather a further development of their core values towards what is more sustainable, more inclusive, more justifiable and what can be authentically globalised. In this sense, we propose the transcendence from the principle of *Personality*, which focuses on the "person-being" of man and on using social incentives to balance the interests of both the individual and the community, to the principle of *Human Dignity.* Again, from the principle of *Solidarity*, which should emphasise the mutual acknowledgement, empathy and support members of a human community owe each other, to the principle of *Corporate Social Responsibility (Eco–Eco Approach),* which must remind the human level of the society, that it is an inextricable part of the generality of nature; and from the principle of *Subsidiarity*, which emphasises self-responsibility as a fundamental right and dissuades unnecessary infringement, especially by oppressive social and political powers, to the principle of *Economic and Social Justice*, where leadership must play a responsible role in providing the prerequisite platform for the *agent's* subsidiary participation (Küpper 2017). In other words, the inherent challenges of globalisation to the global mandate towards sustainability must be addressed *en route* to the integration of the above principles, which, according to Vogt, must not just be themes and concepts but applicable ethical and political programmes to be made relevant, practicable and revolutionary in the people's and societies' ways of life (Ibid., p. 22). The sustainability principle should, therefore, supplement the interpersonal character of the classical principles and introduce in them a sense of objectivity, such that the normative relevance of the principles would be sustainable for both present and future capitals of nature. Lack of proper transcendence from the classical principle, as well as avowed deficiency in the development of sustainability principle, leave globalisation, according to Stiglitz, with varied discontent (Ibid., *Globalisation and its Discontents Revisited*, p. xx).

3.4.2 The Challenging "Sings of the Time" in a Globalised Modern World

For effective development and integration of sustainability principles in the social order, there is the need to observe, analyse and deploy corresponding actions in our present world order so as to better handle the reality of all the empirical factors that influence and affect both the world and the human agents. Vogt designates these factors as the *signs of the time* in a global modern world. Although this term has its origin in catholic social teachings, particularly from the documents of the Vatican Council II, its central idea has become increasingly relevant not only for the Christian world but also for the new wave of the search for a sustainable world process—be it economic, social, political, religious, cultural or ecological. Vogt's position is, thus, made clear in his opinion that a theology of the *signs of the time* should understand this present time prophetically as God's call and turn itself to the "Joys and hopes, griefs and anxieties of the humanity today, especially the poor and the afflicted in all forms" (GS 1). It should, therefore, prove its commitment to the socio-political challenges of our time and situation. Again, the ecological concerns should be given a theological focus as an existential question to the Christian commitment to the liberation and salvation of all humanity and the totality of Creatures (Rom. 8). This is because it challenges the idea of creation as blessed habitat for all creatures and demands from the church to witness to and advocate the liberating presence of God, so as to find solutions to the exigent questions of our time and enable the right orientation towards the *integral call of man* (GS 11) (Ibid., *Prinzip Nachhaltigkeit: Ein Entwurf aus Theologisch-Ethischer Perspektive*, p. 40).

The central message in the appeal to pay attention to the "signs of the time" is to unite the efforts of all and sundry in working out sustainable approaches within institutions, relationships, and organisations, as well as within beliefs and ideologies for our world, which would guarantee better and more integrative societies of both the human capital and other natural givens. This expression appeared four times in the documents of the Vatican II Council, calling for a more sustainable and ecumenical approach to one another in a society that is becoming increasingly complex and diversified (Vatican II Council 1964); for a renewed effort in the respect for religious freedom, which most constitutions and international documents have recognised as a civil right, but which some unjust governmental and political powers violate, thereby creating instabilities which are banes of sustainable human societies (Vatican II Council 1965a); for a more sustainable and cooperative relationship between the ministers of the gospel—the priests and the laity—through unreserved respect and promotion of the dignity of the laity as well as "[…] recognising their experience and competence in the different areas of human activity" (Vatican II Council 1965b); and for a broader approach in scrutinising our present world order, which has profoundly entered a new stage of critical history, and in the words of the Council:

> Never has the human race enjoyed such an abundance of wealth, resources and economic power, and yet a huge proportion of the world's citizens are still tormented by hunger and poverty, while countless numbers suffer from total illiteracy. Never before has man had so keen an understanding of freedom, yet at the same time new forms of social and psychological slavery make their appearance. Although the world of today has a very vivid awareness of its unity and of how one man depends on another in needful solidarity, it is most grievously torn into opposing camps by conflicting forces. For political, social, economic, racial and ideological disputes still continue bitterly, and with them the peril of a war which would reduce everything to ashes. True, there is a growing exchange of ideas, but the very words by which key concepts are expressed take on quite different meanings in diverse ideological systems. Finally, man painstakingly searches for a better world, without a corresponding spiritual advancement. (Vatican II Council 1965c)

These realities of our time above need not be ignored since they challenge the very efforts of our time in creating a world that should be more sustainable. Vogt, therefore, sees relevance in placing emphasis (though not limited to this) on the threefold Christian ethical approach to social issues, namely: **See—Judge—Act**. In other words, the numerous signs of our time, which call for a practicable approach towards the enthronement of sustainability principle as the new social principle, need our world's critical observation, sincere evaluation and corresponding actions. Taken as an approach, this threefold pattern becomes a method, the advantage of which serves to create a more vivid framework that unites all plausible research with both individual and collective approaches, as well as helps to integrate the diversifying circumstances of our time and experiences. In Vogt's expression, it should be a clear framework for the differentiated reception of empirical research and situation analysis (Ibid., *Prinzip Nachhaltigkeit: Ein Entwurf aus Theologisch-Ethischer Perspektive*, p. 41). Paul Michael Zulehner, also referenced by Vogt, would later develop this threefold approach with his three socio-theological concepts, namely *Kairologie*—paying critical attention to the social situations and the signs of the time; *Kriteriologie*—the conscious and determined initiation of action plans and strategies within the socio-scientific levels; *and Praxeologie*—the practical evaluation and application of the numerous objectives, through methodical observation as well as tracking and professional studies to prove their relevance and suitability in dealing with the challenges and signs of our time (Ostheimer 2008).

Theologically and according to Vogt, both the proper recognition of the signs of the times and the application of the right tools in discerning and interpreting them should never be considered secondary in the general Christian approach. They rather form the very salvific approach to the right response to God's purpose of creation. They form the very foundation of justice and love. To this end, the rejection of the same would amount to stark alienation of the purpose of creation and the disdain of the creator himself (Ibid., *Prinzip Nachhaltigkeit: Ein Entwurf aus Theologisch-Ethischer Perspektive*, p. 42). Therefore, the *Kairos* and the challenges they pose for our time socially, economically, politically, ecologically, psychologically and spiritually build strongly the foundations of the call for the social principle that would provide the needed criteria and implement the corresponding actions. The ecological situation of our world is, as at the time of our research, unanimously considered a very critical sign of the time that impacts climate change. Thus, a new social principle should be able to remind humanity that there is now a need for limits, moderation and more sensible dealings with the natural capitals that coexist with humanity. For Vogt, this critical sign of the time –climate change—makes it clear to humanity that the future is shaky and slippery in the hands of our modern society. There is, therefore, an urgent need to give attention to the demands placed by intrinsic natural limits within creation itself and with regard to human actions and behaviour towards the natural capitals. A form of *rational, intellectual and behavioural humility* is demanded (Ibid).

3.4.2.1 The Climate Change Alarm

> Climate change affects every aspect of society, from the health of the global economy to the health of our children. It is about the water in our wells and in our taps. It is about the food on the table. … it is at the core of nearly all the major challenges we face today.
>
> —UN-Secretary General Ban Ki Moon, World Business Summit 2009

Recently, our world has come to live in an era where it can no longer ignore the warnings and recommendations of the same scientific knowledge that has almost placed the planet Earth and beyond in the hands and control of humanity. The above statement is credited to the former Secretary General of the United Nations, Ban Ki Moon, and it vividly describes the situation in which our present world finds itself, or rather has created for itself. The present politics of climate change present camps of divergent, dissident and conflicting opinions, making climate change one of the most debated policies of human history. The political and economic approach to its have brought together policy experts, government and non-governmental parastatals, individuals and groups, leaders of States and organisations, etc.,—all in the effort to pay attention to such reality and recognise it as one of the signs of our time. Vogt sees the new wave of climate change and the challenges it imposes as forming a new era of poverty and drastic socio-ecological change, which impact both human and societal lives. Therefore, it is urgent to see this reality as not just hypothetical anymore but rather as a verified empirical fact of our world and living conditions (Ibid., p. 45).

In various nascent efforts, debates and deliberations, it has become quite clear that the present world must face the critical option of changing course to stem the tide of the drifting humanity, the activities of which have contributed to placing the planet Earth gradually on diverse red lines. Such red lines have their overt and adverse socio-political and socio-economic effects, and such negative effects must be seen as the signs of our time that need the urgent attention of our world, its leaders and policymakers. Vogt identified four of these negative effects, namely: Endangering food security and water supply, spread of diseases, climate change as the cause of war and people's flight away from their land, and monetary costs (Ibid., pp. 47–48). All these put adverse pressure on the life of our world in general and societies in particular.

That the security of food and water supply has become deeply endangered is far from being a myth given the reality of our present moment. Food and water, as essential components of humanity's basic needs, face today's extreme climate conditions that threaten agriculture in general, vegetation, regeneration, crop and animal husbandry, etc. The most affected are the developing economies where mechanisation is still at its lowest ebb and poverty remains like cancer that threatens to eat away all human capabilities and opportunities. The spread of diseases of all kinds takes its turn within the general network of climate change since most negative climatic changes bring with them physical, chemical as well as emotional and psychical challenges, which affect both humans and animals. Imposed conditions of flight from one's fatherland due to situations of drought, flood, hunger and unhygienic environment, as well as the constant struggle for and over-clouded scramble for resources and living conditions in immigration lands, have recorded numerous cases of war and conflict in our present dispensation. In the end, the heavy financial burdens evident on both local and international budgets in the fight against extreme climatic conditions should bring our present world to a point of re-thinking and consequently redefining our current general approach to the natural world—indeed, a challenging sign of our time.

3.4.2.2 Loss of Sustainable Economic Solidarity: The Enthronement of Economism?

Vogt made reference to what he called the "Post-Rio-Rhetoric" paradigm in which our age is gradually losing the sense of sustainable economic solidarity. The effect is the serial enthronement of *economism,* which, in a sense more demonstrative to our focus, is the overemphasis placed on economic factors, the general development theories of which are counter-intuitive to economic ethics. Vogt identified such paradigm (taking the environmental and ecological politics of our time as an example) in all its forms as economic extremism, which takes as a priority an economy that booms, yet it has at the same time no regard for the environment and ecological protection (Ibid., p. 61)—an economy that leads to corporate social *"irresponsibility"*.

The modern global market and the system of operation within it have led to a new form of global capitalism. Such capitalism gives rise to a form of economism that has created an unfortunate trend whereby abject poverty threatens to exterminate the majority of the folk, with only a few having true access to the means of production and wealth. The economic solidarity of the present time experiences danger through an unjust globalisation ideology, which practises domination of the market by the privileged wealthy and developed economies against the developing and underdeveloped economies. The globalised market, therefore, serves as a viable instrument of renewed exploitation in the hands of its proponents, and consequently, it has become a focal point where the need for a sustainability principle is becoming urgent. For Vogt, the present status of the global market is already a phenomenon, the social effects of which can no longer be denied and discarded. To the majority of our world's population, it has remained oppressive, divisive and exploitative (Ibid., p. 62). This economy-based sign of the time, therefore, has diverted and influenced the modern economic paradigm so drastically, such that the effects have become, among others, the destructive challenges of sustainability, justice and responsible leadership at all levels. It is an influential sign of our time which, to a greater percentage, is not to be regarded as a natural fate in our world but a compulsive socio-economic behaviour of our world, with negative influences on both humans and society (Ibid., p. 61).

The vision towards general combat against persistent poverty in our world order has recently become the pre-occupation of the UN and other non-governmental organisations. For such efforts to be strategic and impactful, the principle of *solidarity* must stand out to lead the way towards sustainability. The Pontifical Council for Justice and Peace made a synthesis of what should be the core ideas of such a principle, thus:

> Solidarity must be seen above all in its value as a moral virtue that determines the order of institutions. On the basis of this principle the "structures of sin" [417] that dominate relationships between individuals and peoples must be overcome. They must be purified and transformed into structures of solidarity through the creation or appropriate modification of laws, market regulations, and juridical systems [....] Solidarity is also an authentic moral virtue, not a "feeling of vague compassion or shallow distress at the misfortunes of so many people, both near and far. On the contrary, it is a firm and persevering determination to commit oneself to the common good. That is to say to the good of all and of each individual, because we are all really responsible for all" [418]. Solidarity rises to the rank of fundamental social virtue since it places itself in the sphere of justice. It is a virtue directed par excellence to the common good, and is found in "a commitment to the good of one's neighbour with the readiness, in the Gospel sense, to 'lose oneself' for the sake of the other instead of exploiting him, and to 'serve him' instead of oppressing him for one's own advantage (cf. Mt 10:40–42, 20:25; Mk 10:42–45; Lk 22:25–27)" [419]. (Paricipatory Democracy Cultural Initiative, Inc 2018)

In the final analysis, the required sustainability principle would be able to globalise solidarity and speak the language of the common good and social justice. Only in this form could our world experience an authentic globalisation process in all ramifications.

3.4.2.3 Endangered Peace: Conflict of Interests and Resource Politics

Peace everywhere in our world today is endangered. People and governments everywhere strive in constant search for stability, cooperation and more lasting association. Conflict of interest exists everywhere and the scarce resources of our world set people, countries and opinions against one another. They remain, therefore, the signs of our time that are constantly in need of the right principles to create a more sustainable and stable society. Peace may not be the total absence of war, but the experiences of our world show that the path towards the institutionalisation of peace can never be sustainable without principles that ensure justice and respect for human rights and life. The need to see peace constantly being endangered in our present world as a burning sign of time has become demanding. Since the experiences of the two great wars and the subsequent cold wars, threats to peace and security of lives and properties have translated to varied new forms, especially since the beginning of the twenty-first century. Vogt identified the new waves and trends of violence and terror as the "*Menetekel*" (Daniel 5, 25–28) of our time, in other words, the handwriting on the wall, the signs on the walls of our experiences—the signs of our time (Ibid., *Prinzip Nachhaltigkeit: Ein Entwurf aus Theologisch-Ethischer Perspektive*, p. 68). The asymmetric and sporadic nature of this new development impacts our present world and society with both emotional and psychological imbalances, which make everyone "a suspect" against one another.

Our present society lives in constant tension as to what the next moment would bring or where the next terrorist activities would be carried out. Now and then, we are confronted with tactical and organised political killings and mob actions; former partner countries in treaties, the purposes of which are to retain peace and respect for one another, are gradually withdrawing from them and thereby betraying their political pacts; the new languages and their overt impacts are trade wars, sanctions, withdrawal from the agreement, accusations and counter accusations of terror sponsorships, tax and tariff sanctions and impositions, power delinquencies among political leaders, desperation in so many lives leading to the current migration crises, etc.

The natural capitals of our world—the natural resources—are gifts of nature and nurture, but the selfish human approach to them often leads to conflicts, corruption, enmities and wars. The enthronement of divisions and greed due to struggles for the natural capitals must be seen as a burning sign of our time, which is tearing nations apart, misleading so many leaders, destroying so many international relationships and devastating so many countries due to extreme exploitation of natural resources (anthropocentrism). This is dragging our world's ecology to a redline, exploiting and destroying so many economies and leading to the institutionalisation of extreme corruption in so many systems and structures of our modern world. These call for proper observation, concrete ideological confrontation and empirical actions. Vogt, therefore, sees the need to rediscover the workable principle that would be capable of leading our world to sustainable peace—a constructive part towards recovery that demands the principles of justice, intercultural and interreligious dialogues and understanding, preventive conflict management, the stabilisation of our world's international institutions, as well as the discovery of a distributive approach, which would work to deescalate the tensions and frictions associated with resource greed and covetousness (Ibid., p. 70).

The remedial principle of sustainability must help to combat the four problematic "Ds" associated with our modern social-environmental phenomena, namely: the problems of *Deposition*, through which toxic industrial elements are unguardedly made atmospheric, thereby causing pollution and endangering the lives of living organisms; *Degradation*, which is man-made devastation of the environment and the ecosystem resulting in forced loss of land, aquatic capitals, local economies and means of livelihood, as well as forced migration of people; *Disaster,* due to extreme climatic conditions and changes; and *Destabilisation*, which hits the general social living conditions of people through extreme deficiency in the basic means of livelihood (Ibid., pp. 72–73).

3.4.3 The Principle of Sustainability: Making a Case with Nigeria

While twenty-first century Nigeria may not be written off as a failed economic and political State, the pace and state of general development, both economic and human, have been deemed both nationally and internationally as incongruent with the heavy endowments of nature that she possesses. Making a case with its general economic and political structures, therefore, one accuses Nigeria of being bereft of the right sustainability principle, which consequently has rubbed her of authentic human and socio-economic development since after her independence from the colonial masters. This is a plausible evaluation not only in the Nigerian case but also in the general situation in Africa. The case study with Africa reveals a fundamental problem with the internal structures of her member states, which Osinibi calls "The Tragedy of Africa's Continued Retrogression". He opined:

> It has been observed that at the time of decolonisation in the 1950's and 1960's, the level of economic development in most of Asia was at par with that of Africa. For instance, four decades ago, the per capita income of South Korea was comparable with that of Sudan in Africa. However, since the 1960's, South Korea has achieved an incredible record of growth to become one of the twenty-six richest countries in the world and was able to join the trillion-dollar club of world economies in 2004, while the Sudan is still one of the thirty-three least developed countries in sub-Sahara Africa [….] This is in spite of the fact that African countries have relatively larger endowments of natural resources that Asian countries. While Asian countries have engaged in high rates of savings, investment in education, capital accumulation, sound macroeconomic management practices, relatively open trade policies, dynamic agricultural sectors, the maintenance of relatively equitable income distribution, and political credibility, African countries, with the exception of a few, have done the exact opposite […] The main obstacle to African development is the corrupt autocratic leadership and incompetent government that most African countries have been saddled with since independence. (Olusegun et al. 2013)

The state of affairs in Nigeria at the time of our research would confirm Osinibi's tragic observation and give Nigeria away as a stillborn in matters of development. Her unfortunate history and experiences of political woes and instability deepened her involvement with tragic developmental regress, such that the acclaimed *giant of Africa* appears to be a toothless Bulldog so weak as to revitalise and vivify her economy and the general human development. Two inevitable societal woes, which are consequences of failed sustainable societal principles—according to Markus Vogt—would also be seen plaguing Nigeria, namely Socioeconomic enslavement and Resource conflicts and politics.

3.4.3.1 Socio-Economic Enslavement: A Politico-Ideological Chauvinism in Nigeria

The general exploitation of African natural resources by both internal and external means has been a key factor shaping ideas in international relations, international markets and globalisation in general. However, the structure of trade agreements, with its consequent imbalances and dominance, has refused to combat the concept of enslavement in the general world and economic forums. Consequently, the general scene of forced economic and political dependencies has served to shape the world international forum with a type of chauvinism, which is heavily overt in the many African States as well as in their leaders. Such chauvinism in the Nigerian case remains her internal political and ideological socio-economic enslavement of the masses. One of the avowed keys to the reported South Korean developmental strands was "[…] the maintenance of relatively equitable income distribution" (Ibid). The Nigerian case presents an unjust concentration of the factors of production only in the hands of a few, who, on the other hand, *instrumentalise* the human agents as means towards the maximisation of supplies and gains. The consequences are obvious: exploitative labour, unimaginable and unjustifiable income disparities, unstable and agitating labour force, political and economic instability and confrontations, etc.

The spree of socio-economic enslavement in Nigeria has put her up in the international community as a dastardly corrupt and hijacked nation. Thus, the super-rich minority stops at nothing to maintain their exploitative positions, while the majority are reduced to mere means and objects of planned man's inhumanity to fellow humans. Writing about such enslavement and intractable inequity, Ubani-Ebere observed:

> The sudden death of empathy for decency, honesty, truthfulness, kindness, patriotism, equality, justice and respect for the law created the opportunity that transformed Nigeria into a vineyard of moral delinquency, inhumanity, affliction, abduction, bomb attacks, kidnapping, infringement, injustice, indiscriminate and targeted killing. As a result, Nigeria has become a hub and fierce conduits of exploitation, crimes, extortion, bribery, frauds, nepotism, crookedness, rackets, shadiness and skimming which are the serious symptoms of a sick society that squeezes the poorest of the poor and citizens that have no connections with the upper echelon of Nigeria. (Ubani-Ebere 2017)

Within this orgy of ganged enslavement, one sees an organised political and ideological chauvinism of the few against the common good and welfare of the masses. The lament of the Nigerian President, Muhammadu Buhari, during the 25th Nigerian Economic Summit, held on the 7th–8th Oct. 2019, showed the persistence of economic slavery within the country even in the twenty-first century. For the president, it is still lamentable that the wealth of the country has remained in the hands and control of only a numbered few, who come from only four or five States—including the Federal Capital Territory—out of the thirty-six States of the Federation (Iyinolaka 2019). But the question remains: Does the clue of leaders within the cabinet of the president himself have a concrete and justifiable political and economic agenda to combat the said enslavement and enthrone social and distributive justice? Among the elected representatives of the people themselves exists the scandalous trend of covetousness through budget padding, obnoxious allowances, horrifying salaries and unbridled embezzlement. While the common men and women on the streets of Nigeria struggle for the least to make a daily living, their elected representatives dine and wine in unprecedented galore of corruptly amassed affluence. Obi has this to say with regard to the representatives of the people:

> A referendum on the National Assembly has become imperative because Nigeria's rather fragile structure of democratic government might collapse from the pressure of the cost of government...Now the National Assembly is nothing but a business enterprise and the primary objective of members is to make money for themselves...the authoritative Economist Magazine, has reported that the lowest paid Nigerian legislator earns 116 times of the income per capita, the proportion is 60 in Kenya, 30 in Ghana and 18 in Indonesia. There is nothing to compare with that degree of disparity anywhere in the world. (Obi 2017)

The illustration is in the table below according to the statistics by Obi (Ibid, p. 187) (Table 3.2).

The staggering figure of $961.215.00 is not just imaginary but can be proved with the following most current statistics showing remunerations, salaries and allowances, using the case of the Nigerian senators as an example: Salary—₦9.000.000 ($22.004) per annum; Running allowance—₦163.000.000 ($398.533) per annum; Constituency Projects—₦200.000.000 ($488.998) per annum; Severance Gratuity—about ₦7.043.000 ($17.220) per annum; Furniture Allowance—₦7.045.000 ($17.225) per annum; Motor Vehicles Allowance—₦9.094.000 ($17.235). These remunerations are officially legislated and approved within the Nigerian system. However, the system of inflating these amounts and the subsequent pocketing and privatisation of portions allotted for public and community goods, services and development by the representatives of the people themselves, amount to a structural sin that continues to rob the Nigerian state of both sustainable and people-oriented development. This is the exact evil of political hooliganism embedded in the Nigerian system such that, despite the huge natural capitals within the Nigerian soil and the revenue accruing from them, Nigeria still has blurred prospects, or better put, a defective vision for inclusive development and socio-economic transformation.

Table 3.2 Annual pay of the representatives of the people in selected countries of the world in comparison with Nigeria

No	Country	Annual pay (US$)
1	Sri Lanka	5,100.00
2	India	11,200.00
3	Malaysia	25,300.00
4	Thailand	43,800.00
5	Spain	43,900.00
6	Ghana	46,500.00
7	Saudi Arabia	64,000.00
8	Indonesia	65,800.00
9	Kenya	74,500.00
10	France	85,900.00
11	Sweden	99,300.00
12	South Africa	104,000.00
13	Britain	105,400.00
14	New Zealand	112,500.00
15	Israel	114,800.00
16	Germany	119,500.00
17	Ireland	120,400.00
18	Hong Kong	130,700.00
19	Japan	149,700.00
20	Canada	154,000.00
21	Singapore	154,000.00
22	Brazil	157,600.00
23	United States	174,000.00
24	Italy	182,000.00
25	Nigeria	961,215.00 (2022)

Illustration Mine

3.4.3.2 Resource Conflicts and Politics

One of the most daring banes of Nigeria's path towards sustainability has been her battle with resource conflicts and politics, added to the ever-controversial Land Use Act 1978, the enactment of which portended tensions and frictions for the future and structure of Nigeria as a geopolitical entity. To date, the issues surrounding resource control and the exploitative politics of the policymakers have remained burning in the nation's life and have, in several cases, led to bloodletting and the consequent destruction of lives and properties. The Land Use Act 1978 prima facie was meant to address both economic and political needs, namely:

> The Nigerian Land Use Act 1978 is the principal legislation that regulates contemporary land tenure in Nigeria. Upon its enactment, the law brought about radical, if not revolutionary,

> changes in the erstwhile land tenure system in the country. The law was aimed, among other things, at reducing unequal access to land and land resources, a situation that had caused a great deal of hardship to the citizenry. Massive and unfettered access to land and land resources by the citizens could stimulate the needed economic growth an economy that depends heavily on agriculture and mineral resources. The Land Use Act was equally targeted at reducing the high cost of land required for industrial estates and mechanised agriculture. For these reasons, the law appeared to nationalise land when it placed it in the hands of the government as a custodian, to hold in trust and administer for the use and common benefit of all Nigerians. However, after more than three decades of the operation of the law, it is apparent that most of the problems it sought to cure have resurfaced and certain provisions of the law have themselves worked hardship on the citizens and tended to impede economic development, which the Act initially sought to stimulate. (Nwocha 2016)

While we may not deny outright the benefits that may be associated with such a law if justifiably applied, it is obvious, as Anyanwu observed, that "[...] the Land Use Act was enacted to satisfy the need for larger areas of land for agriculture and non-agricultural purposes, end racketeering and the unending litigations in land transactions due to rising demand for land; checkmate traditional land ownership that had constituted a barrier to national development programmes; [...] carter for the need for sustained security of rights to land in matters of duration, compensation and alienation of rights in land and sharpen government's sensitivity to a system in which only the rich, powerful and influential owned" (Ibid). But the Nigerian system rather twisted the law to mean a weapon of exploitation in the hands of her corrupt leaders and their international *bedfellows*. Consequently, the government's power to regulate the use of land and ensure equitable access to the natural capitals (resources) serves only the needs of the greedy political class and their powerful corroborators. In other words, what is meant to serve the common good now exploits rather the citizenry. In so many cases, the uncontrolled grab of resources by the feudal political lords destroys even the local economy of the people and makes them beggars and aliens in their home country.

The underground politics that is overt in the whole situation with connections to international players leave so many well-intentioned minds filled with questions about why such extremism happens in Nigeria. Even in cases where public projects have been awarded, one discovers, along the line, the illegal padding of the whole project and subsequent misappropriation of the money. In the end, the general body politic is left with unsustainable institutions and unfinished infrastructures, which stand as the bane of Nigeria's stagnant development. This, unfortunately, keeps the Nigeria of today constantly tensed with passions for agitation and leads to sporadic violent reactions. Calls for revolution with ethnic undertones are heard, threatening the land. Hence, Nigeria must know that "The same Revolutionary situation: social inequalities, financial difficulties of the State to meet its obligations to the population, high public debt profile, heavy tax burden on only a section of the population etc., that pushed the people of France to revolt in 1789 against the established political order and *Ancient Regime*'s political and social structures, prevails (and is becoming endemic) in Nigeria today" (Akrasi 2017).

Evaluation

The need to pay attention to the plaguing situations threatening the existence of Nigeria as a geopolitical entity makes Vogt's thesis on the sustainability principle as the new social principle urgent in the Nigerian case. The various signs of the time manifest in her social, political, economic and ecological challenges ring together the alarming bell for concrete responses that would effectively stem the tide of a drifting society. Above all, Nigeria must be ready to live the truth of her active membership and participation in the planned agenda of the United Nations, with the concrete aspiration to realise the prospects of the Development Goals as models and criteria for people-oriented development. These prospects and standards would be dealt with in the discourse that follows.

3.5 The Global Goals and the Planned Prospects

The global sustainable development goals remain the most explicit blueprint, through the united efforts of the world leaders, to give the development concepts of our world concrete and holistic orientation and approach for the period of 2015–2030. They represent planned and proposed steps to make our world more sustainable through a common agenda, the application of which lies heavily on the efforts of individual *member states*. They are not in themselves transformers of political, economic or ecological institutions. Rather, they carry necessary means and instruments designated for functional governmental and non-governmental ideologies for more effective and more reliable development institutions. Both the concepts and applications associated with them revolve within the three fields of "[…] economic well-being, social development and environmental sustainability and regeneration—(and) they collectively represent a set of goals for poverty reduction, embodying a multidimensional conception of poverty" (Black and White 2003). These goals are seventeen in number, taken from the various extensive areas that deeply affect the world, in general, and human agents and nature, in particular. The structure of the Sustainable Development Goals (SDGs) (2015) gives a sense of integration. It means that concerted efforts to achieve success in any one of the seventeen areas invariably lead to or prepare the foundation for success in other areas.

It is to be noted that the Sustainable Development Goals in its present draft and structure is a further and expanded work, which takes its fundaments from the Millennium Development Goals, which the 191 member states of the United Nations agreed in the year 2000 to achieve before 2015. It was all in the effort to evaluate, re-evaluate and make the ideologies behind the goals more inclusive and responsive. In general, and rightly so, whether the proposed development goals have recorded plausible successes and responses to the development demands plaguing our world remains a constant debate that sets, to date, opinions and intellectual minds in conflict with one another. The extent of acceptance, application and achievement within individual member states is also explicitly variable. The United Nations, as a body, uses the

Human Development Indicator as a measuring and monitoring instrument to report member states involvement and efforts in making the prospects of the goals bear on the concrete lives of their citizens and life situations. At present, our research specimen, Nigeria, using a recently available indicator in 2018, ranks 157th out of 193 member states that make up the United Nations (United Nations 2018).

It becomes simply indicative, therefore, that Nigeria as an entity performs below average (fluctuating within the ranks between 157 and 163, 2022 estimate) if, after over 60 years of Independence as a sovereign state, a whopping 71.7% of her employed force still battle the harsh challenge of a minimum decent life. The question of her unemployed population remains a growing and persisting challenge. Based on this, we now try to articulate the core ideas within the individual goals and place them alongside the concrete situations found within Nigeria's involvement in the general development of the country and its people.

3.5.1 The 17 Sustainable Development Goals: An Appraisal Within the Nigerian State

Source Globalance, Ethics Handbook for a Balanced World Post-Covid (Globalance 2020)

Goal 1: End Poverty in All Its Forms Everywhere

The aim of the first goal encompasses all efforts to liberate the affected parts of our world, especially where extreme poverty exists and where people are robbed of their natural rights to economic resources. Special focus would be on the LDCs nations—Landlocked Developing Countries, which are inevitably disadvantaged due to natural, environmental and geographical constraints. The challenges that the first goal sets out to address are summarised in the most recent situations and experiences thus (The United Nations 2019):

GOAL 1

More than 700 million people, or 10% of the world population, still live in extreme poverty. Surviving on less than US$1.90 a day; Having a job does not guarantee a decent living. In fact, 8 per cent of employed workers and their families worldwide lived in extreme poverty in 2018; Globally, there are 122 women aged 25 to 34 living in extreme poverty for every 100 men of the same age group; The majority of people living on less than $1.90 a day live in sub-Saharan Africa; High poverty rates are often found in small, fragile and conflict-affected countries; Poverty affects children disproportionately. One out of five children live in extreme poverty; As of 2018, 55% of the world's population have no access to social protection; In 2018, only 41% of women giving birth received maternity cash benefits, etc.

Development Indicator—Goal 1—Specimen: Nigeria

The challenges of Goal 1 and the targets it has set out still have a long way to go in the Nigerian case. The most recent development indicator shows that the most populous nation in Africa—Nigeria, harbours a greater number of the world's poor. The average Nigerian youth begins life with the fierce struggle to wriggle out of the existing poverty into which he or she was born and is invariably condemned to grapple with. Within such struggle, many are lost in the crowd; perspectives are blurred, vision and innovation die away due to lack of the corporate social and economic institutions that should absolve the talents; human capital and manpower are drained due to flight from hunger and forced emigration, etc. The Nigerian situation is a case of poverty taken as the stifling of the development of capabilities in human agents. While some concerted efforts are being made by some individual groups, activists and opposition groups, the Nigerian problem is, unfortunately, that of an in-built political and economic slavery, with which the greedy and never-too-old political elite minority has held the country in bondage.

Goal 2: End Hunger, Achieve Food Security and Improved Nutrition, and Promote Sustainable Agriculture

Food security and the ability of a country to sustain itself and its human capital nutritionally remain a vital drive to economy and well-being. The second goal seeks to recognise the growing situations of hunger and malnourished persons, especially women and children, alongside the constant waste and discard of food and other material goods in the more developed part of our common globe. The most current core challenges Goal 2 sets out to address are summarised thus (Ibid):

GOAL 2

An estimated 821 million people were undernourished in 2017; The majority of the world's hungry people live in developing countries, where 12.9 per cent of the population is undernourished; Sub-Saharan Africa remains the region with the highest prevalence of hunger, with the rate increasing from 20.7 per cent in 2014 to 23.2 per cent in 2017; 149 million children under 5 years of age—22 per cent of the global under-5 population—were still chronically undernourished in 2018; Agriculture is the single largest employer in the world, providing livelihoods for 40 per cent of today's global population. It is the largest source of income and jobs for poor rural households; Since the 1900s, some 75 per cent of crop diversity has been lost from farmers' fields, etc.

Development Indicator—Goal 2—Specimen: Nigeria

The reality of hunger and malnutrition is evident in the current Nigerian social system, and this has been made worse by Nigeria's almost total loss of her agricultural sector. This was and has been the result of her over-dependence on practically only one sector of her economy—the oil and gas sector. Heavy reliance on the importation of food and other daily products has more or less negative socio-economic effects, namely the high cost of basic commodities, risk of inflation, stagnation of indigenous efforts and economy, self-inflicted increase in mortality rate, etc. The development indicator in the Nigerian case faults the systematic and gradual drift of Nigeria towards socio-economic capitulation, given her inaction in repositioning the agro-sector.

Goal 3: Ensure Healthy Lives and Promote Well-Being for All at All Ages

The third goal recognises the increasing mortality rate around the globe, especially in developing and underdeveloped countries, due to the poor or total absence of health insurance and health institutions. The high records of persistent health challenges and incessant deaths in the most affected areas of Africa, the Middle East, Asia and other parts of the world challenge modern scientific innovations in the general medical field. However, the real problem is the case of failed policies of the policymakers within the affected member states in making the health and general well-being of the citizenry a demand. The third goal, therefore, sees the need for a united effort to combat poor health, diseases and general ignorance about preventive measures, knowing that an affected part can invariably endanger the life of the whole. Some of the key observations of Goal 3 are as follows:

GOAL 3

17,000 fewer children die each day than in 1990, but more than five million children still die before their fifth birthday each year; Since 2000, measles vaccines have averted nearly 15.6 million deaths; Despite determined global progress, an increasing proportion of child deaths are in Sub-Saharan Africa and Southern Asia. Four out of every five deaths of children under age five occur in these regions; Children born into poverty are almost twice as likely to die before the age of five as those from wealthier families; Children of educated mothers—even mothers with only primary schooling—are more likely to survive than children of mothers with no education; Maternal Health – Maternal mortality has fallen by 37% since 2000; But maternal mortality ratio – the proportion of mothers that do not survive childbirth compared to those who do – in developing regions is still 14 times higher than in the developed regions; etc.

Development Indicator—Goal 3—Specimen: Nigeria

Perhaps one of the greatest blows dealt to Nigeria and its teaming population has been the inability to design and implement a functional national health insurance system that would be able to meet acceptable standards within the international community. This failure was articulated by the President of the Nigerian Medical Association, Prof. Mike Ogirima, thus: "Non-implementation of the National Health Act 2014, which provides for not less than 1% Consolidated Revenue Fund as basic healthcare provision fund has further worsened the travail of the health sector, particularly at the grassroots, where the greater burden of the health disease resides. It has also added to the financial burden of the citizens in their quest to seek quality health care, which in most cases is non-existent" (Ojo 2016).

Furthermore, there is a need for Nigeria to invest in integrating the prospects of "reproductive health" into its national scheme. Granted that "[…] Good sexual and reproductive health is a state of complete physical, mental and social well-being in all matters relating to the reproductive system. It implies that people can have a satisfying and safe sex life, the capability to reproduce, and the freedom to decide if, when, and how often to do so [….]" (The United Nations Population Fund (UNFPA) 2019). Africa generally must plan a regulatory reproductive health system to address overpopulation and invariably stem the tide with which the majority of her population and families degenerate to abject poverty. An indigenous family planning system must accompany her process of development while at the same time safeguarding the citizens' freedom and rights to reproductive health. Furthermore, because women take the critical centre stage in the issue of reproductive health, the promotion of gender equality can never be extricated from the sustainable development agenda for Africa and particularly for Nigeria. Nigerian women, especially those still being exploited by some extreme cultural and religious beliefs and practices, must constitute the nucleus of Goal 3 if procedures for sustainable development could become practicable at all. This issue of gender equality will be discussed further in Goal 5.

The impasse that Goal 3's negligence has created has made medical trips outside the country a lucrative business for the countries of Europe, America, and Asia (India). These countries provide health services for the teeming number of Nigerians who, as the case may be, could afford the financial pressures of such trips. A majority of the citizenry is compelled to grapple with the non-performing state or the underperformance of typical Nigerian health institutions. The development indicator still finds Nigeria wanting in this regard.

Goal 4: Ensure Inclusive and Equitable Quality Education and Promote Lifelong Learning Opportunities for All

Goal 4 is a proposal for all member states to make education a priority. With such priority, it behoves the international community, in cooperation with individual member states, to see that the basic standards of learning are made equitable for all, of course, with respect to people's cultural and perhaps religious backgrounds. The inclusive nature of education for all should be made to be attained such that no part would be disadvantaged in terms of the educational basics for an authentic life, for the promotion of both national and international relationships, and for equitable access to life opportunities and individual self-actualisation.

The challenging situations that confront Goal 4 are presented thus.

GOAL 4

Enrolment in primary education in developing countries has reached 91 per cent but 57 million primary age children remain out of school; More than half of children that have not enrolled in school live in sub-Saharan Africa; An estimated 50 per cent of out-of-school children of primary school age live in conflict-affected areas; 617 million youth worldwide lack basic mathematics and literacy skills, etc.

Development Indicator—Goal 4—Specimen: Nigeria

The educational system in Nigeria is associated with a history of fluctuating policies and policy implementation. Its conventionality, according to the constitution of the land, is structured to meet the widely accepted standards in the international community. But the greatest bane of the Nigerian educational system, especially after her independence from the colonial masters, has been the gradual negligence of infrastructure, poor funding, low-quality personnel and misappropriation of the institution's funds. Some years after independence, the citizenry had a short-lived experience of free education from primary to secondary levels, which became functional due to the relatively well-managed resources and revenue of the country. But with the incessant experiences of political instability, sporadic military coups and consequent military rules, corruption was enthroned and made a way of life. In this form, the standard of education experienced a massive fall, which, up to the present, has left the educational institution struggling. Although basic primary education is taken officially to be free and compulsory in the country, the corrupt practices in the public schools have rubbed the institution of the benefits of such policy for the citizenry. UNICEF estimates that "In Nigeria, about 10.5 million children are not in school even though primary education is officially free and compulsory" (United Nations Children's Fund, UNICEF 2022). Added to the situation is the loss of trust in the public and government schools by many with regard to moral and character building. The effect has become the multiplicity of private schools belonging to both individuals and religious organisations. On the one hand, such schools can educate and impart morals and character. However, they are not solutions to the economic burdens of families and guardians of Nigerian pupils and students since their financial demands still pose challenges for the average and poor families of the Nigerian society.

Goal 5: Achieve Gender Equality and Empower All Women and Girls

The issue of gender equality and empowerment of women and girls has gained critical attention in our time more than ever in the history of human society. Thus, the purpose of Goal 5 is to propose the possibility of a more sustainable world if the societies of our world could jettison gender molestation and exclusion. The actual focus of Goal 5, therefore, is the empowerment of all and not the preference for or enthronement of a selected section of the human race. As experience has shown, the greatest challenges to Goal 5 are the disparities seen in cultures, belief systems and religious conventions. While some societies have put in place social structures and policies that work to bridge the gap of inequality, many others still reserve extreme practices that keep excluding the feminine folk and subjecting them to mere objects in the hands of the users. Below are some of the key achievements and challenges associated with Goal 5 and central to the problems remain the denial of equal opportunity and access to employment and education, imposed genital mutilation, as well as forced early marriage (Child marriage):

GOAL 5

Globally, 750 million women and girls were married before the age of 18 and at least 200 million women and girls in 30 countries have undergone FGM; The rates of girls between 15-19 who are subjected to FGM (female genital mutilation) in the 30 countries where the practice is concentrated have dropped from 1 in 2 girls in 2000 to 1 in 3 girls by 2017; In 18 countries, husbands can legally prevent their wives from working; in 39 countries, daughters and sons do not have equal inheritance rights; and 49 countries lack laws protecting women from domestic violence; One in five women and girls, including 19 per cent of women and girls aged 15 to 49, have experienced physical and/or sexual violence by an intimate partner within the last 12 months. Yet, 49 countries have no laws that specifically protect women from such violence; While women have made important inroads into political office across the world, their representation in national parliaments at 23.7 per cent is still far from parity; In 46 countries, women now hold more than 30 per cent of seats in national parliament in at least one chamber; Only 52 per cent of women married or in a union freely make their own decisions about sexual relations, contraceptive use and health care; etc.

Development Indicator—Goal 5—Specimen: Nigeria

While there are recognisable efforts for the empowerment of women and girls in Nigeria, there still exist many cultural and religious obstacles towards commendable implementation. While the openness to programmes and policies of empowerment of women and girls may have gained relative acceptance in the southern part of the country, the push for general acceptance in the North has met an embattled situation that gave rise to the terror group known today as "Boko Haram". The group's revolutionary idea is to combat *Western education (ideologies) and the empowerment of women.* This has caused Nigeria for almost two decades now the loss of lives and properties as well as abuse and kidnapping of young girls from the streets, homes and schools. The practice of child marriage also has a general cultural and religious acceptance in the North. Therefore, the general development indicator for Goal 5 is not yet a *success-for-all* programme in Nigeria but rather a social project under development in the south and controversy in the North. But the unified condemnation of and fight against the fundamentalist group, Boko Haram, may later allow deeper integration of the goal in the wider social system and structure of the nation.

Goal 6: Ensure Availability and Sustainable Management of Water and Sanitation for All

Water and sanitation have basic roles to play in the socio-economic lives of a nation. Human and environmental history is replete with the consequences when the two elements are impeded, be it due to drought, lack of clean drinking water, contamination of water due to industrial and chemical activities, poor refuse disposal, the practice of open defecation as well as zero or poor waste management. The numerous cases of deaths and disease outbreaks, especially in developing and underdeveloped parts of our world, show statistical links to poor management of water and sanitation systems—one being a correlation with the other. In other words, deficiencies within one factor elicit imbalances in the other away as a direct consequence. Identifiable water and sanitary challenges to Goal 6 include the following:

GOAL 6

1 in 4 healthcare facilities lacks basic water services; 3 in 10 people lack access to safely managed drinking water services and 6 in 10 people lack access to safely managed sanitation facilities; At least 892 million people continue to practice open defecation; Women and girls are responsible for water collection in 80 per cent of households without access to water on premises; Between 1990 and 2015, the proportion of the global population using an improved drinking water source has increased from 76 per cent to 90 per cent; Water scarcity affects more than 40 per cent of the global population and is projected to rise. Over 1.7 billion people are currently living in river basins where water use exceeds recharge; 2.4 billion people lack access to basic sanitation services, such as toilets or latrines; More than 80 per cent of wastewater resulting from human activities is discharged into rivers or sea without any pollution removal; Each day, nearly 1,000 children die due to preventable water and sanitation-related diarrheal diseases; Approximately 70 per cent of all water abstracted from rivers, lakes and aquifers is used for irrigation; Floods and other water-related disasters account for 70 per cent of all deaths related to natural disasters, etc.

Development Indicator—Goal 6—Specimen: Nigeria

There was a quasi-system of household and public water supply in the Nigerian social infrastructural system within the colonial period and some years after the independence. During those periods, there existed public services of water cooperation, especially within the urban areas of the society. Some years after independence, this cooperation went defunct, leaving only vestiges of obsolete and dilapidating equipment due to inexperience, corruption, embezzlement and general mismanagement of the cooperation by those who took over after independence. Today, Nigeria, with its teeming population, has no unified and well-structured system of water supply. The wealthy States have some privatised structures in place, but they show incessant disruptions and malfunctions. Wealthy individuals and families, which, of course, make up the minority, resort to the use of private boreholes for the supply of water for their own families and perhaps for the nearby neighbours. The majority of the citizenry resort to water fetching from streams and rivers, where these are not already contaminated through oil exploration or other chemical abuse. The aftermath is clear in the Nigerian environmental and sanitation status, which gives the greater portion of the country away as one of the dirtiest in Africa and the world. The appalling scenes of refuse dumps along the streets and the residential and public areas are overt testimonies of Nigeria's poor regard for a clean and healthy environment. Goal 6, therefore, needs a strategised policy that would first change people's approach and attitude towards the environment and health.

Goal 7: Ensure Access to Affordable, Reliable, Sustainable, and Modern Energy for All

The issue of the modern transition to clean and sustainable energy has become a need in our world. The danger of the harm already done to the human planet due to overdependence on fossil fuel and energy added to its consequent carbon emissions has awoken both scientific and anthropocentric clarion calls for the human world to make urgent alternatives possible. Goal 7 is the call for globalised and united efforts to save our world from further destruction and the protracted dereliction of man himself. While the developed and advanced economies battle to reduce emissions due to industrial activities, the less developed economies, especially the oil reserved areas, must battle a probable loss of ecology and ecosystem due to extreme and uncontrolled exploration of raw materials; in other words, the factors of production. The challenges are articulated thus:

GOAL 7

13% of the global population still lacks access to modern electricity; 3 billion people rely on wood, coal, charcoal or animal waste for cooking and heating; Energy is the dominant contributor to climate change, accounting for around 60 per cent of total global greenhouse gas emissions; Indoor air pollution from using combustible fuels for household energy caused 4.3 million deaths in 2012, with women and girls accounting for 6 out of every 10 of these; The share of renewable energy in final energy consumption has reached 17.5% in 2015.

Development Indicator—Goal 7—Specimen: Nigeria

The presence of fossil fuel energy in commercial quantity in Nigeria has long blurred Nigeria's vision towards alternative and available energy transitions. As an oil-producing giant in the continent of Africa, Nigeria plays down on all innovative technologies for clean and more sustainable energy. As analysed in chapter two of our work, this lack of transition keeps Nigeria as a nation in bondage and has kept her available alternative energy sources unexplored. Instead, Nigeria has relied heavily on extreme fossil fuel exploration and has remained adamant in the face of the horror of environmental woes it has caused and is still causing in the affected areas, which are homes to millions of her citizens. The apparent situation of independent energy generation through fossil fuel combustible generators has made Nigeria also the graveyard to so many helpless citizens who have become victims of generator fumes as the cause of their suffocation and death. Goal 7 was explicit with the statistics of the affected victims. In general, Goal 7 is an emergency in the Nigerian case and still awaits the implementation of the right policy and approach.

Goal 8: Promote Sustained, Inclusive and Sustainable Economic Growth, Full and Productive Employment and Decent Work for All

Goal 8 recognises the inequality that is apparent in our world and within the concept of globalisation. It also recognises the variable situations within the individual member states, with special emphasis on the pay gap existent between the genders—men and women. On the other hand, Goal 8 works as a proposed target that places a demand on States to work towards inclusive economies that would be able to bridge the ever-widening gap between the rich and the poor, as well as reduce the rising unemployment rate, especially among the youths. It is a target as well which proposes, in all its ramifications, the need for decent working conditions for all, contrary to all conditions that exploit the human person and see him or her not as an end in himself or herself but only as a means and an object to be used. The main targets are as follows:

GOAL 8

The global unemployment rate in 2017 was 5.6%, down from 6.4% in 2000; Globally, 61% of all workers were engaged in informal employment in 2016. Excluding the agricultural sector, 51% of all workers fell into this employment category; Men earn 12.5% more than women in 40 out of 45 countries with data; The global gender pay gap stands at 23 per cent globally and without decisive action, it will take another 68 years to achieve equal pay. Women's labour force participation rate is 63 per cent while that of men is 94 per cent; Despite their increasing presence in public life, women continue to do 2.6 times the unpaid care and domestic work that men do; 470 million jobs are needed globally for new entrants to the labour market between 2016 and 2030.

Development Indicator—Goal 8—Specimen: Nigeria

The reality of economic growth in Nigeria after her independence would not be denied on the one hand. On the other hand, it is necessary to emphasise that Nigeria's economic growth is inversely proportional to its population growth. This has placed the percentage of the employed labour force below 40% and the recent report of the National Bureau of Statistics shows that 23.1% of the available workforce is unemployed (The Federal Republic of Nigeria 2020). It all means that concerted efforts are needed to stabilise the economy and sustain population growth. Within our research period, the issue of an acceptable decent minimum wage for jobs in Nigeria has continued to set the government and Labour against each other. Economists and analysts see the whole situation as needing reform and restructuring. For one, to make Goal 8 a reality in Nigeria, the cost of governance and the uncontrolled access to wealth by her politicians must be nipped in the bud.

Goal 9: Build Resilient Infrastructure, Promote Inclusive and Sustainable Industrialisation and Foster Innovation

Sustainable infrastructure and institutions, industrialisation and a sense of innovation are catalysts to sustainable growth and development. The target of Goal 9 is to unite alliances that would be capable of pushing up development, especially in developing and underdeveloped economies, while at the same time uniting efforts in fighting for cooperation at the international level that would necessitate inclusive socio-economic participation. Excerpts from the basic targets are as follows:

GOAL 9

Basic infrastructure like roads, information and communication technologies, sanitation, electrical power and water remains scarce in many developing countries; 16% of the global population does not have access to mobile broadband networks; For many African countries, particularly the lower-income countries, the existent constraints regarding infrastructure affect firm productivity by around 40 per cent; The global share of manufacturing value added in GDP increased from 15.2% in 2005 to 16.3% in 2017, driven by the fast growth of manufacturing in Asia; Industrialization's job multiplication effect has a positive impact on society. Every job in manufacturing creates 2.2 jobs in other sectors; Small and medium-sized enterprises that engage in industrial processing and manufacturing are the most critical for the early stages of industrialization and are typically the largest job creators. They make up over 90 per cent of business worldwide and account for between 50-60 per cent of employment; etc.

Development Indicator—Goal 9—Specimen: Nigeria

Nigeria possesses many colonial and post-colonial socio-economic and political infrastructure. However, with the development indicator, Nigeria needs a reformed attitude towards infrastructural politics. Many social services and institutions became dilapidated and some extinct within some years after independence due to negligence and poor maintenance. The pace of industrialisation in Nigeria is not as should be expected due to related factors and, chief among them, energy and power supply. This same factor places high pressure on production, hinders prospects for innovation and scares investors away from the economy. In the main, Goal 9 has high prospects in Nigeria, given her natural endowments and human capital.

Goal 10: Reduce Inequality Within and Among Countries

Inequality within and among countries remains persistent in the international community and it is as well the bane of modern efforts towards globalisation. Goal 10, therefore, has the prospect of uniting efforts of the international community in bridging the extreme socio-economic gaps among nations and peoples. The main focus is on the plights of economic and social injustices that exist in the least and developing economies. Concrete observations are:

GOAL 10

In 2016, over 64.4% of products exported by the least developed countries to world markets faced zero tariffs, an increase of 20% since 2010; Evidence from developing countries shows that children in the poorest 20 per cent of the populations are still up to three times more likely to die before their fifth birthday than children in the richest quintiles; Social protection has been significantly extended globally, yet persons with disabilities are up to five times more likely than average to incur catastrophic health expenditures; Despite overall declines in maternal mortality in most developing countries, women in rural areas are still up to three times more likely to die while giving birth than women living in urban centres; Up to 30 per cent of income inequality is due to inequality within households, including between women and men. Women are also more likely than men to live below 50 per cent of the median income, etc.

Development Indicator—Goal 10—Specimen: Nigeria

Nigeria itself is yet to be free from varied socio-economic inequality embedded in her structure: the widening gap between the rich and the extremely poor; the failure of the State to make a functional and inclusive social health system a reality; inaction towards investment in and transformation of the health institutions, as well as empowerment and protection of local economies. The socio-economic burdens placed on individual households and families due to mismanagement and misappropriation of the common good contribute to the inequality within, and this is responsible for so many related factors that weaken the general structure and make it porous to constant instability—social, economic and political. There are available laws and proposed policies capable of combating the adverse effects, but the policies of the majority of her successive powerholders since independence have been corrupt and weak.

Goal 11: Make Cities and Human Settlements Inclusive, Safe, Resilient and Sustainable

Goal 11 focuses on the basic need for shelter and the important role that security and the general protection of lives and properties play in any approach to development. Emphasis is on the exploding transformation of human settlements into quasi and megacities due to economic, commercial and industrial factors. On the other hand, Goal 11 recognises the direct consequences of such transformation, especially concerning energy consumption and ambient air pollution due to industrial activities. Urbanisation, as implicit in Goal 11, must be structural if it is to be sustainable. It must also be inclusive and affordable if our world would ever want to bridge the gap between the *haves* and the *have-nots*. The salient observations of Goal 11 are thus:

GOAL 11

Half of humanity – 3.5 billion people – lives in cities today and 5 billion people are projected to live in cities by 2030; 95 per cent of urban expansion in the next decades will take place in developing world; 883 million people live in slums today and most them are found in Eastern and South-Eastern Asia; The world's cities occupy just 3 per cent of the Earth's land, but account for 60-80 per cent of energy consumption and 75 per cent of carbon emissions; Rapid urbanization is exerting pressure on fresh water supplies, sewage, the living environment, and public health; As of 2016, 90% of urban dwellers have been breathing unsafe air, resulting in 4.2 million deaths due to ambient air pollution. More than half of the global urban population were exposed to air pollution levels at least 2.5 times higher than the safety standard.

Development Indicator—Goal 11—Specimen: Nigeria

The Nigerian system still develops the real meaning of urbanisation and the basic need for development plans. Difficulties arise when and where the acceptable development plan has been hijacked by corrupt political thugs who choose to allot themselves undue and illegal privileges out of the public good. To do this, they embark on all sorts of nepotism, bribery and corruption so as to maximise profit. The consequences are illegal citing of buildings—residential as well as commercial; inflation of basic acquisition rent and the consequent exclusion of the financially weak majority; unapproved architectural plans; extortion, etc. Improper and poorly structured waste management remains the bane of many of Nigeria's urban settlements. Improvement is still only sporadic. The general community consciousness and awakening of interest within the Nigerian setting need re-orientation and revitalisation.

Goal 12: Ensure Sustainable Consumption and Production Patterns

The gradual degeneration to a near culture of waste amidst the growing situation of hunger and impoverishment of millions of our world's populations is the focus of Goal 12. The Goal seeks the partnership of policymakers in both government and non-governmental organisations to combat human exploitative grab and impact on nature, as well as to tell our world that the present system of consumption is gradually pushing our age to the red lines. The target would be a new orientation towards a responsible approach to nature and the need for people to live a more responsible lifestyle. The main areas of concern are energy, water and food, and the key factors in Goal 12 are the need for conservation, moderation, regeneration, substitution, subsistence, protection, stewardship and equitable balance between "***housekeeping***" (economics) and "***nature-keeping***" (ecology). Observable facts within the focus of Goal 12 are as follows:

GOAL 12

Should the global population reach 9.6 billion by 2050, the equivalent of almost three planets could be required to provide the natural resources needed to sustain current lifestyles; With rises in the use of non-metallic minerals within infrastructure and construction, there has been significant improvement in the material standard of living. The per capita "material footprint" of developing countries increased from 5 metric tons in 2000 to 9 metric tons in 2017; 93% of the world's 250 largest companies are now reporting on sustainability; **Water** – Less than 3 per cent of the world's water is fresh (drinkable), of which 2.5 per cent is frozen in the Antarctica, Arctic and glaciers. Humanity must therefore rely on 0.5 per cent for all of man's ecosystem's and freshwater needs; Humankind is polluting water in rivers and lakes faster than nature can recycle and purify; More than 1 billion people still do not have access to fresh water; **Energy** – If people worldwide switched to energy efficient lightbulbs, the world would save US$120 billion annually.... **Food** – While substantial environmental impacts from food occur in the production phase (agriculture, food processing), households influence these impacts through their dietary choices and habits.... Land degradation, declining soil fertility, unsustainable water use, overfishing and marine environment degradation are all lessening the ability of the natural resource base to supply food; The food sector accounts for around 30 per cent of the world's total energy consumption and accounts for around 22 per cent of total Greenhouse Gas emissions, etc.

Development Indicator—Goal 12—Specimen: Nigeria

Nigeria, as a developing nation and home of valuable natural resources, has peculiar challenges with resource production. With reference to the consumption of energy, food and water, the Nigerian case is still the case of underproduction of energy—at the moment, still fighting to stabilise a meagre sum of 6000 megawatts for a population of over 200 million people. Household and public water supply lie dormant among the list of the dividends of Nigeria's democracy. In the Nigerian case, it is about the need for sustainable food production, provision and availability for the citizenry, not just sustainable consumption. There would be a need to re-energise and fortify production in the agricultural sector to reduce importation and provide affordability within the indigenous markets. The struggle by millions of the populace for a decent daily living is still a challenge to the Nigerian polity.

Goal 13: Take Urgent Action to Combat Climate Change and Its Impacts

Goal 13 is the global call on our age to unite efforts, principles, policies and actions to tackle the reality of climate change. The key focus is to curb extreme dependence on fossil energy so as to combat global warming and its impact on the planet Earth. The prospect is a responsible radical shift to renewable energy possibilities and a revolutionary approach to cubing and reducing drastically extreme industrial emissions to the environment. The key observations are as follows:

GOAL 13

As of April 2018, 175 parties had ratified the Paris Agreement and 168 parties had communicated their first nationally determined contributions to the UN framework convention on Climate Change Secretariat; As of April 2018, 10 developing countries had successfully completed and submitted their first iteration of their national adaptation plans for responding to climate change; Developed country parties continue to make progress towards the goal of jointly mobilizing $100 billion annually by 2020 for mitigation actions; Thanks to the Intergovernmental Panel on Climate Change we know; From 1880 to 2012, average global temperature increased by 0.85°C – To put this into perspective, for each 1 degree of temperature increase, grain yields decline by about 5 per cent. . .Oceans have warmed, the amounts of snow and ice have diminished and sea level has risen – From 1901 to 2010, the global average sea level rose by 19 cm as oceans expanded due to warming and ice melted. . .. Given current concentrations and on-going emissions of greenhouse gases, it is likely that by the end of this century, the increase in global temperature will exceed 1.5°C compared to 1850 to 1900 for all but one scenario – The world's oceans will warm and ice melt will continue. Average sea level rise is predicted as 24 – 30cm by 2065 and 40-63cm by 2100, etc.

Development Indicator—Goal 13—Specimen: Nigeria

Nigeria is a signatory to the most recent partnership of nations *en route* to fighting the growing challenges of climate change. The greatest challenge confronting Nigeria in this direction is in connection with her energy, oil and gas sectors of the economy. The dual problems of the unending story of gas flaring in the Niger Delta and the increasing air and environmental poisoning due to compulsive private power generation through combustible power generating sets make the focus of Goal 13 emergent in Nigeria. While the situation is not the worst for Nigeria compared to the more industrialised areas of the world, it is still needful for Nigeria to integrate into her policies scientific, technological and anthropological approaches that would harmonise and form an alliance towards effective combat against climate change.

Goal 14: Conserve and Sustainably Use the Oceans, Seas and Marine Resources for Sustainable Development

The marine resources of our world are essential components that serve human, scientific, technological and biological purposes. Goal 14 uses the word conservation to emphasise the need for global action towards protecting the marine capitals of our world from extreme and adverse human activities. In other words, cubing the wave at which human socio-economic, scientific and technological activities impact marine life remains one of the prerequisites for sustainable marine life and economy. The fundamental observations of Goal 14 are as follows:

GOAL 14

Oceans cover three quarters of the Earth's surface, contain 97 per cent of the Earth's water, and represent 99 per cent of the living space on the planet by volume; Over three billion people depend on marine and coastal biodiversity for their livelihoods; Globally, the market value of marine and coastal resources and industries is estimated at $3 trillion per year or about 5 per cent of global GDP; Oceans contain nearly 200,000 identified species, but actual numbers may lie in the millions.... Marine fisheries directly or indirectly employ over 200 million people; Subsidies for fishing are contributing to the rapid depletion of many fish species and are preventing efforts to save and restore global fisheries and related jobs, causing ocean fisheries to generate US$50 billion less per year than they could; Open Ocean sites show current levels of acidity have increased by 26 per cent since the start of the Industrial Revolution; Coastal waters are deteriorating due to pollution and eutrophication. Without concerted efforts, coastal eutrophication is expected to increase in 20 percent of large marine ecosystems by 2050.

Development Indicator—Goal 14—Specimen: Nigeria

The intended prospects of Goal 14 have the urgent need for integration into the Nigerian system of resource exploration. Nigeria's crude oil exploration is both offshore and on-shore. Offshore exploration has led to various marine oil spillages and the hazards they pose to marine life. The impact on the social and community life of the citizenry is obvious since the threat to marine life and the consequent migration of marine economic capitals impact heavily on the local economy of the people, who, as a consequence, must battle with fish migration, water poisoning and other health hazards. The on-shore area of oil exploration has imposed both environmental hazards and poisoning of drinking water sources due to the spilling of oil and other chemicals being injected into the soil. In all, therefore, the development indicator in Nigeria meets a Nigerian system full of rhetoric on how to remedy the situation and make the prospects of Goal 14 a reality, but at the same time, fails to make concrete impacts to salvage the extent of the devastation.

Goal 15: Protect, Restore and Promote Sustainable Use of Terrestrial Ecosystems, Sustainably Manage Forests, Combat Desertification, and Halt and Reverse Land Degradation and Halt Biodiversity Loss

Conscious of the role the ecosystem plays in the general symbiosis among living organisms, Goal 15 focuses on the need for a more responsible human approach to economic trees and the general use of land. Extreme and unregulated felling of trees, as well as the current incessant desertification of the forests and farmlands due to wildfire disasters caused by extreme weather conditions due to climate change, build the nucleus of concerns for Goal 15. The targeted prospect is the unity of efforts on the part of world leaders and stakeholders to protect biodiversity and all endangered species—plants and animals—through functional policies and laws that would guard human commercial and economic activities. The key factors in Goal 15 are:

GOAL 15

Forests – Around 1.6 billion people depend on forests for their livelihood, including 70 million indigenous people; Between 2010 and 2015, the world lost 3.3 million hectares of forest areas. Poor rural women depend on common pool resources and are especially affected by their depletion; **Desertification** – 2.6 billion people depend directly on agriculture, but 52 per cent of the land used for agriculture is moderately or severely affected by soil degradation; Within one year, 20 million tons of grain could have been grown; 74 per cent of the poor are directly affected by land degradation globally; **Biodiversity** – Illicit poaching and trafficking of wildlife continues to thwart conservation efforts, with nearly 7,000 species of animals and plants reported in illegal trade involving 120 countries; Fish provide 20 per cent of animal protein to about 3 billion people.... Over 80 per cent of the human diet is provided by plant. Only three cereal crops – rice, maize and wheat – provide 60 per cent of energy intake; As many as 80 per cent of people living in rural areas in developing countries rely on traditional plant-based medicines for basic healthcare; Micro-organisms and invertebrates are key to ecosystem services, but their contributions are still poorly known and rarely acknowledged, etc.

Development Indicator—Goal 15—Specimen: Nigeria

According to World Bank 1995 report, the Niger Delta area of Nigeria holds the third largest mangrove forest in the world and the largest in Africa (Ugochukwu and Ertel 2008). This part of the Nigerian state is home to a rich concentration of *flora* and *fauna.* The tidal actions of the Atlantic Ocean provide creeks of water as well that make fishing lucrative and support the local economy. However, the whole area and the general biodiversity stand deeply endangered due to crude oil exploration and its negative impacts. Implementation of Goal 15 in the Nigerian situation would mean applying the political will that makes functional remediation possible. A variety of approaches have been tried out before now. Nevertheless, each of them has always met the Nigerian greatest negative syndrome, namely corruption. It remains, therefore, a persistent political and economic dilemma if the nation fails to strike the balance between a *progressive economy and sustainable ecology*.

Goal 16: Promote Peaceful and Inclusive Societies for Sustainable Development, Provide Access to Justice for All and Build Effective, Accountable and Inclusive Institutions at All Levels

Development can never be realistic and sustainable without the necessary structural, infrastructural and institutional elements, which serve as catalysts for the prospects and principles that go into development rhetoric. Such elements must also be designed to meet all principles of social justice; in other words, they should be inclusive, effective and accountable. This is exactly the focus of Goal 16. Its prospect is embedded in the call for respect for the rule of law and conscious resolve for and united efforts towards an inclusive socio-economic society. Observations are as follows:

GOAL 16

Among the institutions most affected by corruption are the judiciary and police; Corruption, bribery, theft and tax evasion cost some US $1.26 trillion for developing countries per year; this amount of money could be used to lift those who are living on less than $1.25 a day above $1.25 for at least six years…only 46% of Sub-Saharan Africa have had their births registered; Approximately 28.5 million primary school age who are out of school live in conflict-affected areas; The rule of law and development have a significant interrelation and are mutually reinforcing, making it essential for sustainable development at the national and international level; Violence against children affects more than 1 billion children around the world and costs societies up to US$ 7 trillion a year; 50% of the world's children experience violence every year; 1 in 3 internet users worldwide is a child and 800 million of them use social media. Any child can become a victim of online violence; Child online sexual abuse reports to NCMEC has grown from 1 million in 2014 to 45 million in 2018; etc.

Development Indicator—Goal 16—Specimen: Nigeria

With the development indicator, it is to be discovered that Nigeria has many forms of inequalities that challenge the general structure: the widening gap between the rich and poor; the imposed oligarchy from the political class; the oppressive nepotism in public services; the never-ending influence of the *godfathers* and self-acclaimed *kingmakers*; etc. These forms of inequalities kept the nation in constant rift within her general structure shortly after independence. These have caused national wars and rivalries as well as impacted economic recessions. To date, there exist varied forms of division within the general citizenry—each struggling for inclusion and emancipation. While the nation grapples with the reality of her population boom, the oppressive rulers and deceptive policymakers manipulate the whole system to keep themselves unruly in power and persistently exploit the masses economically. The proposed prospects of Goal 16, therefore, still await fruition within the Nigerian state, and nothing else would be more effective than the masses' resolve to carry on organised non-violent struggles and revolutions against all forms of social bondage existent in the general system.

Goal 17: Strengthen the Means of Implementation and Revitalise the Global Partnership for Sustainable Development

Goal 17 recognises, on the one hand, the need to follow-up on proposed policies and prospects of development through forms of empowerment for the proposed means towards implementation. On the other hand, the measure of economic strength, which remains differential among the nations of the world, makes it invariably necessary to form partnerships among the nations for a united approach to sustainable societies. In such partnerships, the weaker economies would be helped to achieve stronger stability through the relational efforts of the stronger economies. In such partnerships, the resolve to approach development more inclusively and holistically would place responsibility on the more privileged nations to help combat the challenges of the less privileged ones, especially those naturally landlocked from socio-economic activities due to geographical factors. Observable facts in Goal 17 are as follows:

GOAL 17

Official development assistance stood at $146.6 billion in 2017. This represents a decrease of 0.6 per cent in real terms over 2016; 79 per cent of imports from developing countries enter developed countries duty-free; The debt burden on developing countries remains stable at about 3 per cent of export revenue; The number of Internet users in Africa almost doubled in the past four years; 30 per cent of the world's youth are digital natives, active online for at least five years; But more four billion people do not use the Internet, and 90 per cent of them are from the developing world, etc.

Development Indicator—Goal 17—Specimen: Nigeria

The mainstream process towards strengthening the means of implementation of sustainable development goals, especially for developing and undeveloped economies, remains finance. Nigeria, in varied forms, has also been one of the beneficiaries of such empowerment. The development indicator, with a focus on Nigeria and her socio-economic and political situation, is rather critical. This is because Nigeria has voted so many millions and billions of dollars into developmental and infrastructural projects yet remains poorly graded on the international barometer. The country is a known debtor in the international community and relations. However, such borrowings have failed to lift Nigeria out of persistent poverty and economic instability due to poor governance and corruption. Perhaps, the general international rhetoric of "*Help for Africa*", added to the antecedents of colonialism, has remained the ideological and psychological bane of making development sustainable in the global South. But Goal 17 chose the right word and approach with the introduction of the word Partnership, which gives the sense of the union and share of responsibility between and among agents and institutions in a mutual relationship with one another. The concept of *help*, in the context within which we make a critique of the Nigerian situation, already creates the feeling of subordination and unconditional dependence on the "provider of help". In consequence, therefore, many of the developing and underdeveloped economies, mainly in sub-Saharan Africa, have failed to translate decades of "help for Africa" into concrete and effective means for socio-economic transformation. Thus, the focus of Goal 17 is not just a mere financial investment to *aid* the weaker economies but rather the establishment of forms of partnership that place responsibility on and demand responsibility from the *relating-partners*. That is an urgent demand in the approach to the Nigerian case.

3.5.2 New Strategic Management for the SDGs: Breaking the Chains of Colonial Development Thoughts

The SDG agenda is, in the context of our contemporary world, both an integral and complex project. Therefore, there is a need for a designed strategic management of the project in its holistic, inclusive and participative form. Till now, the fiercest criticism of this project pertains to the governance and politics of leaders and actors surrounding the project. In the opinion of Sachs, the very structure of the United

Nations reserves the power to design qualitative ideas and policies for purposeful governance. However, the practical flow of those ideas into daily real life and experiences has been known to be constantly deficient. With reference to Agenda 2030, Vogt observed that one of the greatest obstacles has been disunity among actors in the governance of the agenda. Thus, the declamatory self-commitment of all governments to the full implementation of the 2030 Agenda always suffers the blatant contradiction in the hands of the same governments, who most often pursue their individual selfish interests to the detriment of the common agenda. This can be classified as a mode of "simulative democracy" or—somewhat more soberly—as a political dilemma in the field of tension between national and international challenges (Vogt 2021).

Our research, therefore, sees the need to emphasise the shift of narrative, which must develop alongside practical strategic management approaches to close the gap between rhetoric and actuality. There exist already many theoretical ideas, proposals and political promises, yet our world remains trapped in the same obsolete model of doing things. The new strategic management for the sustainable development goals should be a turnaround away from the *colonial* form of development thoughts to a *post-colonial* market, political and socio-economic liberalisation ideology, which builds more on the ideas of inclusion, participation, solidarity, subsidiarity and sustainability. Without this transition, the status of the world's international relations would remain alien to the contemporary urgency of development concepts that would be able to engage indigenous languages, worldviews and cultural identities (Ibid., p. 529). We would elaborate on this in the concluding chapter with reference to Pope Francis's appeal for a cultural revolution. However, the character and mode of the strategic management must be *remedially post-colonial and de-colonial*. It would integrate, in a reformed manner, the three constitutive aspects of development, namely chronopolitical, geopolitical and civil-political aspects.

3.5.2.1 Post-Colonial Development Thought as Chronopolitical

To be *chronopolitical*, the new strategic management of the development process of our world must make a shift from a mere embrace of control, power, natural resources or profit maximisation to a system of administering development policies to reflect the temporal conditions of human existence (Chronopolitics et al. 2012), in other words, to reading and managing the signs of the times and conditions within which both the agents and means of development narratives live. The basic idea here is a focus on those efforts that would bring contemporary politics to a linear model of development policies, where the various actors receive a form of relative but simultaneous participation rights, which provides an equal platform for relationships at all development levels.

As a thought model, chronopolitics is fundamentally rooted in the general belief in progress—a form of Christian eschatology of development (cf. Ibid., *Christliche Umweltethik, Grundlagen und zentrale Herausforderungen,* p. 530)—not really based on the concept of the hereafter, but rather the idea of development progress towards fulfilment (*German: Vollendung*) that is to be achieved through participative relational progress of the world's societies in respect of common good. Colonial development thoughts, therefore, contradict chronopolitics. This is due to its character of *allochronism*, whereby the disparity seen among the contemporary actors in the development governance singles out in man-made categorisations some sections of the "other" and designates them with concepts connected to forms of binary temporal attributes that are remote from the user's present (worldview, actuality). To this effect, contemporary development concepts and policies still battle the (negative) *bloc ideologies*, where ascriptions hewed out of previous historical experiences of our globe now aid the development categorisation of societies into "[...] modern, backward, primitive (or) advanced [....]" (Chronopolitics 2013). The imposed forms of derogatory bloc ideologies have become complex and synergistic in the hands of actors and groups of actors always working against any proposed form of decentralised development ideas and, to this end, only an ethically based and applied post-colonial chronopolitics could attempt a change. Its main characteristic would be to bring the agent of development himself or herself to his or her present world, placing him or her under an equitable platform of approach where only the category of humanity becomes the criterium.

3.5.2.2 Post-Colonial Development Thought as Geopolitical

While post-colonial chronopolitics would try to prove that development politics should, as a demand, play the politics of time change and transition, geopolitics demands review and redefinition of the contemporary geopolitical approach to development. One of the greatest banes to our world's development process and approach is the geopolitical ideological impositions which have contributed to splitting our world into the categories of *Donors* and *Recipients*. In this way, the international relationships among various actors and policymakers have always proved to be the encounter between the *sole provider* and the *sole consumer*, giving rise to superiority-inferiority complexes. Such a form of development politics has been described as a relic of colonial experiences (cf. Ibid., *Christliche Umweltethik, Grundlagen und zentrale Herausforderungen,* p. 530), which today has become geopolitically systemic and implicit in the contemporary imperial forms of commitment to global markets and economic growth.

Until more recently, the concrete result became the fact that development ideologies were regarded as the sole properties of the then-colonial masters, which must guide and shape the approaches of the colonised territories, even with disregard to unique cultures, identities and worldviews. Against these development odds, post-colonial development thought, which is understood as new strategic management for a restructured development geopolitics, looks at the Sustainable Development Goals as a world agenda for relative but simultaneous development of humanity and not just a designed project for the emancipation of a designated section of the world where nothing good comes from except hunger, disease, crime, illiteracy, high mortality rates, etc. Our research sees such geopolitical fixations as *persistent imperial impersonations* ravaging the so-called underdeveloped and developing economies.

3.5.2.3 Post-Colonial Development Thought as Civilisation Policy

The concept of *indigenous people* has gradually proved to be fundamental to the general development process of our world. Given that the core colonial ideology was dominated by socio-political, socio-economic and even socio-cultural impositions, authentic civilisation policy eluded decades of encounters and policy implementations, especially in the lands and soils that provided habitation for visiting cultures, civilisations and ideologies. The modern approach to the development process, therefore, is a witness to the gap created by such non-engagement, non-dialogue and, much later, the non-indigenisation and/or purification of colonial and imperial experiences. To this effect, a post-colonial development thought taken as a civilisation policy would be a bridge for such a gap, whereby the liberalisation and globalisation of the socio-economic tracks of our world should better speak the language of fair play, equity, non-dominance, healthy competition, just negotiations, etc. It means that when post-colonial development policies approach the SDGs with a civilisation spirit, development polity itself would be shaped through the culture, language and identities of the people for whom it is meant. A reversal creates opportunities for detractions, oppressions and impositions, and, following the opinion of Sachs, such reverse situations may only be preparing development processes for their inevitable collapse (Sachs 2021).

Furthermore, the envisaged civilisation policy in approach to the SDGs for development processes is to be understood as a socialisation paradigm, in other words, a culturally based project of dialogue and, following the opinion of Michael Ley, a dialogue towards the socialisation of civilisation (*Zivilisierung der Zivilisation*) (Lay 2005). The very act of socialising the development goals is a way of inculturation and enculturation. Strategic management as inculturation for the SDGs is first a recognition of the cultural diversity of the actors within the development process. Secondly, it would mean the preparedness and the determination to appreciate uniqueness and respect boundaries. Again, strategic management as enculturation for the SDGs is the diffusion of the development plans and policies through the indigenous and cultural agents of transmission—the family/parents, the educational institutions, peer groups, and now almost inevitable, through socio-culturally driven social media. This approach would help to save the SDGs governance from the rejection it suffers as a project, especially in the minds of stakeholders in underdeveloped or developing economies, namely that the development concept has been long dominated by Eurocentrism. Consequently, all social, economic and cultural forms of Europe, added to the colonial standards bequeathed to the respective former colonies, were all projected as decisive norms for global progress and development projects (cf. Ibid., *Christliche Umweltethik, Grundlagen und zentrale Herausforderungen*, p. 531). This negative feeling has contributed to one of the obstacles to SDGs implementation and governance in many countries and administrations.

However, our research suggests that constant bemoaning of the development governance model would perpetually stifle the needed aspirations for possible transformative projects. The development strategies proposed in the SDGs should always be read as objective blueprints capable of dialogue with every culture, convention and worldview. An independent working group in connection with the Brundtland project observed:

> The 2030 Agenda for Sustainable Development and the Paris Agreement on climate change that was adopted in the same year, are tangible proof of the benefits of multilateralism and the indispensable role that the United Nations can play to find global solutions to global challenges. Their implementation offers a pathway to a world where poverty, inequality and conflict will not blight the life chances of millions of people who are currently denied the opportunity to enjoy their fundamental rights and freedoms. (Ibid., p. 530)

Administrations, institutions and governments are unanimous in acknowledging that there are general constraints and conditionalities associated with the SDGs project. Nevertheless, the best approach that our research would eventually rely on would be *to consider constraints as inherent challenges that can be confronted with positive, indigenised and structuralised approaches*. Rejection and withdrawal would be counterproductive. Participation and critical indigenisation of the known development principles would pave the way for a relatively more stable global community.

Evaluation: From Development Polemics to Concrete Actions

Despite our projection of a hopeful and workable future of our world of today, varied opinions show discontent with our modern polemics with development policies. The modern world is a witness to conferences, where heated debates and planned actions are often further stifled by the failure of so many governments, leaders and institutions to translate policies into concrete results. In the face of such massive indolence—from the government down to the individual human agents—there is the need for a socio-cultural project that would legitimise the normative orientations needed for proper actions that would translate policies to people-oriented facts. In other words, there is an urgent need for sustainability ethics which would provide the normative guidelines for the proper approach to development.

3.6 Sustainability Ethics: Towards a Functional Socio-Ethical Approach to Development

To propose ethics for sustainability is to seek what should guide competence towards the fundamental values and attitudes, which social policies and economics must apply to bring actions to positive orientations. Such positive orientations are the instruments for empowerment and transformation. They are drivers for holistic change towards socio-economic and socio-political integrity for all. They must be competent enough

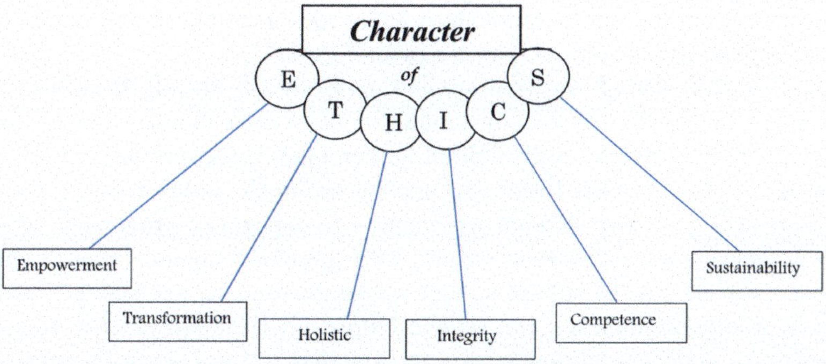

Fig. 3.5 The nature and character of ethics. *Text adapted from Christoph Stückelberger, Globethics.net President and Founder Globethics.net Newsletter No.1/2018. *Illustration mine*

to achieve sustainability. This is, according to Stückelberger, the basic character of ethics as represented below (Fig. 3.5).

3.6.1 Ethics: Its Objectives and Goals

Ethics, in general, is both a prescriptive and normative approach to the question of the moral choices we make or have to make to promote authentic life. In other words, it is a moral tool of analysis applied in relation to the human agents, their decisions, actions or omissions. Decisions and actions that enhance our humanity help to make our lives authentic. The opposite makes us inauthentic. The objective and goal of ethics, therefore, is to help establish principles that would help make authentic life possible and practicable.

In Christian parlance, such a state of authentic life is intended with the ancient Hebrew concept of *shalom* or peace—wholeness or health—which is fundamentally the will of the creator for the harmonious existence of the self in relation to others and the whole of creation (Kansas State University 2019). But this state of harmonious relationship does not exist automatically, nor is it a natural characteristic of human society. It is only a possibility and an ideal, which could be realised through combined related factors. Thus, the question about why we do not find ourselves living automatically in this peaceful, paradisiacal situation receives only one answer, namely: *because we have made and we still make bad moral choices* (cf. Ibid). On the other hand, for the Greeks, ethics should aim at the enthronement of the state of *Eudaimonia,* meaning *flourishing or happiness*. However, in contrast to the Hebrew approach, the source of such a eudemonic state is nothing outside human reason—not the will of God or the Creator. In other words, the human agents have the power to reason and apply ways and means capable of establishing Eudaimonia. The fundamental goal of ethics, therefore, "is defined as the search to identify those human choices, which will promote authentic humanity. And authentic humanity is defined as humans living in the right relationships with self, others and the world. In such a world, humans flourish and find their happiness" (Ibid).

It must be emphasised that such ethical harmony could only have the chance to exist when the need for common ground is given concrete attention. Efforts to establish a common ground in the field of human knowledge have been made in the past and have led to various forms and approaches in the field of ethics. Concrete examples are (1) *The sentimentalist approach*—An appeal to common human moral feelings or sentiments rather than reason. This approach assumes that the human sentiments of kindness, friendliness, honesty, gentleness, etc., would be chosen by most over cruelty, dishonesty, etc. (David Hume 1711–1776); (2) *The rationalist approach*—An appeal to human reason to establish a "categorical imperative", in other words, what we have the duty to do, no matter what the consequences are. This "categorical imperative" included a demand to act only on principles that could be made universal. Also, treat people as ends in themselves and never as means to our individual ends. (Immanuel Kant 1724–1804); (3) *The utilitarian approach*—An attempt to base ethics on utility or "the greatest happiness of the greatest number" defined in terms of pleasures and pain common to all human beings. (Jeremy Bentham 1748–1832 and John Stuart Mill 1806–1873); *The contractarian approach*—An appeal made to a social contract that all reasonable persons, despite their differing values, could agree upon. (Thomas Hobbs 1588–1679, John Locke 1632–1704, Jean-Jacques Rousseau 1712–1778 and John Rawls 1921–2002) (cf. Ibid).

The proponents of the natural law theory would later hold human nature and human reason as the two components that can build the basis for the establishment of common ground, where people can discover their true humanity together and establish principles that would help it to flourish. These two components were and remain fundamental to the core concept of ethics. As seen in Aristotle, they are the two assets, namely (a) the rational asset, which aids the human person in discerning the natural world and its structures, and (b) the human natural asset, which empowers the ability to strive for something and triggers the corresponding action towards achieving the goal. For Aristotle, therefore, the human person is at best when he or she is able to actualise these two assets, leading to an ideal life, in other words, life in the commonwealth—an actualisation of the ethics of life (Heimbach-Steins 2004).

Our research uses Christian social ethics to confront the dilemmas encountered with theories and the applications of sustainable development concepts. In this way, therefore, it proposes the sustainability principle as a fundament for establishing the ideals of human life in general and applying those ideals for a harmonious society. In doing this, Christian social ethics must be competent enough to, first, show that its method and rational reflections are philosophic enough to confront the societal realities; secondly, it must be able to demonstrate accountably its place and character in theological justifications; and thirdly, it must prove its competence with regard to the analysis and ethical assessment of societal institutions and processes and dynamics (cf. Ibid., *Christliche Sozialethik, Ein Lehrbuch, Band 1, Grundlagen*, p. 8).

3.6.2 The Role of Christian Social Ethics as a Research Tool

As a discipline, social ethics develops ethically based social principles, which it uses as tools to engage the social world on all its levels—social, political, economic and cultural. The objective is twofold, namely (a) to make the church and faith both ideologically and concretely present in respective social situations and experiences and (b) to develop, implement and live the faith authentically within the social realities of our world's time and spaces (Sander 2013). Christian social ethics, therefore, should develop as a method, not just a reflection on the pure principles of ethics and morals but rather the bridging of such principles with the social realities through practical and action-based normativity and orientations. The method means that the developed principles remain only relevant when they become principles for *something* and not for *nothing*. In other words, the objectivity of the ethical social principles in Christian social ethics must be competent enough to engage the social realities of both humanity and the world that engage them. They should be competent enough to function as buildup structural principles for the society (Nell-Breuning) and/or be able to provide the open society with the appropriate socio-scientific grammar (Vogt) in the effort towards developing befitting and sustaining structure (Heimbach-Steins 2022).

Furthermore, the general effort would be to develop criteria and strategies that would aid the Christian social teaching which is neither "[…] a (mere) bundle of practical instructions for the solution of social questions nor a skilful selection of certain findings of modern sociology useful for Christian social training, but "an integral component of the Christian doctrine of man" (Mater et Magistra)" (Höffner 1997). The reference here is to the whole man, the imago Dei, the human person (Mmadụ), who "[…] is related to the interpersonal other and the community" (Ibid). The tool, in other words, the principle of Christian social ethics, therefore,

> Over against a widespread of supranaturalism […], emphasises that even after the Fall there is an order of common social life grounded in the social nature of man and thus divinely willed. This 'social order, its restoration, and its perfection according to the saving plan of the good news' (*Quadragesimo anno*), the 'formation of it in the light of Christian teaching' (*Mater et Magistra*), is the object of Christian social teaching. (Since) God has not left the fallen age to his adversary. (Ibid)

This nature and character of the social principles taken as tools for Christian social ethics should, therefore, function under three categories, namely as tools of analysis, organisation and justification. These categories would accompany the Christian social mandate and its ethics within the known conditions of the human person (Mmadụ) in his or her inter- and intra-relational self within the community of existence, where *social circumstances* could divert and pervert the will of the creator for creation itself, namely eternal salvation. The image of *eternal salvation,* in the understanding of our research, is not identifiable with some form of salvific exoteric ecstasy but rather with the prime objective of Christian eschatology, which begins in an integrative process of a life lived in goodness, justice, love, social well-being and social sustainability. It is the hungry person who is fed now; the thirsty whose thirst is assuaged; the stranger who is loved and sheltered; the naked and vulnerable who is cared for; the sick and wounded who is socially, economically and psychologically bandaged and cured; the unjustly oppressed whose cries and pains must be heard and responded to, etc. (cf. Mt. 25: 25–36). Höffner painted another insightful picture of this truth and deserves an extensive reference here:

> An unsettling importance with respect to our salvation attaches to social conditions as a result of our humbling dependence on the given milieu because 'men are often diverted from doing good and spurred toward evil by the social circumstances in which they live and are immersed from their birth' (Gaudium et spes, 25). The perversion of the divinely willed order 'is now such as to put in the way of vast numbers of mankind most serious obstacles which prevent them from caring for the one thing necessary; namely, their eternal salvation' (Quadregesimo anno, 130). Conditions adverse to salvation, such as the levels of poverty in some developing countries, are scandals that cry out for redress, and not only in the form of critique and alms, but also in the form of a new social order that accords with the principles of Christian social teaching. The hard and passionate struggle against poverty, hunger, disease, misery and need is a Christian duty. Premature resignation would not be a Christian surrender to the will of God, but a fatalistic quietism that would bring the reproach of being 'an opium for the People' upon the Christian faith. (Ibid)

The social principles as ***analytical*** *(Analysekategorien)* refer to their competence as tools for identifying social structures and systems calling for change or replacement (cf. Ibid., *Sozialprinzipien*, p. 185). In the area of technology and scientific innovations, the principles help to analyse and suggest relevant *exnovations* for more *sustainable innovations*. The general approach here is through a planned, concrete and focussed perception of social reality within which the observer tries to reflect and evaluate a social principle taken as a measuring tool. The social principle becomes the reference point upon which a probable problematic situation or a given fact would be critically and ethically evaluated. To this end, the analytical function provides a gateway for a mutual diffusion of action and orientation between socio-scientific and socio-ethical approaches to social challenges (Ibid). The expected outcome would be the transformative insight into the systems and behaviour of our modern societies and the gaining of better orientational knowledge for better ethical approaches (Ibid) (Fig. 3.6).

Secondly, the ***organisational*** *(Ordnungskategorien)* competence of the social principles should offer criteria for the purposeful administration of all proposed

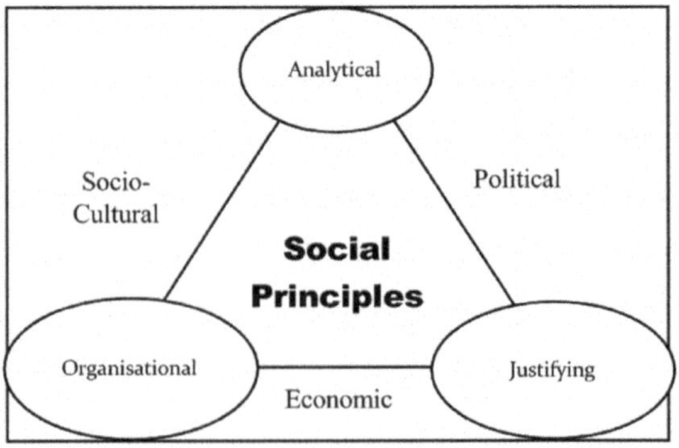

Fig. 3.6 Functional character of the social principles. *Illustration mine*

social changes, which must not be taken just as situations associated with fate but rather as realities that require concrete political and economic decision-making (Ibid). They should introduce corresponding frameworks for developing norms that should aid the formation and evaluation of existent or proposed social institutions. Here, the organisational character of the social principles would eventually create a mutual diffusion of knowledge, especially between social ethics and other practical domains of law and politics (cf. Ibid). The objective would be to apply ethical questions in both general and particular social and institutional givens for the purpose of making social norms and legislations human and adaptable.

Thirdly, the developed criteria of the social principles should be as critical as they are ***justifying*** *(Begründungszusammenhangs)*. However, the competence to offer justification is not to take the status of command and imposition but rather a platform for further dialogues, modifications, implementation and re-implementation. For this reason, the certification and the conviction that the principles prove competent enough to justify opens, therefore, the continuous dialogue between CSE and social philosophy (Ibid., p. 186). In this way, the social principles keep their analytic and organisational competencies constantly renewable, debatable, verifiable, as well as innovative.

These three categories give Christian social ethics its functional character, and, as Heimbach-Steins opined, it would be insightful to get the formation of the social principles strictly identifiable with the social changes they are meant to provide criteria of approach. Consequently, the consciousness of the context of formulation helps to purify better how to formulate and for what reason is there to do so; it would help to direct properly the philosophical and theological tools to better dialogue with and engage the ever-changing and challenging dynamics of our modern societies. In the end, the whole project of developing and applying criteria would mean the systematised efforts towards making social structures and institutions more sustainable, more subsistent and more suitable for the central agent of all analytical, organisational and justification-based decision-making, the human person, the Mmadụ. A further elaboration under "sustainability ethics" is to give a more precise reflection on the core social principles that lead our research on its way to contributing and participating in scholarly, literary and practical applications of measure towards the contemporary project of sustainable development discourses, processes and movements.

3.6.3 Sustainability Ethics: General Understanding

Sustainability ethics used as a concept and in our context calls for the need to rediscover those normative values that guide and place obligations on the human agent within the general network of interaction in our world. In other words, sustainability ethics focuses on the need to place importance on those ethical values that give the right orientation not only towards what we seek to sustain but also on how and in what ways the general decisions and approaches of the human agents work to impact the general network of socio-economic, political and ecological systems. It is the focus on those factors that serve as the platforms for justice, equality, the common good and respect for both human agents and nature itself. This has become increasingly relevant for our time since the authentic spirit of sustainability "...can only be ignited by addressing such profound issues like the meaning of work, the purpose of life and the relevance of ethics and values. Otherwise, we are destined to confront the looming crisis, the signals of which are already amply evident in many countries across the globe along all fronts—social, economic, cultural, ecological and spiritual" (Mukherjee and Stückelberger 2014b). In the final analysis, our research finds the following core principles fundamental in any sustainability project that is worth the name.

3.6.4 The Normative Value Orientations: Investing in the Core Values for Sustainability

The general social structure is a network of factors that must work legitimately to ensure stability and harmony. However, the experiences of human societies show incessant conflicts and crises, which lead to the conviction that certain objective principles and values must be contractually obligated, first, to empower the social structure and, secondly, to define ideals and values that must place responsibility on contracting subjects and within the general system. The relevance is for the ideal social values that would be capable of offering authentic legislation and legitimisation without being plagued by the social vicissitudes, which are predominantly man-made. Nevertheless, such principles must be resilient enough to combat the realities of our world order and its social challenges (cf. Ibid., *Christliche Sozialethik: Ein Lehrbuch*, p. 263). These core values in the approach to sustainability we would now examine below.

3.6.4.1 Personality (Personhood)

The human person has the pride of place in every known human culture. The person stands as the central force around which all other aspects of nature revolve. The person lives in and within nature and remains an endogenous part of it. The famous expression about the human being taken as the measure of all things is credited to Protagoras of Abdera and it remains foundational for any ontology of natural existence. Therefore, presenting personality as a core concept for sustainability ethics is to emphasise that all approaches to the dynamics of development must focus on the human person as both the means and the end in itself. As a means, the person remains the driving force and actor towards achieving sustainability. As an end, all normative orientations must be morally plausible and must not violate the sanctity, in other words, the dignity of the human person—both in his or her person or in relation to the givens outside of the human person. The person in his or her personality, in other words, in his indispensability, remains the defining social principle or force upon whom and in reference to whom all normative orientations revolve (cf. Ibid., p. 265).

The indispensability of the *person* for sustainability ethics should and must not be replaced with any other means. Otherwise, this personhood suffers "objectification", which violates his dignity. This personality, therefore, for Kant should never be put up for a price or compromised in all social dynamics at the centre of which the *person* stands as the pivot. Thus, Kant, on the human person, made a distinction between Price and Dignity. For him, every other thing outside of the person can be priced, in other words, such things could be compensated with something equivalent. But the human person has dignity, which is an end in itself (cf. Ibid). It follows that all social structures must be designed in such a way that the *person* retains his or her place as a capable and functional *subject* within the structure and never as an *object* to be commercialised.

Therefore, in emphasising the need for normative guidelines for achievable sustainability, sustainability ethics refer to "personhood" as the point, which is centripetal for all social structures. For this purpose, the Pastoral Constitution, Gaudium et Spes made it clear:

> Man's social nature makes it evident that the progress of the human person and the advance of society itself hinge on one another. For the beginning, the subject and the goal of all social institutions is and must be the human person which for its part and by its very nature stands completely in need of social life. Since this social life is not something added on to man, through his dealings with others, through reciprocal duties, and through fraternal dialogue he develops all his gifts and is able to rise to his destiny […] (it then means that) […] the social order and its development must invariably work to the benefit of the human person if the disposition of affairs is to be subordinate to the personal realm and not contrariwise, as the Lord indicated when He said that the Sabbath was made for man, and not man for the Sabbath. (Vatican II Council 1965d)

It means that since the person is an integral part of the natural order and, in this way, reciprocally related to it, he remains a moral being (subject) capable of moral decisions and self-reflection (self-transcendence) both unto himself and towards the natural realities that are necessary parts of his social existence. Such capability, therefore, places responsibility on him to settle for principles that can only ensure sustainable development of himself and the world around him. Thus, for the sustainable principle to function, the person must be given his or her rightful position as the moral force capable of moral actions (Fig. 3.7).

3.6.4.2 Common Good

Desire and respect for the common good is a normative responsibility that establishes justice between the individual and the society in which he lives. The concept itself encompasses such attributes as Association, Charity, Civil Rights, Collaboration, Community, Democracy, Goodwill, Equality (Equity), Justice, Responsibility and

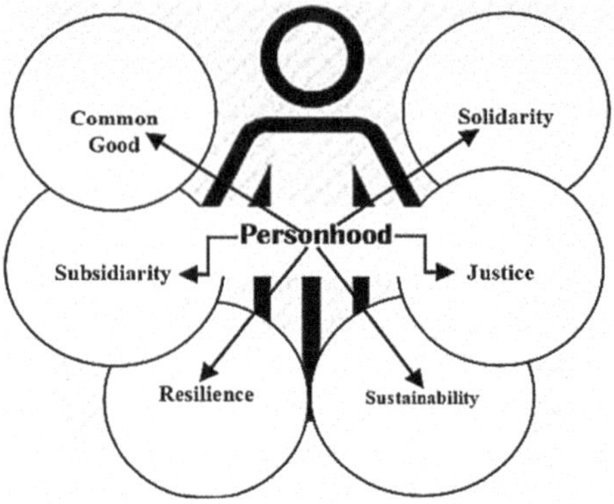

Fig. 3.7 The centrality of the human person within the normative social values. *Illustration mine.* The person remains the moral *SUBJECT* to whom, from whom and around whom all the normative orientations take their meaning and relevance

Sharing. The development of this concept within its legal, utilitarian and social levels is found spanning through the thoughts of thinkers like Plato, who, in his political view on society, proposed, as an ideal, a harmonious unity of the society, where each of its parts is allowed to flourish but never at the expense of others. However, the realisation of such societal order, for Plato, can never be attained without virtue (without ethics) (Nwosu 2016).

Moral responsibility, in reference to the common good, is intrinsically binding and helps to give both the individual and the society the orientation towards common values and objectives. At the same time, the two agents—individual and society—can never function disparagingly. Instead, they must build a symbiotic ethical union so as to keep the general structure functional. The sense of common good establishes the moral basics through norms and institutions working together to find the common ground to accommodate the diverse approaches to values, objectives and schemes of life that the members of a particular society and culture represent (cf. Ibid., *Christliche Sozialethik: Ein Lehrbuch*, p. 271). Opinions have allusions as well to the nature of the common good as the construct of both normative a priori and a posteriori sources. A priori, in the sense of ideal objective goods, perceptible to the human mind but not dependent on the agreement between the society and its members; a posteriori, in the sense of the interplay of historical dynamics and empirical facts determinable through both individual and collective interests oriented towards the good. The union of these dual motivational sources makes the sense of the common good a normative duty for a sustainable human society. For Plato, this builds up the character and the basic objectives of all political and societal alliances, through which the needs, interests and happiness of all citizens could be equitably realisable (cf. Ibid).

With the current focus on climate change and the discussions surrounding it, the theme of the common good deserves more extensive interpretation involving the consequences, which the need to protect the climate places as a duty on states, societies and the people. Discourses on the common good, therefore, call for a renewed reflection on the focus of international law, which, as the signs of the time demand, must change ideas and narratives to better suit the contemporary paths towards transformation. One of these practical paths, following the opinion of Epiney in his *Gerechtigkeit im Umweltvölkerrecht*, should be to treat international law not just as a contract for the right to *coexistence* (Koexistenzrecht), but rather as a shared norm that places responsibility on all and guarantees the right to *cooperate existence, the right of cooperation* (Kooperationsrecht). This broadened understanding of the nature of the common good is to be taken as a new global social contract (Völkervertrag) (cf. Ibid., *Christliche Umweltethik, Grundlagen und zentrale Herausforderungen*, p. 250). The encyclical Laudato Si speaks in this context about responsibilities that are not just common but differentiated with respect to cultures, worldviews, social circumstances and uniqueness (Francis, Encyclical Letter Laudato Si 2015). Thus, the encyclical refers to the need for global cooperation in perceiving the difficulties faced by human societies that are victims of varied forms of ecological degradation, which impose forceful migrations and emigrations (cf. Ibid., no. 25).

Furthermore, Eva Horn gave an insightful reflection and contribution in her *Das Denken im Anthropozän and die Enzyklika Laudato Si* on the importance of shifting the narrative in the approach to the common good, which suggests that it has become striking to emphasise the cooperative common good of both human and non-human creatures. It should be a model built not just on the principle of appropriation, distribution or redistribution, which aligns with the current capitalist economic system operative in its welfare state variants and which turns things and people into goods, resources and commodities. This modern approach should rather treat the common good as aiming at a network of sharing and coexistence, not just the market and its power of instrumentalising. The common good should mean an economy of sharing that does not just aim at the exchange relationships and is not based on monetisation and fairness. The binding vectors are immaterial goods such as belonging, identity, feelings, etc. It is based on the use and care of the so-called *commons*, in other words, goods that belong to everyone, namely the public spaces, bodies of water, meadows and forests. Furthermore, and in the language of today, it is based on open-source software, or simply put, the *common good knowledge,* which should be made available free of charge (Horn 2018).

3.6.4.3 Solidarity

For a sustainable human world, the sense behind the concept of solidarity must be made a normative principle. In solidarity, the consciousness that the individual is an integral part of the whole remains central. It is to be understood as an identification process in which the individual persons, the society and the natural order reciprocally relate with each other, with the human person being the uniting force. In such relation, therefore, the individual identifies himself or herself with the aptitude of the group and remains an ardent part of the general commitment to responsibility (cf. Ibid., *Christliche Sozialethik: Ein Lehrbuch*, p. 271).

The sense of solidarity is both feeling and action. On the one hand, the feeling of being part of a larger group and, on the other hand, the motivation to act in the best interest of the group while at the same time preserving the uniqueness of the individual members. The commitment of the individuals to functional solidarity is a kind of contract, which places responsibility on each for the sake of the common good. John Paul II gave a substantial description of what the nature of authentic solidarity should be, thus:

> (Solidarity) is not a feeling of vague compassion or shallow distress at the misfortunes of so many people, both near and far. On the contrary, it is a firm and persevering determination to commit oneself to the common good; that is to say to the good of all and of each individual, because we are all really responsible for all. (John Paul 1987)

As a normative orientation towards sustainability, therefore, the individual and the society, in relation to the natural order, work to establish a more equitable world; a human relation to nature that is more regenerative and renewable, as well as a society that sees its members as ends in themselves and never as objects to be manipulated for inordinate gains. The reciprocity that is in-built lubricates the relationship and makes every part an active and relevant contributor—*one for all and all for one* (cf. Ibid., *Christliche Sozialethik: Ein Lehrbuch,* p. 284).

Taken from empirical social experiences, there is no social structure that could properly thrive without an organised and institutionalised order and ideals, which must work harmoniously to keep the societal structures stable. One of the requirements is the principle of solidarity, which uses criteria to unite both agents and ideologies in a symbiotic way. On the other hand, the fact that solidarity groups or concepts could also exhibit elements of individual or group egoism, discrimination and other forms of negative class consciousness makes it important to emphasise here the needed character of any solidarity concept that should be considered applicable and acceptable. Thus, for solidarity to play its normative role towards sustainability, it must have orientation towards 1) **Universality (objectivity)**—it means that solidarity must strip itself of all forms of purposeful discrimination, selfishness and oppressive exclusion; 2) **Spontaneity**—the readiness to do the right and the noble at the first sight; 3) **Human Dignity**—promotion and protection of the dignity of the human person remains always its planned objective; 4) **Re-Integrating**—in this sense, the principle of solidarity reaches out even to the guilty and delinquent of the society through the process of reintegration and rehabilitation, and with the conviction that there is always the possibility for a second chance and a new beginning; 5) **Flexibility and Reliability**—it means that solidarity as principle is never to be reduced to and understood only as individual or group moral attitude to be held on to, without the virtue of being open and flexible to changes and other positive influences. Its openness makes it more reliable and gives it a greater chance to permeate the general social structure through its institutions and organisations (cf. Ibid., pp. 289–291).

In all, therefore, when the principle of solidarity fulfils these characteristics, its authentic character as a normative principle would be able to give society the orientation needed for a sustainable human world both for the present and for future generations.

3.6.4.4 Subsidiarity

Subsidiarity as a concept takes from its Latin form *subsidium,* meaning literary—support from a reserved position (*German: Hilfe aus der Reservestellung*) (cf. Ibid., p. 293). An action that is subsidiary, therefore, is invoked to support the mainstream activity, especially within organised smaller groups, and not to undermine or obliterate it. In this form, the subsidiarity principle is an appeal against all forms of state and governmental policies, which tend to subvert the capability of the citizens themselves to exercise their proper rights and freedom in a social structure. The encyclical Quadragesimo Anno of Pope Pius XI, 1931, was the origin of this concept and it defined the idea behind the concept in no. 79 thus:

> As history abundantly proves, it is true that on account of changed conditions many things which were done by small associations in former times cannot be done now save by large associations. Still, that most weighty principle, which cannot be set aside or changed, remains fixed and unshaken in social philosophy: Just as it is gravely wrong to take from individuals what they can accomplish by their own initiative and industry and give it to the community, so also it is an injustice and at the same time a grave evil and disturbance of right order to assign to a greater and higher association what lesser and subordinate organizations can do. For every social activity ought of its very nature to furnish help to the members of the body social, and never destroy and absorb them. (Pius 1931)

As a social principle and one of the normative orientations needed for a sustainable human society, the principle of subsidiarity must establish respect for individual capabilities but at the same time emphasise the subsidiary function of the socio-political institution, which has the sole responsibility to subsidise those affected areas, where the individual or the subordinate groups would inevitable, not out of negligence, lack the power to transcend. In this way, the whole system and structure would be able to provide the citizens with the opportunity to be active and participatory rather than passive. Such a form of interaction makes for a sustainable social order. Thus, in this sense, subsidiarity has four authentic characteristics, namely: (a) It emphasises the autonomy of the moral subject—the individual, who can make decisions and take responsibilities; (b) It must be concerned with the need for the promotion of competence. It ensures that an individual is not, in the first instance, detracted from the competence to initiative and creative efforts, especially by other social and political powers above him or her. Subordinated matters, which individuals in a social order could easily and competently handle, should not be denied to them; (c) It emphasises the autonomy of the individual and the subordinate order of the society while, at the same time, placing responsibility on the higher order of the society to do its part in giving subvention to those areas, where the individual and subordinate capabilities prove naturally incompetent. In this form, the assistance of the higher order must always have the character of being an aid, not the absorption of individual capabilities. In other words, it should take the form of *a socio-relational subsidy (mine)*; (d) Its principle emphasises the need for the higher societal order that provides subvention and intervention at a given time to step aside once successful and functional empowerment of the subordinate has been attained. In other words, an unnecessary persistence in the form of intervention would eventually degenerate into domination, infringement and, in some extreme cases, oppression (cf. Ibid., *Christliche Sozialethik: Ein Lehrbuch*, pp. 295–297).

3.6.4.5 Justice

Justice as a normative orientation defines the foundation upon which relationships between individual persons, groups and organisations should be based on the one hand and, on the other hand, between and among the different structures of the society. As a general concept, it establishes, first, the normative measures required for the mutual relationship between the individual and the socio-institutional structures; secondly, it presents itself as a virtue that should be found naturally in the attitude of the human person as a moral being (cf. Ibid., p. 316). The early antecedent theories of justice all found their roots in the works of Aristotle (Nicomachean Ethics) and Thomas Aquinas (Summa Theologiae). These works laid the foundations for the classical formulation of the basic constituents of Justice, namely: *Iustitia legalis* (legal justice), *Iustitia commutativa* (commutative justice) and *Iustitia distributiva* (distributive justice). These forms present only the three different modes of approach to the application of justice as a normative principle. However, with regard to criteria, they represent the basic principles of the relationship of individual persons within the collective whole on the one hand. On the other hand, they are taken as standards in the interpersonal relationships of individual persons or social groups (cf. Ibid., p. 319).

To serve as a normative principle for sustainable social order, therefore, the sense of justice must be made a virtue and its objective must be the promotion of the common good. This sense of justice for a well-ordered society, according to John Rawls, should be the general assumption by all—assumptions that must be recognised as standards (Rawls 1999). The well-ordered society becomes then the society that could boast of maintaining a relatively sustainable life and sustainable social institutions, made possible by the guiding principles decided upon by its polity. Rawls articulated this thus:

> Now a well-ordered society is also regulated by its public conception of justice. This fact implies that its members have a strong and normally effective desire to act as the principles of justice require. Since a well-ordered society endures over time, its conception of justice is presumably stable: that is, when institutions are just (as defined by this conception), those taking part in these arrangements acquire the corresponding sense of justice and desire to do their part in maintaining them. (Ibid)

Each desiring his or her part in promoting social justice is, therefore, the demand for distributive and sustainable socio-political and economic structures of the human society—both for the individual lifestyle and for the harmonious participation of each in community life. This, from common sense, is the ethics of participation, interaction and association needed for sustainable and inclusive social order.

3.6.4.6 Sustainability as an Attitude

For it to qualify as a norm for the social order, the sense of sustainability must be made an attitude, especially with regard to man's relation to nature and his environment. An attitude is a feeling, a passion towards a necessity. Such a necessary cause for the development of an attitude in response must and should be for the benefit of the society, for the *moral good,* which should benefit society in general. This nature-man relationship establishes the link, which provides the normative guidelines in man's attitude and approach to the natural capitals. In other words, man orientates himself in his actions and decisions as an integral part of his world, which lies in his hands to make or mar. With such an attitude, the whole social structures—politics, economics, science, technology, ecology—all work harmoniously in achieving a corporate, social and responsible network of interaction for sustainable growth and development.

Four cardinal approaches in mind must serve as the basis upon which this attitude must be developed as a norm, namely: (1) the integration of the operative social structures—ecological, economic and social—in a relational system for reciprocal effects; (2) the orientation of every human attitude and approach to understanding "stewardship" as a moral obligation to nature and environment. This will invariably teach the society the need to respect the *planetary boundaries* for protection, maintenance and regeneration of the natural order; (3) the call for solidarity as a social norm that awakes active participation of all to respond to the acute needs of the developing economies and human conditions, thereby developing functional strategies to eliminate poverty and man-made inequitable life situations; (4) the promotion of the sense of intergenerational justice, which reminds the present generation of its moral obligation to make the earth safe and more habitable even for the future generation (cf. Ibid).

Developing an attitude towards sustainability remains one of the laws of social life which, according to Gaudium et Spes, the creator has written into man's moral and spiritual nature (Vatican II Council 1965d). It is an attitude towards improvement and change, an attitude that brings the community of mankind in harmony with the nature it has to maintain and promote. Such promotion of nature means recognising the autonomy reserved in those natural capitals—in their laws and values. Human activities, therefore, can decipher and use them, however, by proper laws and order since all magnified powers of humanity bereft of all proper attitudes towards sustainability invariably threaten the race itself (cf. Ibid., nos. 36, 37). Making the sustainability principle an attitude, therefore, is to answer the question of *retinity*, which demands that all levels of human approaches and decision-making must qualify to "... account for the interrelatedness of economy, ecology and the social sphere..." (Fränzle et al. 2010).

3.6.4.7 Resilience as a Virtue

To achieve sustainability as a social principle, resilience must be lived out as a moral obligation, which awakes in both individuals and the society itself the ardent determination always to *bounce back*—the ability to recover, reinvigorate, reinvest or transcend a given situation (usually an adverse situation) and transform for the better. As a normative capability, being resilient means yielding to the demands of the human mind, informed by the experiences and situations of life, by confronting the challenging situations of our life with corresponding and transformative measures. For "… The more we understand about the situation we find ourselves in and the options we have for dealing with it, the more likely it is that we will be able to deal with the situation at hand" (Richardson 2019).

In resilience, it is not enough just to be aware of the emerging problem or challenge and thus reduce reaction to a mere feeling of rejection, discontent and disapproval. There must be a concrete transition from reaction to planned action and volition, drawing attention to personal, social, cultural, economic or ecological resources available for the combat (Vogt and Schneider 2020). The orientation towards planned actions remains the ideal for resistance, regeneration and reconfiguration (Lepore and Ravenson 2006) amidst the known tensions that exist between stability, adaptation and transformation in the discourse on resilience. Consequently, these factors make different normative and cognitive demands which need integration in order to arrive at the normative targets (cf. Ibid., *Ökologische Sozialethik: Kompass für eine nachhaltige Transformation der Gesellschaft*). For Vogt and Schneider, resilience should be identical to a pragmatic variance of sustainability. Within such pragmatic variance, the required normative orientation spindles between three dimensions of volitional approaches, namely persistence, adaptability and transformability. These help to build resilience by translating into coping capacities (Bewältigungspotentiale), adaptive capacities (Anpassungspotentiale) and transformative capacities (Transformationspotentiale) (cf. Ibid). It all concerns the capacity to determine the right response for corresponding transition and transformation. Further differentiation specifics would talk about engineering resilience, ecological resilience and evolutionary adaptive resilience.

However, within the scope of our approach, to demand resilience as a virtue, therefore, is to emphasise the need for the human society—both individually and collectively—to develop a strong mind and character in the approach to the challenging situations of our world—be it socio-political, socio-economic, socio-ecological or socio-cultural. What it means is that the whole structure of a given society must be resiliently integrated for the common purpose of achieving a stable social order. Such stability is the requisite criterion for sustainability and resilience should be made its norm. Being a normative orientation, therefore, resilience resides in the ability of the human agents to give such power of the mind and character the concrete attention it needs, through effective plans of action. The SDGs demonstrated this attitude in the projected action plans of six among its seventeen development goals, namely: Goal

1—on the resilience of the poor; Goal 2—on resilient agricultural practices; Goal 9—on resilient and sustainable social infrastructures; Goal 11—on resilience in approach to both natural and man-made disasters; Goal 13—on resilience towards adaptation in climate-related hazards; and Goal 14—on the resilience towards protecting marine and coastal ecosystems (cf. Ibid., *Resilience*).

3.7 "*Abundantia Vitae in Legatione Christi*" (Jn. 10:10): Towards Social Sustainability and the Freedom of Mmadụ

Given that our research is done in the area of Christian social ethics, it is pertinent to anchor our strategic approach to the human person, who is the agent and focus of all concrete sustainable and development processes, on the teaching-, learning- and belief-source of Christian knowledge, faith and morals, namely the Christ. *Abundantia Vitae in Legatione Christi,* meaning *the abundance of life in the mission of Christ*, adapted from the words of John's Gospel, "I came that they may have life and have it in abundance",[4] depicts the mission statement of Christ himself inspired by his participation in the social situations of his time and milieu. It is a proclamation of the determination to launch a revolutionary attack on the status quo and chart a course for transitions and transformations. It announces the dawn of time to investigate the root causes of the chains that are binding the human agents and destabilise the human-nature relations and connections within the community of existence. It sets the tone for plans to liberate the society from the structures that keep it detracted, corrupt and malfunctioning. It reiterated the sense of assurance in the prophecy of Isaiah (Is. 61:1–3), in which Christ himself made a mission statement the very moment he revealed himself and entered the daily human life experiences, namely that the spirit is upon him; that he is sent on a mission; a mission to identify with the poor and downtrodden, liberate the societal chains of inhibition and proclaim freedom (the year of the Lord's favour) for the human person and for his or her capabilities. The anticipated year of the Lord's favour should not be a fixed or pre-assigned literal year. It is rather a realisable space of time that begins the very moment there is the decision to *do "the good"* and *deal with "the disorderly"*. It begins the very moment a society sets itself on course to instigate order and justice, as well as institutionalise respect for the human agents and nature itself. It is a space of time, which itself remains a process towards (relative) peace, in other words, sustainability. To do this, the master himself took up a plan which gave direction and orientation to his actions and this plan contains the objectives of his *mission*, the applied *method* for the transformation and the liberating *message* of his work (Fig. 3.8).

---

[4] Cf. As translated from the German version, *Die Bibel, Einheitsübersetzung der Heiligen Schrift*. Gesamtausgabe, Stuttgart 2016.

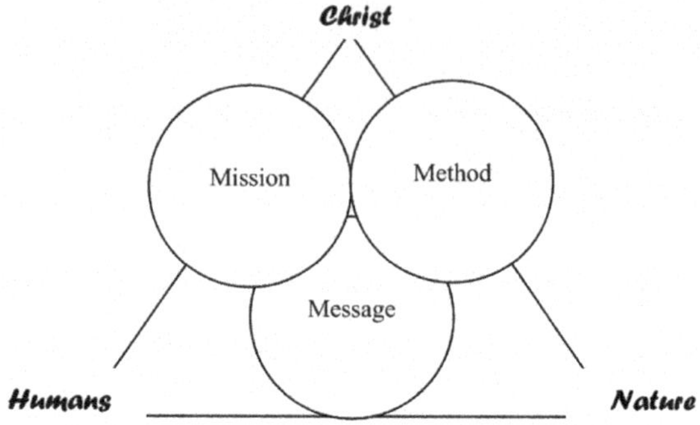

Fig. 3.8 Schematical overlapping of the mission, method and message of Christ. *Illustration mine*

3.7.1 The Mission of Christ

The mission was distinct and direct. It was the mission to bring change and transformation. The central focus was the human person, the *imago Dei* (image of God). It was a mission designed to meet and engage the human person within his or her daily life experiences and in relation to other creatures and natural realities surrounding those experiences within a unified whole—the community of existence. As mentioned above, with reference to the prophecy of Isaiah, it was a mission towards recovery that must begin with observation and identification of the structural problems, namely the debased human status, plights of the poor, oppression, as well as social and institutional abuses. Furthermore, the mission's anticipated effects, eventual changes and transformations were never automated or pre-programmed occurrences. The *abundancia vitae in legatione Christi*—the fullness of life, well-being in life, sustainable life or living—was a mission that also required methods, strategies and approaches in order to restore and liberate Mmadụ and sustain him or her in well-being.

3.7.2 The Method of Christ

Describing the mission of Christ as being strategic makes the statement predicated on planned preparatory and implementation factors: it was clear that the social situation he was to meet needed to be rescued; it was demonstrative that a section of the human agents was under siege due to some forms of political, economic, social and cultural structures that were "*sinful*", in other words, dysfunctional and unsustainable; there existed an atmosphere of anticipation, expectation and the feeling of the need for liberation; the events surrounding his entrance to rescue the situation was a social phenomenon with all the characteristics that are known and associative of it, namely, the combined atmosphere of joy, excitement, inquisitiveness, reservations, the feeling of insecurity by the lovers of the status quo, organised camps of opposition to the mission itself, camps of followers and detractors, etc. These concrete factors need approaches and defined methods that would be able to give the general mission

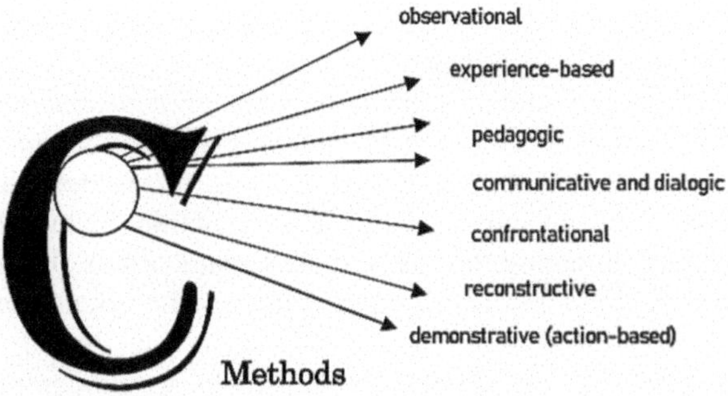

Fig. 3.9 Christological methods towards transformative human and social sustainability. *Illustration mine*

work some direction. Christ's mission was not exempt. He strategically began by identifying himself with the system and structure with the disposition to observe the situations, to evaluate or make necessary judgements about them in order to arrive at concrete and applicable actions as responses to the situations. Thus, his method was ***observational***, in the sense of applying the perceptive powers in identifying the problems, the root causes, the victims and perpetrators, as well as the effects on both humans and the society itself. It was ***experience-based***, taken out of the real situations and oppressive conditions of his audiences.

Furthermore, Christ's method was ***pedagogic***, in the sense of teaching and proclaiming to his listeners and the entire structure of the society the objectives of his mission, which was to liberate the ignorant and the oppressed and chart a sustainable course for the freedom of the human person. It was a method of ***communication*** and ***dialogue***. He was listening, asking questions and dialoguing with all without exception or exclusion (cf. Lk. 2:46). His method was sporadically and situationally ***confrontational***, not in the sense of imposing norms or traditions, but rather in the bid to send a clear message to the social elites who are bent on keeping the system and people exploited and oppressed to their own selfish advantages (cf. Lk. 13:32). Christ's method was ***reconstructive*** since he made it clear that his mission was not to destroy the law and norms make for order in the human society, but to uplift them and give them modes of sustainable application (cf. Matt. 5:17). His method was ***demonstrative***. His various teachings were accompanied by his exemplary actions and volitions, which his mission of change and transformation required. He was always demonstrative in demanding concrete and concerted actions from social and political actors who must translate theories and principles into practical plans and strategies (Fig. 3.9).

3.7.3 The Message of Christ

Christ's message remains distinct, namely that his mission is fulfilled and the principles are laid down—*consummatum est*, "It is finished". It means that the way is prepared, the foundation is laid and a price has been paid for prospective gains, restorations and transformations. However, the consummated mission is not identical to a programmed and automated "*perfectus mundi*", a perfect world. It is rather a consummated mission in the sense of "*pro mundo in processu*"—a world in a process, a world on a course, a world in progress. Thus, the opposition, oppression, suffering and death encountered within the mission never spoke the final word in consummating the mission. The real symbol of consummation that initiates the new beginning of actions and volitions (the orientation for action), in the Christian context, is the *empty tomb* (the resurrection, Jn. 20) and the subsequent *commissioning of actors* (Matt. 28:19–20), who would continue a mission of teaching and applying the laid down principles for sustainable human and social transformations. Our research tries to bring its own contribution and proposals within the body of actors with a foundational focus on the human person in his or her transcendent, immanent and relational self within the community of existence.

The above-reflected Christological application of our approach helps position our research's focal elements to prove their relevance in both contemporary formal intellectual spaces and the Christian socio-ethical and moral theories. It means that Christian social reflections, beginning with the examples laid down by the master, Christ, can never be relevant outside the daily experiences of the common human-nature relationship. Sustainability, therefore, is an applied concept to emphasise the integral process of our world, which must remain on course in providing well-being for human agents and in protecting and respecting the non-human capital, nature and its boundaries. The status and centrality of the human agent would be examined and reflected upon in our next chapter.

Evaluation

In this third chapter, the main focus of our research has been on the varied approaches to the development of our theme, using precedents and antecedents to explain the stages, experiences and situations of our world today. In our research opinion, the movement has never been permanently smooth. Rather, the whole historical development processes of our human world have always known both turbulence and relative progress. However, every human epoch has its attitude and approach towards development—socially, politically, economically, ecologically and culturally. The only common denominator remains the fact that all indices of development have the human person as both the agent and the subject from whom the dynamics of development take off and with whom they reflect their objectives. It means that all development approaches and policies stand to have consequences on Mmadụ, the human person and the world around him. The central question is: To what extent have the known dynamics of development of our world helped to give meaning to the dignity of the human person and therefore work to make the person, the *Mmadụ* and his world sustainable, both for the present and future generations? The next chapter will be based on a critical and philosophico-theological focus on the human person—the Mmadụ. Our vantage point would be a cultural and contextual analysis of the person within the worldview of the Igbos of the eastern part of Nigeria. The purpose would be to analyse the deep socio-cultural view of the human person as a moral subject within an intimate interaction with the nature around him. This would provide the basic tools to evaluate the place of the person in the general dynamics of development within the Nigerian state, as well as explore the reasons behind the shaky politics of development within the Nigerian experience.

References

Akrasi I (2017) Lessons from history. The French revolution and the total transformation of France. In: Nwangwu PU (ed) We the people. Building a new democracy in Nigeria as a model for Africa. American Congressional Press, Texas, pp 268–274

Balaman ŞY (2019) Decision-making for biomass-based production chains: the basic concepts and methodologies. Academic Press, London, p 77

Benedict XVI (2009) Encyclical letter caritatis in veritate. On Integral Human Development in Charity and Truth, Rome

Black R, White H (eds) (2003) Targeting development: critical perspectives on the millennium development goals. Taylor & Francis, London, p 3

Chronopolitics ID (2012) A theory of social rhythm. In: Bal M, De Vries H (eds) The future and its enemies: in defence of political hope. Standford University Press, New York, pp 77–89

Chronopolitics KI (2013) A conceptual matrix. Prog Hum Geogr 37(5):673–690. https://doi.org/10.1177/0309132512472094

De Janvry A, Sadoulet E (2016) Development economics: theory and practice. Routledge, New York, p 67

Francis (2015) Encyclical letter Laudato Si. On Care for Our Common Home, Rome

Francis, Encyclical Letter Laudato Si (2015) On care for our common home, Rome

Fränzle O, Kappen L, Blume H, Dierssen K (2010) Ecosystem organisation of a complex landscape. long-term research in the Bornhöved Lake District, Germany. Springer, Berlin, p 320

Globalance SC (2020) Ethics handbook for a balanced world post-Covid. Globethics.net Focus Series No. 57. Globethics, Geneva, p 238

Globalisation and Livelihood Options of People Living in Poverty, GLOPP (2019) The three dimensions of sustainable development. https://glopp.ch/A2/en/html/resear_area_present_1_3.html. Accessed 3 June 2019

Heimbach-Steins M (ed) (2004) Christliche Sozialethik, Ein Lehrbuch, Band 1, Grundlagen. Verlag Friedrich Pustet, Regensburg, p 23

Heimbach-Steins M (2022) Sozialprinzipien. In Heimbach-Steins M, Becka M, Frühbauer JJ, Kruip G (eds) Christliche Sozialethik, Grundlagen—Kontexte—Themen, Ein Lehr-und Studienbuch. Verlag Friedrich Pustet, Regensburg, pp 170–186

Höffner J (1997) Christian social teaching (as translated and digitalised). Ordo Socialis, Academic association for the promotion of Christian social teaching, 2nd edn. Cologne, p 15. https://ordosocialis.de/wp-content/uploads/ecoenga4neu.pdf. Accessed 10 Dec 2019

Horn E (2018) Leben in einer beschädigten Welt. Das Denken im Anthropozän und die Enzyklika Laudato Si. In: Bertelmann B, Heidel K (eds) Leben im Anthropozän. Oekom Verlag, München, pp 65–78

IEEE EARTHZINE (2012) The seven critical issues at Rio+20. A publication of the oceanic engineering society. https://earthzine.org › Articles. Accessed 4 Oct 2019

Iyinolaka O (2019) The 25th Nigerian economic summit. Nigeria 2050, shifting gears. Nigerian Economic Summit Group. https://nesgroup.org/blog/The-25th-Nigerian-Economic-Summit:-Shifting-Gears-to-a-Prosperous-Nation. Accessed 11 Nov 2020

John Paul II (1987) Encyclical letter Solicitudo rei Socialis. On the Twentieth Anniversary of Populorum Progressio, Rome

Kansas State University (2019) Ethical decision making. Establishing a common ground. https://www.k-state.edu/assessment/resources/ethics/carlin09.pdf. Accessed 10 Nov 2019
Kukah MH (2012) Nigeria as an emerging democracy. The dilemma and the promise. In: Daily trust, Nigerian daily, September 2, 2012. https://dailytrust.com/nigeria-as-an-emerging-democracy-the-dilemma-and-the-promise-i/. Accessed 10 Feb 2020
Küpper A (2017) Die Ordnungsethik der Katholischen Soziallehre. In: Kirche und Gesellschaft, Nr. 436 der KSZ. J.P. Bachem Medien GmbH, Köln, pp 12–13
Lay M (2005) Zivilsationspolitik. Verlag Königshausen & Neumann, Würzburg, Zur Theorie einer Weltökumene, p 7
Lepore SJ, Ravenson TA (2006) Relationship between resilience and posttraumatic growth: recovering, resistance and reconfiguration. In: Calhoun LG, Tedeschi RG (eds) Handbook of posttraumatic growth, research and practice. Routledge, Mahwah, NY, pp 24–46
Malthus TR et al (2018) An essay on the principle of population. In: Stimson SC (ed) The 1803 edition. Yale University Press, pp 18–21. http://www.jestor.org/stable/j.ctv1bvnf95. Accessed 3 Feb 2019
Mukherjee S, Stückelberger C (eds) (2014a) Sustainability ethics: ecology, economics and ethics. Globethics, Geneva, p 302
Mukherjee S, Stückelberger C (eds) (2014b) Sustainability ethics: ecology, economics and ethics. Globethics, Geneva, pp 18–19
Nkrumah K (1974) Neo-colonialism. The last stage of imperialism. Panaf, London, p 6
Nwocha ME (2016) Impact of the Nigerian land use act on economic development in the country. Acta Universitatis Danubius Administratio 8(2):117–128
Nwosu CF (2016) Ethical approach to abuse of common good and human rights in Nigeria, the polity, oil, violence and Niger Delta. AV Akademikerverlag, Belin, p 84
Obi L (2017) Corruption in the national assembly: abuses in budget approvals, confirmation hearing, oversight duties and constituency projects. In: Nwangwu PU (ed) We the people: building a new democracy in Nigeria as a model for Africa. American Congressional Press, Texas, pp 173–192
Ojo J (2016) The non-implementation of Nigeria's national health act. Punch Nigeria, Nigerian Daily. https://punchng.com/non-implementation-nigerias-national-health-act/. Accessed 23 Oct 2019
Ojomo PA (2011) Environmental ethics: an African understanding. Afr J Environ Sci Technol 5(8):572–578
Olusegun OM (2013) The political economy of rising Asian interests in Africa. In: Falola T, Achberger J (eds) The political economy of development and underdevelopment in Africa. Routledge, New York, pp 213–227
Ostheimer J (2008) Zeichen der Zeit Lesen. Erkenntnistheoretische Bedingungen einer praktisch-theologischen Gegenwartsanalyse. W. Kohlhammer, Stuttgart, pp 33–34
Paricipatory Democray Cultural Initiative, Inc (2018) The principle of solidarity. Pontifical Council for Justice and Peace, p 1189. https://democraciaparticipativa.net/documentos-data-a-referenda/documents-in-english/christian-social-thoght/14691-the-principle-of-solidarity.html. Accessed 26 April 2019
Paul VI (1967) Encyclical letter populorum progressio. On the Development of Peoples, Rome
Pius XI (1931) Encyclical letter quadragesimo anno. On Reconstruction of the Social Order, Rome
Pruzan P (2014) The future is in our hands. Conceptual, economic, ecological and spiritual challenges to sustainable development. In: Mukherjee S, Stückelberger C (eds) Sustainability ethics, ecology, economics and ethics. Globethics, Geneva, pp 297–316
Rapley J (2007) Understanding development. Theory and practice in the third world. Rienner Publishers, New York, p 1
Rawls J (1999) The theory of justice. Harvard University Press, Harvard, pp 397–398
Richardson K (2019) Resilience. In: The sustainable development goals: a global, transdisciplinary vision for the future. University of Copenhagen, Copenhagen. https://www.coursera.org/learn/global-sustainable-development. Accessed 13 March 2019

Sachs W (2021) Papst versus UNO sustainable development goals and Laudato Si: Abgesang auf das Entwicklungszeitalter? In: Vogt M, Umweltethik C (eds) Grundlagen und zentrale Herausforderungen. Verlag Herder, Freiburg im Breisgau, p 530

Sachs W (2022) Papst versus UNO sustainable development goals und Laudato si': Abgesang auf das Entwicklungszeitalter? In: Budrich Journals, PERIPHERIE, pp 245–260. https://doi.org/10.3224/peripherie.v38i2.06

Sander H (2013) Sozialethik, theologisch: Der locus theologicus alienus Menschenrechte. In: Markus V (ed) Theologie der Sozialethik. Verlag Herder, Freiburg im Breisgau, pp 252–278

Schönborn G, Berlin C, Pinzone M, Hanisch C, Georgoulis K, Lanz M (2019) Why social sustainability counts. The impact of corporate social sustainability culture on financial success. J Sustain Product Consump 17:1–10

Sen A (2013) The ends and means of sustainability. J Hum Develop Capabil 14:6–20. https://doi.org/10.1080/19452829.2012.747492

Siep L (2016) Konkrete Ethik. Grundlagen der Natur-und Kulturethik. Frankfurt am Main: Suhrkamp Verlag, Frankfurt am Main, pp 288–289

Stiglitz JE (2017) Globalisation and its discontents revisited. Penguin, p xii

Suh R (2015) Climate change has met its match in the will of a United world. NRDC President, Natural Resources Defence Council. In: NRDC issue brief, "the paris agreement on climate change", IB: 17-II-A. http://www.nrdc.org/media/2015/151212.asp, https://www.nrdc.org/sites/default/files/paris-agreement-climate-change-2017-ib.pdf. Accessed 11 April 2019

Sulabh International (2002) The world summit on sustainable development, Johannesburg, 2002. Sulabh International Social Service Organisation, New Delhi. http://www.sulabhinternational.org/the-world-summit-on-sustainable-development-johannesburg-2002/. Accessed 11 April 2019

The European Bank for Reconstruction and Development (2019) Economic inclusion strategy (EIS) 2017–2021. https://www.ebrd.com/ebrd-economic-inclusion-strategy.pdf. Accessed 3 June 2019

The Federal Republic of Nigeria (2020) Labour force statistics, Volume 1. Unemployment and underemployment report. National Bureau of Statistics Q4 2017–Q3. https://www.nigerianstat.gov.ng/pdfuploads/Q2_2020_Unemployment_Report.pdf. Accessed 15 Dec 202

The United Nations (2019) Take action for the sustainable development goals. https://www.un.org/sustainabledevelopment/sustainable-development-goals/. Accessed 17 Oct 2019

The United Nations Population Fund (UNFPA) (2019) Sexual and reproductive health. https://www.unfpa.org/sexual-reproductive-health. Accessed 23 Oct 2019

Ubani-Ebere O (2017) The roots and foundations of corruption in Nigeria: corrective tools for mitigation. In: Nwangwu PU (ed) We the people: building a new democracy in Nigeria as a model for Africa. American Congressional Press, Texas, pp 211–232

Ugochukwu C, Ertel J (2008) Negative impacts of oil exploration on biodiversity management in the Niger Delta Area of Nigeria. J Impact Assess Project Appraisal 26(2):139–147

United Nations (1964) United Nations conference on trade and development, UNCTAD. Prosperity for all. Geneva. https://unctad.org/. Accessed 11 April 2019

United Nations (1972) Report of the United Nations conference on the human environment. Stockholm, p 3. https://www.un.org/en/conferences/environment/stockholm. Accessed 10 Oct 2019

United Nations (1981) Report of the world commission on environment and development. Our common future. https://sustainabledevelopment.un.org/content/documents/5987our-common-future.pdf. Accessed 10 Oct 2019

United Nations (1992) The rio declaration on environment and development. Rio de Janeiro. https://www.un.org/en/conferences/environment/rio. Accessed 10 Oct 2019

United Nations (1994) Report of the international conference on population and development. Chapter 1, Preamble 1.1, Cairo, 5–13 September 1994. https://www.unfpa.org/sites/default/files/event-pdf/icpd_eng_2.pdf. Accessed 10 Oct 2019

United Nations (1995a) Report of the world summit for social development. Copenhagen, 6–12 March 1995, Annex 1, no. 2. https://www.un.org/development/desa/dspd/world-summit-for-social-development-1995.html. Accessed 10 Oct 2019

United Nations (1995b) Report of the fourth world conference on women. Beijing, 4–15 September 1995, no. 7. https://www.un.org/womenwatch/confer/beijing/reports/. Accessed 10 Oct 2019

United Nations (1997) Kyoto protocol to the United Nations framework convention on climate change. Art. 10(ii)c. Kyoto. https://unfccc.int/kyoto_protocol. Accessed 10 Oct 2019

United Nations (2002) World summit on sustainable development. Johannesburg. https://www.un.org/en/conferences/environment/johannesburg2002. See also: an assessment for IISD. https://www.iisd.org/publications/report/world-summit-sustainable-development-assessment-iisd. Accessed 11 April 2019

United Nations (2011) Fourth United Nations conference on the least developed countries. Istanbul, p 16. https://www.un.org/en/conferences/least-developed-countries/istanbul2011. See also The United Nations conference on trade and development. Prosperity for all. https://unctad.org/. Accessed 11 April 2019

United Nations (2012) Rio+20 United Nations conference on sustainable development. III (56), Rio de Janeiro. https://sustainabledevelopment.un.org/rio20. Accessed 10 Oct 2019

United Nations (2014a) Third United Nations conference on small island developing states. Apia, Samoa. https://www.un.org/en/conferences/small-islands/apia2014. See also Island Voices, Global choices. UN conference on small island developing states

United Nations (2014b) Second United Nations conference on landlocked developing countries, I (1), Vienna. https://www.un.org/ohrlls/content/official-documents-lldcs. Accessed 10 Oct 2019

United Nations (2015a) The third UN world conference on disaster risk reduction (Sendai framework for disaster risk reduction 2015–2030). Sendai. https://www.un.org/sustainabledevelopment/un-world-conference-on-disaster-risk-reduction/. Accessed 11 April 2019

United Nations (2015b) Paris agreement. Paris, France. https://www.un.org/en/climatechange/paris-agreement. Accessed 23 May 2019

United Nations (2015c) Integrating the three dimensions of sustainable development. A framework and tools. Economic and social commission for Asia and Pacific. Greening of economic growth series, 2015 ST/ESCAP/2737. https://repository.unescap.org/handle/20.500.12870/3161. Accessed 3 June 2019

United Nations (2018) Nigeria, human development indicators. United Nations Development Programme, Human Development Reports. https://hdr.undp.org/data-center/specific-country-data. Accessed 17 Oct 2019

United Nations (2019) UNESCO moving forward the 2030 agenda for sustainable development. UNESCO task force on the 2030 agenda for sustainable development, p 7. https://en.unesco.org/creativity/sites/creativity/files/247785en.pdf. Accessed 26 May 2019

United Nations Children's Fund, UNICEF (2022) UNICEF warns of Nigerian education crisis as world celebrates international day of education amid Covid-19 concerns. Press Release. https://www.unicef.org/nigeria/press-releases/unicef-warns-nigerian-education-crisis-world celebrates-international-day-education. Accessed 22 July 2022

United States Catholic Bishops (2009) Economic Justice for all: pastoral letter on catholic social teaching and the U.S. economy. United States Conference of Catholic Bishops, Washington DC, p 88

Vatican II Council (1964) Unitatis Redintegratio, decree on Ecumenism, Paul VI, Rome

Vatican II Council (1965a) Dignitatis Humanae, declaration on religious freedom, Paul VI, Rome

Vatican II Council (1965b) Presbyterorum ordinis, decree on the ministry and life of priests, Paul VI, Rome

Vatican II Council (1965c) Gaudium et Spes. Pastoral constitution on the church in the modern world, Paul VI, Rome

Vatican II Council (1965d) Gaudium et Spes. Pastoral constitution on the Church in the modern World, Paul VI, Rome

Vogt M (2013) Prinzip Nachhaltigkeit, Ein Entwurf aus theologisch-ethischer Perspektive, 3rd edn. Oekom, München, p 16

Vogt M (2021) Christliche Umweltethik. Grundlagen und zentrale Herausforderungen. Verlag Herder, Freiburg im Breisgau, pp 527–528

Vogt M, Schneider M (2020) Ökologische Sozialethik: Kompass für eine nachhaltige Transformation der Gesellschaft. Forschungsprojekt, Bayerischer Forschungsverband, Fit for Change, 2013–2017. https://www.bayfor.org/de/unsere-netzwerke/bayerische-forschungsverbuende/forschungsverbuende/project/forchange/oekologische-sozialethik-als-kompass-fuer-eine-nachhaltige-transformation-der-gesellschaft.html. Accessed 10 March 2020

Wikipedia (2019a) Kyoto protocol. https://en.wikipedia.org/wiki/Kyoto_Protocol. Accessed 10 April 2019

Wikipedia (2019b) Paris agreement. https://en.wikipedia.org/wiki/Paris_Agreement. Accessed 23 May 2019

Wikipedia (2019c) UNESCO. https://de.wikipedia.org/wiki/UNESCO. Accessed 26 May 2019

Wikipedia (2019d) Social sustainability. https://en.wikipedia.org/wiki/Social_Sustainability. Accessed 3 June 2019

Chapter 4
The Socio-Cultural Approach to the Moral Agent of Sustainable Development—The *Mmadụ*

Abstract The goal of all political, economic, social, cultural, and ecological interactions is predominantly to create harmony and allow the human person to work towards achieving his or her full potential within varied and interrelated vectors of opportunities. Therefore, the person remains the agent, the receiver, the object, and the subject of the development dynamics itself. To thwart or attack this goal is to detract the human person of authentic freedom. The Afro-Igbo ethnocultural understanding of the human person, Mmadụ, as a 'distinct being' who is both autonomous and relational sheds light on the need to maintain balance in the relationship that should exist between the two. A functional mutuality would guarantee and improve human dignity, Ùgwú Mmadụ, while the opposite leads to a loss of Ùgwú Mmadụ. Thus, wherever there is an intrigue that causes a divide in the authentic life of the human person—as is always the case in detractive societies—there must be a need for a remedial ethical regeneration towards viable ecological humanity.

Keywords Mmadụ · Autonomy · Relationality · Human dignity · Authentic freedom · Ecological humanity

4.1 Introduction

Our research has tried to examine the development nuances, which border on the general socio-economic and political experiences of our present world but with reference to the Nigerian situation. In doing this, it tries to prove the relevance of focusing on the human person in relation to the entire natural capitals as the only relevant criteria for a sustainable human society. The present chapter would deepen this understanding through a socio-cultural argument to the fact that the focus on *Mmadụ* as a criterion of approach is intrinsically motivated, and it is a core factor within humanity's metaphysical and empirical world views.

C. F. Nwosu, *Systemic Remediation*, Environmental Humanities: Transformation, Governance, Ethics, Law 4, https://doi.org/10.1007/978-3-031-61909-0_4

4.2 The Place of *Mmadụ* as a Moral Agent of Development

Mmadụ—the human person—is a being, a creature with meaning. The inherent understanding that every human life possesses intrinsic value and deserves respectful treatment is a foundational aspect of human dignity—*Ùgwù*. This is ontological to life itself. The various approaches to human life and the obligation to accord this life respect and dignity define a people's normative worldview, which helps to shape both their cosmological and metaphysical quests about life and experiences. The meaningfulness and the dignity of the life of *Mmadụ* make him or her an agent with normative inclinations—a moral agent. Thus, the *Mmadụ*, with a constant search to unmask the world around him, elicits the core element that defines him or her as a person, namely: reason—rationality. That *Mmadụ* is rational stands fundamental in all intercultural interpretations of the human person and the life that he possesses. Nevertheless, there are some peculiarities stemming from people's experiences, beliefs, myths and empirical interactions with their immediate world—the cosmos, which also influence their unique mentalities, approaches and worldviews.

Reason—rationality, therefore, is a natural condition of *Mmadụ* and it is both a defining and driving factor that makes the human person a teleological being—a being whose life is purposeful. He does not just exist. Rather, the human person, Mmadụ, *lives existence*. Mmadụ is a being with "entelechy"—the ability to be involved and the ability towards activity and not passivity generally driven by the power of reasoning. This "*entelechical*" power in the human person makes him or her a being in relation to all other realities outside of the human "self"—nature and all its capitals, both living and non-living. This relational character of the human person, which makes him self-transcendent as well, establishes a natural link that invariably places an obligation on him to observe, learn and follow some set rules and approaches in relation to other existent realities outside of him. The consequence of the rational and healthy relationship in this regard depicts all that goes into development—human, social, political, economic, ecological, etc. When the contrary is the case, the relational character of Mmadụ himself remains distorted and elicits chains of malfunction, which invariably influence the general life within him and outside of him. In such a situation, one speaks of underdevelopment, dehumanisation, dysfunction, oppression—both socio-economic and socio-political, etc. Therefore, the authentic place of *Mmadụ* as the moral agent of development remains the true exercise of the innate relational character within and outside his own species, in other words, within humanity and in relation to nature upon which the Mmadụ himself lives and depends. Subsequent dependent subtitles of this chapter would use an intercultural semantic understanding of the human person—Mmadụ—to prove the fundamental understanding of Mmadụ as the centre and the moral agent of development seen within cultures, beliefs, understanding, mores and worldviews in human history. The chapter would also use the three terms, namely "Mmadụ", "human person" and "person" interchangeably.

4.2.1 Mmadụ—*The Human Person—in a Trans-Cultural Worldview: The Western and African Comparative Analysis*

Two cultural worldviews play wide representative roles in the approach to the socio-cultural understanding of who a person is. The schema below shows these two different but relational perspectives:

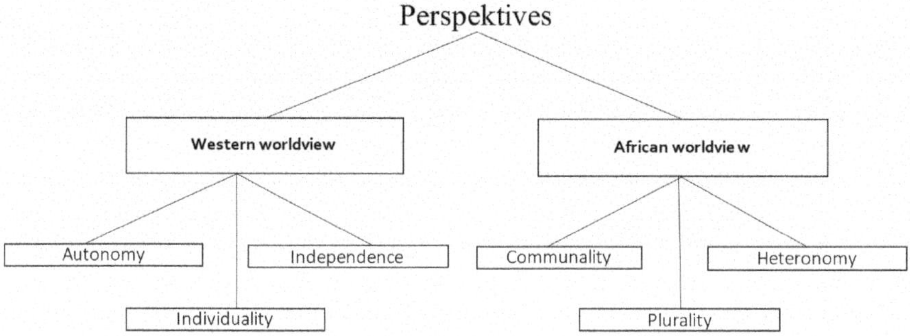

Illustration mine

4.2.1.1 The Western Perspective

From Protagoras' human person as the centre and measure of all things to Thomas Aquinas' ontologically founded and ethically oriented Imago Dei; from Fredrick Nietzsche's Super Man, who must reject all forms of conventional moral values, to Martin Heidegger's human *self* as the actual entity to be analysed; from Ludwig Wittgenstein's identification of language as the human person's only weapon to expand his world to the cultural ontology of the human person, exhaustively defined only in relation to the other beings outside of him, and attributed to the thoughts and worldviews of both non-literary and later literary African cultures, especially when literacy is viewed in its conventional and formal sense—the *Mmadụ* remains the focus around which all cosmological dynamics of existence revolve. A trans-cultural view of this human person, therefore, is a comparative approach to who the person is in relation to the world within which he or she lives.

For more clarity, we focus on the opinions of Boethius and Aquinas as complementary representations of Western thought on the human person. The Igbo from Nigeria, with reference to other core thought-elements from other tribes within Nigeria, as well as the Bantu comparative thought line, which streamlines other African thoughts and cultures, remains our focus for the African approach to understanding the human person. In Western thought, following Boethius, *Mmadụ* is "[...] *rationalis naturae individuae substantiae*" (Iroegbu 2022). This is translated as: A *person is an individual substance of a rational nature*. The definition contains three core concepts about the human person, namely substantiality, individuality and rationality. Seen as a substance, *Mmadụ*, therefore, is a being that subsists in itself and this subsistence defines the essence of Mmadụ's individuality, upon which and through which the rational powers depend and are exercised. Thus, the human person as a fundament subsists on a nature characterised by self-will and autonomy. Such attributes define an independent existence that is concrete and substantial—and this substance is either the first or second substance.

The individuality of *Mmadụ* as a core factor of his essence is located solely within the concept of the first substance and can only be abstracted from within the concept of the second substance. Hence, the core of the essence of *Mmadụ* is defined with the first substance, in the sense of a concrete reality existing on its own, a reality that is self-subsistent—a person, an individual, a German person, a Nigerian person, a man, a woman. The second substance refers to the inhering mental abstractions—the "abstracts" of the first substance like personhood or personality, *germanness* or *nigerianness*, humanity, etc. Furthermore, *Mmadụ* subsists as one of the core characteristics of his existence, the power of cogitation—rationality. Thus, with Bothius' classical definition, *Mmadụ* is a singular being who can use the power of reason to engage his world as well as influence and shape his life. His existence depends on a series of cognitive dynamics that provide him with the tools to confront himself and other realities outside of him. Simply put, and in line with Boethius, a person, *Mmadụ,* "[...] is a type of existing separate being that can think" (Ibid., p. 69). In other words, the capability of thinking in Mmadụ makes him different from all known realities that have no power to exercise such cognitive function.

Mmadụ, therefore, in the Western experience, is a concrete, *substantial* reality of its own, who can subsist in his *individual* self and with the *rational* power to influence both himself and the realities around him. In pure Western polemics, a person is a rational individuality. However, a comparative approach to the nature of the human person would often view this Western position as acceptable but not inclusively exhaustible. It is acceptable, on the one hand, in as much as it identifies Mmadụ as a singular reality of its own with the power of cognition. On the other hand, it is non-inclusive since the definition played down the relevance that other existing realities have in relation to Mmadụ, as well as the role they play in the general understanding of the human person. Contemporary Western philosophers, anthropologists, scientists, psychologists and theologians would work to complement this fundamental position by introducing some correlational concepts. Such a manner of correlation is to be seen in Thomas Aquinas.

Aquinas, following Boethius as the representative of the subsequent group of thinkers with correlational ideas, found Boethius' definition plausible, though with a seemingly conflicting choice of words, especially with regard to the human person being an *individual substance.* Aquinas may have found the expression tautological since the very concept of a substance is always predicated on an actuality that is singular, particular or individual. To attempt more clarity, therefore, Aquinas preferred the two terms "absolute and relational" as basics in understanding the "*kpim*"[1] of the human person. The individuality and separateness of the person as a concrete substance in Boethius are viewed in Aquinas as the absolute constitution of a person. In other words, and following Aquinas, Mmadụ is separate, exists differently, is rational and thus, absolutely constituted (Ibid, *The Human Person in the Western and African Traditions: A Comparative Analysis*, p. 70). Thus, Aquinas defines a person as "[…] *subsistens distinctum in natura rationali*" (Ibid.), meaning, in effect, that a person (Mmadụ) is a distinct subsistent of a rational nature. Following this definition, Aquinas thought about two choices of words that would better review and reinterpret Boethius' use of the *individual substance*, namely, *distinct* and *subsistent*. Hence, Mmadụ, in Aquinas' approach, is a distinct being, one with the certainty of existence, a being that has a concrete nature, separate and individual. Secondly, the Thomistic Mmadụ subsists on its own (first substance) even when all accidents (second substance) are taken away. In other words, the subsistent Mmadụ is a being whose existence is absolute, and this absolute nature is the essence—a being as an "another", a being distinct and separate from other beings. However, one would rightly admit that the distinction between Boethius and Thomas Aquinas, in this first instance, can only be identified in the choice of words they used. Ideologically, both views are like the two sides of a coin, for "[…] an individual is always a distinct being and a distinct being is equally an individual. Furthermore, substance and subsistence are *grosso modo* synonyms. The substance of a being is that which subsists when accidents are taken away. Subsistent is, therefore, substance" (Ibid., p. 71).

Nevertheless, Aquinas' concept and his choice of the word "subsistent" led the way to much later comparative and correlational works in understanding the human person, basically as an individual and separate being, but who subsists on its own even within the reality of other beings or orders of existence outside of him. Therefore, Aquinas would subsequently talk about the ever-enduring part of the human person—the soul, which defines his personality and makes his constitution absolute. Hence, "… This spiritual part of the human being allows him to possess abstract knowledge that is universal. In that universality is also embedded a *relationality* to other beings, including God, the world and other created beings" (Ibid., p. 72). This means that the real constituent of what a person is embodies both the rational and relational aspects—a process that is both spiritual and social. This subsequent understanding of the relational aspect of the human person—Mmadụ—would later expose the intellectual Western world to the need for a comparative approach to the philosophy and anthropology of the human person, especially the need to collaborate with other cultures and worldviews in a dialogue of opinions. Such need, and for our research, brings Africa into focus.

[1] Kpim—is a phonological word in the Igbo language, which has flexible linguistic application. Its use is always descriptive and emphasis-based. The meaning is associated with such words and phrases as the core, the focus, the nucleus, the central idea, the basis, the groundwork, the fundament, the foundation, the determinant factor, the central point or fact, etc.

4.2.1.2 The African Perspective

Africa is a multilingual and multiethnic continent, with over 16% of the total population of the world divided into continents. In other words, the second most populous continent after Asia (over 59%). Amidst this diversity, there is a deep and close union of ideas about Mmadụ, which places emphasis on human relational ability as an indispensable factor in the normative constitution of who a person is. We would analyse the concept of the human person among the Bantu tribes as an exhaustive representation of the African approach within the tribes and cultures in the Sub-Saharan African area outside Nigeria. Secondly, the Igbo culture and worldview from the southeastern part of Nigeria, which has already provided the unique concept of *Mmadụ*—the human person, would be employed to comparatively analyse the quintessence of the African view and approach to the human person.

The Bantu tribes and cultures in Sub-Saharan Africa represent about 30% of the population of Africa. The larger groups are found in their millions, spanning countries like Zimbabwe—the Shona (about 12 million); South Africa—the Zulu (about 12 million); Democratic Republic of Congo—the Luba (about 7 million); Tanzania—the Sukuma (about 9 million); as well as Kenya—the Kikuyu (about 8 million). For these extensive African cultures, the human person—Mmadụ—is fundamentally constituted as self-conscious, rational and autonomous. These elements define the absolute nature of the human person and make Mmadụ an individual, distinct and separate from other realities. Taking the Kikuyu language as representative of other Bantu linguistic genres, Mmadụ—the human person—is given the name "Mundu". When we analyse the two-syllable word, it shows that "**Mu**" means *being*; "**undu**" means *reason or rationality*. Thus, "mundu" means a being with reason, a rational being, or a cognitive being. On the one hand, a being with reason, therefore, is a being with purpose and orientation. On the other hand, a being whose existence is purposeful and orientated can only realise the purpose of his or her existence in relation to both the "*self*" and the other "*selves*". In other words, the very essence of Mundu—*Mmadụ*—the human person, takes its meaning not just because of the cognitive powers, which make the human person conscious and autonomous, but more so because of the natural ability of *Mmadụ* to transcend the individual conscious self and subsist in relation to other realities outside of him or her. Put succinctly, the individual, Mmadụ, exists within and lives a shared consciousness.

In the Bantu worldview, therefore, personhood could be lost; in other words, a person, *Mundu, the Mmadụ,* could never fully realise the constituent potentials when his natural conscious self loses or denies the interrelation with other existence outside of the person. Iroegbu would argue:

> The dry, abstract analysis of substance, nature, hypostasis and ultimate transcendence, and so on (with regard to who a person is), totally removes the person from the real world where his roots and development are found. Any authentic analysis of the person must take into account the real world in which the person lives, moves, and enjoys existence. Even metaphysics, the study of being in general, must be metaphysics for real people, i.e., people living in a world of space and time and history. Thus, the culture, language, experience, and

> concrete background of such people should be brought into the explanation of the identity of the person for relevance and real explanation to be achieved. (Ibid, p. 76)

This means that the individuality, rationality or autonomy of *Mmadụ* unite in an individuality capable of subsistence within a communal existence; it is a rationality that is not trapped merely within the conscious self but a cognitive power, which can transcend the self and is capable of translating its powers in relational dynamics towards the external. Again, it is not an autonomy out of nothing but, rather, an autonomy out of something and within a circle of other existing realities. Thus, in the Bantu perspective, the absoluteness and the autonomy of the human person (the *Mmadụ*) could face the danger of being diminished or lost completely under two conditions, namely: (a) complete loss of or distorted interconnection with other diverse beings—in Kiswahili: "**Mtu duni**"; (b) loss of the cognitive power, loss of reason—in Gikuyu: "**kindu**" (Kahiga and Eberl 2013).

While Sub-Saharan African thought acknowledges the Cartesian cogito as the basis of individuality and autonomy, it emphasizes that cognitive power is not isolated but rather develops and operates within a collective framework. In other words, Mmadụ's absolute value as an autonomous being who exists separately and is capable of reasoning would be grossly irrelevant if such core components are interpreted without reference to the *community of existence* outside of and integrative of the person himself or herself. To this effect, the Sub-Saharan Igbo culture and worldview of the human person use the name Mmadụ—*mma-ndụ, mma-si-na-ndụ*, the beauty of life (existence), the core of existence, the custodian of existing realities, an existence within existence, etc., as the ontological name for who a person really is. We will deal with this more explicitly later. Mmadụ, therefore, as an existing substance of a rational nature, becomes meaningful only as a subsisting reality within an existential space and only within this space is the concept and art of *subsisting* comprehensible. This is so because the cognitive powers of Mmadụ are understood among the Igbos as the powers of both self-determination and self-communication. However, self-communication, understood in this context as the ability of the "*human self*" to transcend its conscious self and to be able to recognise other conscious realities outside of him, can only happen within relational dynamics, in other words, within the community of existence. Thus, for Iroegbu, the community,

> […] entails relating to one another at the human-social level. Community undergirds the spirit and practice of giving what we have (self-communication), and receiving what others give in co-operation and mutuality. In the sharing process, each person receives what he needs and gives to others what these others need to live. In a wholesome community, the members make the lives and activities of one another possible and able to flourish [….] All are products of the community. (Ibid, *The Human Person in the Western and African Traditions: Comparative Analysis*, p. 76)

The Igbo cosmology, therefore, accepts that the individual Mmadụ is an individual substance. However, Mmadụ remains a subsisting substance whose individuality is essentially participatory. This participation is the essence, the relevance and the determinant factor about who he or she is. Such *individuality is participability* (cf. Ibid, p. 78). The essence of Mmadụ, the human person, the absoluteness and the distinctness of the individual, needs, at the same time, the ontological language of "Ìgwè bụ Íké" meaning *community (communality) is strength*. For the Igbo, the autonomous Mmadụ has a separate individual existence. However, autonomy subsists and is meaningful only within the heterogeneity built with other realities outside of the human person. This fact is basic to the ideology of the human person within Sub-Saharan Africa. For there can be no authentic discussion about community without individuality and no approach to individuality is exhaustive without reference to the community. Iroegbu describes this as communal dialectics.

From the African perspective, therefore, this communal relationship remains core to understanding the human person. This is in no way to deny the distinctness of Mmadụ from other beings. It is rather to emphasise that the prime constituent of the human person, taken as autonomous, rational and individual, is a relational product that is based on both *union* and *communion* (Fig. 4.1).

The dialectic begins fundamentally with a union made up of two distinct individual existences. The individual product of the union, therefore, owes his or her existence to *something* and never to *nothingness*. In other words, the essence of the individual Mmadụ cannot be nihilistic. Individuality is substantive and dialectical. The individual is from a union and communes "... in the community and the community in the individual. While there is no community without individuals, there is also no individual without the community. Each is inextricably involved in the other" (Ibid, p. 79). This means that the autonomous, rational and individual Mmadụ is a self-embodied being whose essence lies in self-communication and self-transcendence, which strictly flow from the ability to relate and commune with other existing realities

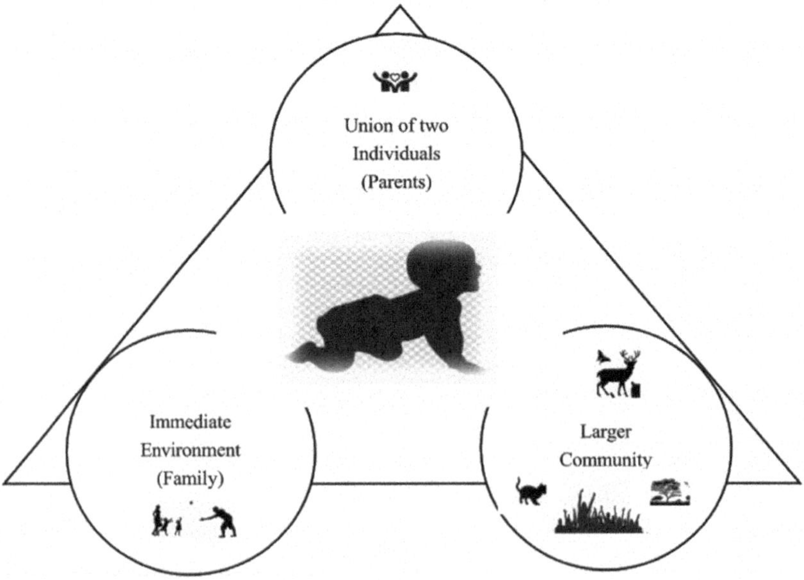

Fig. 4.1 Individual-Collective Relationship to the Concept of Mmadụ

outside the human person. Maurier puts it this way, "Relation is the constituent of the human person: Without it, the human person is isolated and falls into nothingness" (Ibid., p. 81).

The final thesis here and for the purpose of our research is to accept this comparative approach, not as an anthropological dissonance about the *"kpim" of Mmadụ*, but rather as the ontological categories that combine to define the real essence of the human person, namely the Mmadụ as an individual, which is demonstrable in Descartes' idea of "cogito ergo sum", I think, therefore, I exist; and Mmadụ as a relational being, more akin to Ubuntu's idea of "I am because we are":

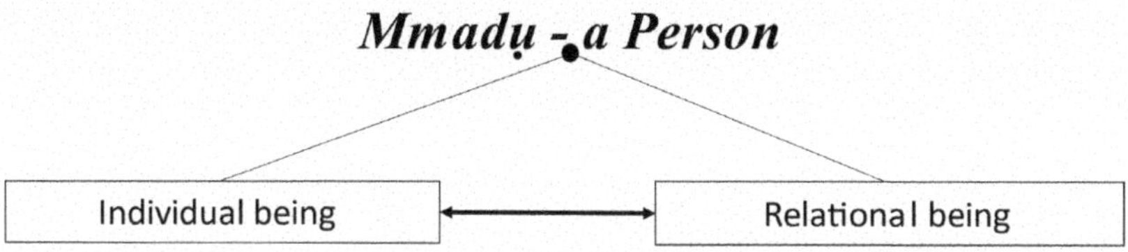

Illustration mine

In this relationship, three essential elements would unite the various approaches to understanding the human person, namely, community, self-(hood) and transcendence. Mmadụ, with autonomy, rationality and individuality, lives the language, mentality, values, worldview and attitudes of the community of existence outside of him. Mmadụ is an individual self that is distinct and singular. This is seen in the rational ability to individual thought, knowledge, will, interest and volition. This selfhood is the autonomous self. Finally, therefore, Mmadụ is self-transcendent, meaning that the human essence is never a stagnant or non-communicable reality. The essence of Mmadụ and existence, in general, is relational. However, such self-transcendence is not robotic or automated. It begins with articulation and the launching of the self towards the external—"It must be a conscious act of the spirit of man—the soul, mind and will" (Ibid., p. 83).

These three basic elements are rooted in a dialectical process that is community-based or communal-based. The community (society), therefore, plays a fundamental role in the transcendence of self for an authentic life and should be able to offer a functional environment for such self-giving to others. The mutuality that exists and runs the entire dialectical process goes into all that defines all forms of socio-political, socio-economic and socio-ecological development processes, which are at the nucleus of Mmadụ and his or her relational entelechy with the external realities. A short semantic of the keyword *Mmadụ* brings the reality home within our context, namely the Nigerian situation and society.

4.3 Contextualising Mmadụ and His Essence in the African-Igbo Cosmology: The Constitutive Elements

Igbo cosmology has a deep regard for realities and the names they bear. Thus, just as the Igbo language is highly phonological, words behind names, concepts and ideas within the culture carry deep meaning. The Mmadụ, the human person, who is both subject and object of development in the context of our research, lends a deep semantic inroad to a possible understanding of what must have gone wrong in the general approach to development, where and how the wrong was constituted and the reason for its persistence.

4.3.1 *Mmadụ: The Semantic Nuances*

Mmadụ basically is a combination of two important Igbo words,

Mma + Ndụ = literary, beauty + life. Thus, Mmadụ = Beauty of life, Beauty in life.

The two words describe the deepest sum of the fundamental elements of existence. Therefore, they have related synonyms construed as explicating the broad fundaments of both human and natural realities. Thus,

Mma + Ndụ = Mmadụ **Synonyms**

↓

Mma ⟶ *Beauty, quintessence, core, nucleus, centre, pivot, hub, goodness, value, order.*

Ndụ ⟶ *Life, existence, reality, regeneration, existential mobility*

The broad interpretation of the two key words—mma + ndụ—allows, therefore, other formulations of the same basic concept about Mmadụ, namely: *Mma—ndụ*: the beauty of life; *Mma—bụ—ndụ*: the reality, substance (individual) that is life; *Mma—na—ndụ*: the quintessence for life; *Mma—ji—ndụ*: the core of life; the nucleus of life; *Mma—si—na—ndụ*: the subsistent in life; subsisting in existence. Ndụ, as a traditional cum cultural concept, is seen as the highest good not only with regard to human life but also within other natural living things, though some sorts of hierarchy in approach and valuation of life exist within the humans in relation to existences often described as the *lower beings*. Thus, Mmadụ as an entity becomes an inviolable symbol (*mma*, the goodness, the *summum bonum*) of this life (Ndụ). As a consequence, the value placed on Ndụ awakes in Mmadụ himself or herself the corresponding responsibility to protect, respect and promote life through values and ideals that work for the improvement of Ndụ—life. Such values and ideals would later be dealt with as factors that harness and promote the capabilities that every rational and normal human being has the right to pursue.

Both concepts, "mma + ndụ" in relation to understanding Mmadụ, bring the anthropocentric Igbo worldview into play. Mmadụ, in this context, becomes the conscious manifestation of vitality, spiritedness and transcendence. Thus, the central belief is that,

> […] Mmadụ stands at the centre of existence; that although all beings receive their being from Chukwu (God), they (even the deities) tend towards Mmadụ. All beings are conceived, more or less, as existing for the sake of Mmadụ. Every activity in the cosmos, according to the Igbo worldview, is directed – directly or indirectly – towards the welfare and preservation of Mmadụ. This form of perception of existence is the first indication of the moral value the Igbo repose in Mmadu. They believe that Mmadụ has an unquantifiable value and is, as such, of high dignity and deserves special respect. However, the concept of this moral value has not always possessed the same meaning in the different historical epochs of the Igbo people. It has continued to evolve. The concept received a major shift in meaning with the advent of Christianity in the nineteenth century. (Oforka 2016)

In line with our comparative approach to Mmadụ, this semantic interpretation within the Igbo worldview holds Mmadụ—the human person, as a distinct being existing among other categories of being—but, most importantly, as a being in relation to other beings and realities in existence. Referring to Mmadụ as *mma,* the nucleus, essence or quintessence, is to emphasise his absoluteness among other existing realities, and this is the basis of his autonomy, rationality and individuality. In other words, the "mma + ndụ", Mmadụ, reserves an existence that is distinct, conscious, self-determining, but also relational. This places Mmadụ at the centre of existence and accords him with dignity and respect, which assumes inviolability, especially within the communal dialectics. The Igbo worldview refers to this as "Ùgwú- Mmadụ"—the dignity of the human person.

The care and concern for Ùgwú- Mmadụ stand, therefore, at the centre of life in general and all cosmological dynamics of life—socio-political, socio-economic, socio-ecological, are geared towards promoting and protecting this dignity—*Ùgwú-Mmadụ.* Thus, Ùgwú- Mmadụ remains the "mma"—the core, the essence, the *kpim,* of what Mmadụ is. This Ùgwú- Mmadụ is both *ontological* and *cosmo-phenomenological.* Ontological, in the sense of being intrinsic in the human person's nature, since Mmadụ possesses the cognitive powers that guarantee his autonomy and individuality among other existing realities—this is the level of Mmadụ's autonomy and inviolability. *Ùgwú-Mmadụ* is also cosmo-phenomenological. This means that it is oriented towards self-transcendence, an orientation towards nature and nurture. This is because Mmadụ possesses both the power of subsisting and a distinct but relational power of being (and participating) in the community of existence. This is the relational-communal level of Mmadụ. However, while Ùgwú-Mmadụ was held autonomous and inviolable, especially in the pre-Christian Igbo worldview, the imbalance created in attitude towards the ontological and cosmo-phenomenological levels of who Mmadụ is and is meant to be, witnessed a preference of approach to Mmadụ, not as the autonomous rational individual, but solely as a generic entity. This gave rise to some practices and approaches that violated the Ùgwú-Mmadụ and sacrificed individual autonomy at the altar of some extreme utilitarian moral principles. Some examples of such violations are the superstitious practice of killing twins, varied forms of human sacrifices, the Òsú caste system, as well as other forms of social and gender dehumanisation and abuse. Oforka opined that the epoch in question:

> [...] was more interested in the moral status of the generic Mmadụ and its preservation as a generic entity. It concentrated more on the defence of the common good and the preservation of the community of Mmadụ. It is believed that this could be essentially achieved by maintaining cosmic equilibrium, which constitutes Igbo ethics – moral good acts help to maintain cosmic order, and the moral bad actions achieve the contrary. The effort to restore this order in the face of moral offences resulted in a quasi-utilitarian morality, which permitted any solution that would benefit the common good, even if it compromised the sanctity of any individual Mmadụ .... Hence, the idea of *Ùgwú-Mmadụ* in the pre-Christian era did not recognise the dignity of the individual Mmadụ as an inalienable and inviolable character. Its concept of dignity could be said to be generic more than appreciating the autonomy and inviolability of the individual Mmadụ. (Ibid., p. xxi)

It is needful here to emphasise that the generic way of approach to Mmadụ is seen as the fundament within African anthropology and cosmology since the individual is strictly defined only in relation to his or her communal affiliations. On the other hand, the approach experienced forms of extremism, which in many cases undermined the very essence of Mmadụ as a distinct moral entity. Perhaps, this worldview, attitude and practice provided for the smooth sailing of the slave trade, which experienced so much lucrativeness within the Igbo societies (by extension, within so many African societies) during this dark period of history. However, a notable reconstruction and socio-cultural *inculturation* happened.

4.3.2 The African-Igbo Mmadụ: The Enculturated *Status*

The encounter between so many African cultures and societies with the Christian missionaries beginning in the fifteenth century and the consequent Christianisation of Africa marked a turning point in the history of the continents involved. In the context of an understanding of the human person and the beginning of formal education, it was the story of the meeting between Western individualised concepts with the African generic approach to Mmadụ. Thus, the Western form of autonomy and individuality of the human person taken as the central thesis in defining Mmadụ met the African preference for communality (commonality, community) as the basis for authentic individuality of the human person. This resulted in a dialogue, which, of course, spanned over four centuries but which eventually witnessed some positive cultural and institutional purifications, modifications and assimilations. Consequently, the African generic approach to the human person, which invariable initiated and permitted some concrete cultural and institutional practices that violated the autonomy and absoluteness of the individual, began to integrate and assimilate more of the belief in the inviolability and sanctity of life of Mmadụ, in other words,

Mmadụ as an individual self, autonomous, rational and self-subsistent. Within the Igbo culture, this resulted in a kind of cultural reformation, which repaired the general approach to the status of Mmadụ and helped to put an end to some barbaric practices. Chief among them were the practices of infanticide of twins and some forms of human sacrifices carried out as requirements by the gods, either as oblation or as appeasement.

However, during this period of cultural and institutional transformation, southeastern Nigeria witnessed a socio-political and economic detractive force in the form of the slave trade (1700–1850, the heaviest period), which inflicted another form of inhumanity in the African societies in general and Nigeria in particular. By the time the slave trade was abolished in 1807, another form of attack on Africa's indigenous and independent worldview played itself out in the practice of colonialism. Nigeria has its own share in the period between 1900 and 1960. Within this period, the political and economic agenda implicit in the colonial blueprint attacked and detracted both the individuality and the transcendence of the African-Igbo Mmadụ. Thus, with the events of colonialism, the Igbo *Mmadụ*, alongside other African men and women, experienced an ideological pressure, which he or she was compelled to grapple with even unto the present age. At the initial stage, the cultural attitude that was fundamentally communal effortlessly gave the harbingers of colonial ideology a smooth landing. However, the host culture would eventually be made to live in a constant rethink of its involvement and hospitality to the guests. What was thought to be an encounter for mutual enrichment would eventually crystallise into forms of superimposition, material exploitation and, in recorded extreme cases, intimidation and violent conflicts, which also claimed lives and properties. Achebe re-lived this concrete experience in his literary choice of words thus: "The white man is very clever. He came quietly and peaceably with his religion. We were amused at his foolishness and allowed him to stay. Now he has won our brothers, and our clan can no longer act like one. He has put a knife on the things that held us together and we have fallen apart" (Achebe 2010).

Our research may not accept totally that the moments of encounter between the African-Igbo Mmadụ and other cultures and civilisations foreign to the indigenous one brought a total deformation of the Igbo worldview that is irreparable. However, the dual experiences of the slave trade and colonialism left a scar, which has become persistent in the African person's struggle for authentic life and emancipation. Such a scar has put a divide on the intrinsic nature of Mmadụ—a scar in the form of a violation of personhood. To date, therefore, the African Mmadụ remains in the struggle to prove his or her identity as a being that is distinct, subsistent and autonomous; he or she must struggle to tell and represent his or her subjective worldview which often is labelled primitive and barbaric, etc. It became an ideology that was designed and sold into history for the political and economic purposes of the designers, and it worked. However, conscious of the harm that was already done, the Igbos of southeastern Nigeria, whose concept "Mmadụ" we have chosen as the focal point in this study, initiated an ideological revolution within the first phase of independent Nigeria so as to retrieve the authentic view about Mmadụ and what the essence should mean. The initial response to this saw the establishment and formal opening of the University of

Nigeria in eastern Nigeria on 7 October 1960, six days after Nigeria's Independence from colonial Britain. Oforka opined:

> Sequel to (the) United Nations declaration on human dignity, on 10 December 1948, the first indigenous university in Nigeria, the University of Nigeria Nsukka, adopted the following motto: 'To restore the dignity of man' [....] The university was founded by Igbo elites. These elites – many of whom were scholars – perceived the period following colonialism as a period of rebuilding the long history of the serial abuse of human dignity in the indignities of slave trade and colonialism [....] The Igbo shared greatly in this brutal history. (Ibid. *Afro-Igbo Mmadu and Thomas Aquinas' Imago Dei: An Intercultural Dialogue on Human Dignity*, p. xix–xx)

"To restore the dignity of man" is, therefore, predicated on the backdrop that the human person—Mmadụ, following the events of some dehumanising pre-Christian cultural practices, the affront of the slave trade as well as the oppression of colonialism, remains wounded within the dynamics of the encounters. Thus, the founding fathers initiated a plan for action to combat the indignities and restate the status of the human person by appealing to the indigenous worldview and educational emancipation. The initial success was the lead recorded in the South with regard to the number of intellectuals produced, as well as the acceptance and importance placed on basic and formal education for the average child looked upon as the leader of tomorrow's Igbo society. However, the forced fusion of different cultural conventions within the Nigerian state, which many consider, to date, an imposition given the event of amalgamation, created and still causes confusion about the authentic identity of Mmadụ. The harm is considered done through a socio-phenomenal attack on the "**mma**"—the beauty, the essence, the fundaments, the prerequisites, which invariably debases "**ndụ**", the life, existence. This is viewed, in effect, as a socio-psychical semantic divide, which keeps the Mmadụ within the Nigerian context, struggling, staggering and in a constant feeling of not being self-actualised.

The final thesis in this context, therefore, is: the *enculturated* view about Mmadụ, namely, the priority placed on Mmadụ as a distinct, rational and autonomous individual, as well as the concurrent generic approach to Mmadụ as a relational being that is both self-subsistent and self-transcendent, provides the contemporary view about Mmadụ in the Nigerian context. But the "*mma*" taken as the quintessence of this human person and within which lies the power of self-transcendence and self-communication remains under attack within the Nigerian socio-political and socio-economic systems, which emit adverse detractive forces (structures of sin) in the forms of behavioural, attitudinal, structural as well as infrastructural machinations that detract the union of "*mma*" and "*ndụ*". This causes identity conflicts and sets the human *self* in a persistent struggle for self-actualisation on the one hand. On the other hand, the person battles forces impeding the power of self-transcendence. In effect, it means that Mmadụ's relational powers, which make Mmadụ central within the community of existence, become impaired, disempowered and dehumanised. This leads exactly to the *loss of ecological humanity*—in the African-Igbo parlance, the "*loss of Ùgwú-Mmadụ*".

4.3.3 *The Critique of* **Ùgwú-Mmadụ** *in the Contemporary Nigerian Situation: A Capability Approach*

The events of inculturation did not introduce a new ontology about the Mmadụ in Africa. It rather helped to chart a course towards the retrieval of an integrative concept about Mmadụ—a concept exhaustive for all that the human person's essence and existence should mean. In other words, Mmadụ's status remains authentic only within the confluence of his *autonomous* self and the *relational* self. The healthy relational dynamics—*ad intra* and *ad extra*—constitute the *Ùgwú-Mmadụ*, human dignity. The loss of Ùgwú-Mmadụ can invariably lead to a loss of the *self*, a loss of the person or personhood. It would represent a situation whereby *mma* undergoes a forced separation from *ndụ*, causing the very existence of Mmadụ to be disoriented and developmentally amorphous. The recovery of Ùgwú-Mmadụ, therefore, is the cardinal point on which the development dynamics for sustainability in Africa, in general, and Nigeria, in particular, should focus. The aim would be to revitalise, reintegrate and reinstate Mmadụ within the general polity as the true moral agent. Nevertheless, for more clarity of approach, Ùgwú-Mmadụ as a concept needs to be understood. How?

Ùgwù is the Afro-Igbo word for dignity, regarded as a normative integrative concept that applies to and is intrinsic to Mmadụ. The Ùgwù of Mmadụ is the attributive essence of who a human person is, namely: that Mmadu is distinct, subsistent, rational and autonomous, and that Mmadu has the power of self-transcendence, which is the basis for his or her relational identity within the community of existence. These intrinsic attributes go into the definition of Ùgwú-Mmadụ, human dignity. Thus, our working thesis in this regard is:

> ***Ùgwú-Mmadụ*** is the worth and essence of Mmadụ, it is the ability of the Mmadụ to live as a distinct, rational autonomous individual, whose relational powers make him as well a concrete subsistent being within the community of existence. The unhindered ability to exercise this power is the core and meaning of the Ùgwù that is intrinsic to Mmadụ. Even in situations where such powers are impaired due to biological and accidental disabilities, Ùgwú-Mmadụ, as an ontological endowment, is not destroyed or detracted. It would rather be complemented relationally and communally in the community of existence (opinion mine)

The complemental relation is at the basis of Igbo everyday life and philosophy articulated and lived out in the concept of "*onye aghana nwanne ya"* (literary, no one should desert or detract his or her brother or sister), meaning as well, be your brother's (sister's) keeper. This ideology expresses the fundamental belief of the Igbos in corporate social equality as a guiding and inspiring principle.

For a non-ambiguous critique of Ùgwú-Mmadụ in the context in which we approach the Nigerian situation, it is necessary to observe and clarify the trend that tends to substitute the concept, human dignity—Ùgwú-Mmadụ, with its correlate "human rights". While the two concepts are related, keeping them distinct in usage is fundamentally necessary. Thus, while human dignity defines the total value and worth of who and what a human person is and should be, the human rights concept contains specifications and guidelines for the protection of human dignity—they are rights drawn from the *Ùgwú*, the dignity of the human person. After the brutal and devastating World War II experience, the ardent search to reconstruct our world and societies eventually gave rise to the formative articulations of human rights laws, culminating in the Universal Declaration of Human Rights (UDHR). Although this articulated document is viewed to date by varied opinions as a Western-dominated framework with a universal mandate, the human rights concept began to receive an overwhelming acceptance that often tends to substitute and replace the distinctness of the concept of human dignity. However, there is always the need to articulate the two concepts distinctly though relatedly. Secondly, the general image given to the 1948 framework as the UDHR must be brought under review and be approached critically. Onwubiko opined:

> As a matter of fact, the UDHR 1948 stands as the climax of the Western articulation of moral norms into what we have today as human rights. Incidentally, human rights have become at the same time a veritable political instrument of the West in its various forms of centralisation. Human rights thought as universal human rights stay under suspicions of imperialism since universalism itself as a concept is normative and can only fit human dignity as shared heritage. But human rights as laws coming from human dignity can only be understood against their historical cultural developments. The chances of a genuine universal interpretation and application can be gleaned only if the rights are left interpretatively open with the view of dialogue and convergence. (Onwubiko 2012)

This work, in line with the above concerns, finds in the UDHR a framework with fundamental values taken from the inalienable principles that should guide Ùgwú-Mmadụ, but the articulation of which was not extensive and intercultural. Therefore, the *Anwendung*, the application, while not limiting or destroying the core concept of human dignity that it defends, must rely heavily on shared cultural heritages and worldviews, on comparative interpretative approaches as well as on constant effort to initiate grounds for authentic and justifiable convergences. On this platform, the present work finds it socio-culturally relevant to identify the problems of development in general and sustainable development in particular within Nigerian settings. As a first premise, our research proposes that the hinge of the general problem lies in the detractive attack on and the disregard of Ùgwú-Mmadụ within the general Nigerian polity. Such an attack is, therefore, directly proportional to an attack on the human rights principles meant to protect the Mmadụ himself or herself, as well as direct policies and policymakers towards people-oriented approaches and decision-making.

A critique of Ùgwú-Mmadụ as both the ground for an evaluative understanding of the developmental status of Nigeria and as the ground for achieving sustainability in modern developmental dynamics serves the purpose of, first, establishing the fact that the Ùgwú of Mmadụ has been abused and detracted and thus prove how this was done or is still being perpetrated. Secondly, a substantive argument can be made that progress towards a sustainable Nigerian society hinges on structurally and systematically revitalising the *Ùgwú-Mmadụ*. This requires overhauling (exnovating) ineffective policies and practices to foster practical, enduring innovations.

4.3.4 The Status of Ùgwú-Mmadụ Within the Socio-Cultural Contemporary Nigeria

We have argued that *Ùgwú* describes an essence that is intrinsic and invariably inalienable. It is the sum of ***mma***—the good, goodness or essence, which is an intrinsic part of a being. Immanuel Kant equates the "mma" with his deontological concept, "the *goodwill*", which for him remains the only intrinsically known "good" that commands categorically. Now we examine the status of this Ùgwú-Mmadụ within the current Nigerian social structure, using as tools of approach the re-examination and re-evaluation of the *capabilities* and *functionings* as the core categories of Ùgwú-Mmadụ and as control principles within the discourse. In doing this, the purpose would be, first, to ascertain the extent to which the general Nigerian system has worked to achieve development as the freedom of *Mmadụ* himself or herself to choose the form and manner of life and participation in a polity for which he or she has the right to. A second purpose is to determine what the detractive elements are, which form the banes of sustainable development and which demand the need for critical *exnovations*, predominantly when the reverse becomes the case. It is an examination of how the capabilities of the autonomous agent, Mmadụ, are being responded to through social systems and structures that should guarantee authentic functionings for sustainable society and development. Thus, our research views sustainable development of the authentic Mmadụ as the totality of the overt and covert effects flowing from all the intrinsic attributes of Ùgwú-Mmadụ categorised under life; bodily health; bodily integrity; senses, imagination and thought; emotions; practical reason; affiliation; play; control over one's environment; and concern for other species. These are attributes of the transcendent Mmadụ, the relational Mmadụ—the Ùgwú-Mmadụ (Fig. 4.2).

The figure below synthesises the core ideas of Sen and Nussbaum by visually representing the collective human capabilities that underpin individual autonomy and the shared societal well-being. Simply put, they are the values that should be harnessed so as to respect and improve Ùgwú-Mmadụ. To concretise our critique of Ùgwú-Mmadụ as the fundament for any practical inroad towards social and

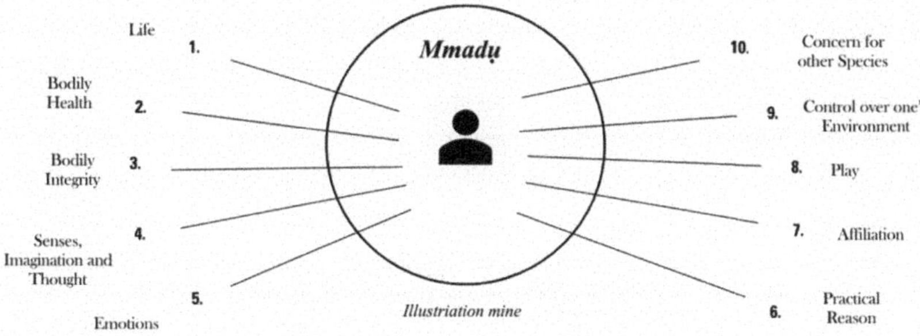

Fig. 4.2 Categories of capabilities as basic attributes of *Ùgwú-Mmadụ*

human development in Nigeria, these values must be examined within the current socio-political, socio-economic and socio-cultural Nigerian society. In doing this, we apply the method of examining concrete areas of political, economic, social and cultural interests. Their verifiable extent of performance within the Nigerian system would be evaluatively scored FAILURE—representing danger, risky or non-performing; DEFICIENT—representing warning, urgent need for action and change or underperformance; or SUCCESS—representing satisfaction, credibility or functional.

1. Category	Means of Evaluation	Social Alert
Life	Social system, health policy and infrastructures, health budget.	FAILURE

***Life*—avoiding escapable morbidity and premature morbidity**: The Igbo cosmology places the greatest value on the capability, life (ndụ) and all that should and must enhance it. Thus, "Ndụ-bụ-isi" remains one of the most populous names in Afro-Igbo culture. It is the *Ndụ* that defines everything a person is and ought to be. Subjected to our means of evaluating this life in the context of our research, namely the social system, health policy, health infrastructures and health budgets, the situation in Nigeria has a red card, in other words, failure. First, the Nigerian social system is predominantly not inclusive. If the **social system** is understood to mean patterns of relationships meant to form a "coherent whole" among individuals, groups and institutions, then the Nigerian divisive system is an assault on the general socio-economic life. The multiethnic character of the Nigerian society, with its consequent ethnic politics, is a burden to the lives of the majority of her citizenry since there exist varied dichotomies, which create the camps of the elite rich versus the disadvantaged poor; the self-acclaimed "born to rule" versus the underprivileged, who must submit their consent even involuntarily; the Islamic North versus the Christian South; the elite Oil Magnates living in their castles and mansions versus the destitute indigenous people of the oil resource areas. Moreover, the general social system feeds on the blood of millions of helpless Nigerians, who, on a daily basis, are frustrated with life. The 2020 EndSARS#[2] revolution in Nigeria and the way so many resorted to the looting of goods and properties in the scramble for what to eat showed an extremely divided and non-inclusive system, which has reduced its citizenry to all categories of *palliative patients.* A functioning social system should make citizens agents of development, not its patients.

The **health policy** in Nigeria has never been a success story since Nigeria's independence. Thus, the life expectancy in Nigeria has remained retarding, with the most

[2] EndSARS# is the designated concept for the 2020 youth revolution against the police and other law enforcement agencies' brutality and inhumanity in the discharge of their duties. While the revolution recorded successes from the beginning, it later became bloody again when a group among the military used firearms against unarmed protesters. It was also alleged that the main ideology of the group that initiated the revolution was later hijacked by some other groups who hid under the umbrella of revolution while carrying out other forms of looting and banditry.

recent statistics showing evidence of about 55.75% for both sexes—male and female. On the other hand, the majority of the health infrastructures and institutions can never qualify for any standard accreditation: they are poorly managed, poorly cited, underfunded, equipped with obsolete medical equipment, infested with fake and expired drugs, substandard medications and poorly staffed. Furthermore, it is alleged that there is a healthcare system packet in the Nigerian system. However, the annual budget has never given corresponding attention to that. Hence, the budget allocations clearly show the inconsistency and indecision of the Nigerian state in defending this category of capability—life.

2. Category	Means of Evaluation	Social Alert
Bodily Health	Poverty rate, access to food and clean drinking water, rate of personal and daily income, the residential system.	FAILURE

***Bodily Health*—being well-sheltered, being adequately nourished, having both bodily and mental health:** Bodily health is a prerequisite for functional and resourceful manpower in any polity. It is a social surplus, which is bound to have a positive multiplier effect within the factors of production, especially on labour. As a social surplus, investing in the means that would enhance and respond to the individual capability to and right to bodily health should be made a requisite agenda in any given political, economic and social system. The means of examining this capability in the current Nigerian situation identifies the poverty rate and the widening gap between the rich and poor as debilitating conditions, which threaten the continuing existence of the whole nation. Amidst her natural resource endowments and the inflow of the petro-dollar, the poverty level in Nigeria has continued to hold the nation within the red lines of the international poverty index. The 2020 multidimensional global monitor and report locate Nigeria as being persistently trapped in low performance and lacking the political will to drive poverty-alleviating policies with the general polity. Placed alongside Ethiopia and Sierra Leone, the socio-economic and socio-political fate that is awaiting Nigeria in the coming decade is shrouded in uncertainty and confusion since the situation, as it is currently known, cannot be identified with progress but rather with unpredictability and indecision. Below is recent concrete statistical evidence (UNDP and OPHI 2020) (Table 4.1).

This second capability, therefore, marred by the situation of poverty among the citizenry—about 28–39.1% (current index)—elicits other negative social factors identified in the lack of varied basic requirements of daily life such as food and clean drinking water, poor sanitation and dehumanising residential conditions, poor income and the rising rate of youth unemployment. Revealing a staggering statistic, Tanko observes that,

> […] For many years, a high unemployment has been a key challenge to the Nigerian government, aggravated by youth restiveness and other vices in all parts of the country, particularly the Niger Delta region […] A National Bureau of Statistics (NBS 2009) report showed that the highest number of the unemployed fell within the 22- to 44-year age bracket, indicating

Table 4.1 The 2020 Global Multidimensional Poverty Index (MPI)

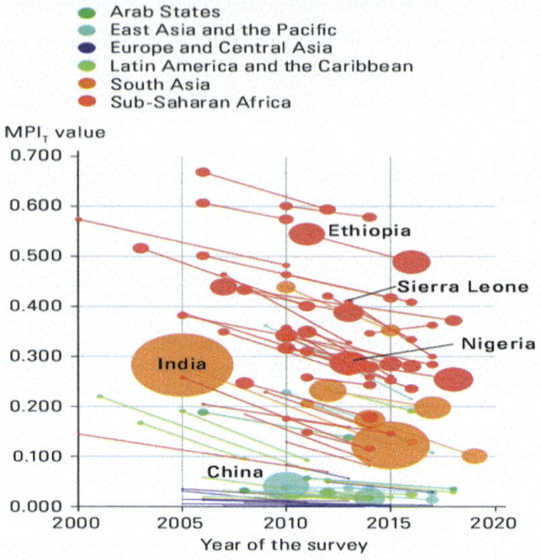

Source Oxford Poverty and Human Development Initiative, OPHI

> that young Nigerians suffered more unemployment challenges than the younger and older folks. Following closely were those between 15 and 25 years, while the elderly, whom the bureau classified into 45- to 64-year group, suffered the least unemployment. (Tanko 2013)

While the Nigerian situation is not the worst in Africa, the level of poverty is considered predominantly a killer factor (Stiglizt 2017) in the general progress of the social system and in the effort to see sustainable institutions manifest within the general polity. No sustainable bodily health could be achieved if people are forcefully plunged into daily experiences that keep life hanging on the fringe of existence.

3. Category	Means of Evaluation	Social Alert
Bodily Integrity	Freedom of movement, security of lives and properties, public mobility system.	FAILURE

Bodily Integrity**—ability to avoid unnecessary or non-beneficial pain, the right to pleasurable experiences, right to secure public mobility:** Integrity conceptually is the quality of being whole and complete. Bodily integrity is the capability of having the reason to live an integrally functional life. This does not mean the total absence of pains and worries. It means the capability to live the necessary pains of life as part of life itself—pains that are an intrinsic part of the ontological meaning of life, not inflicted or man-made physical as well as psychic pains, which could be the results of imposed poverty situations, economic slavery, violent conflicts and wars, dictatorial governance, corruption, etc. When the means of evaluation is applied to

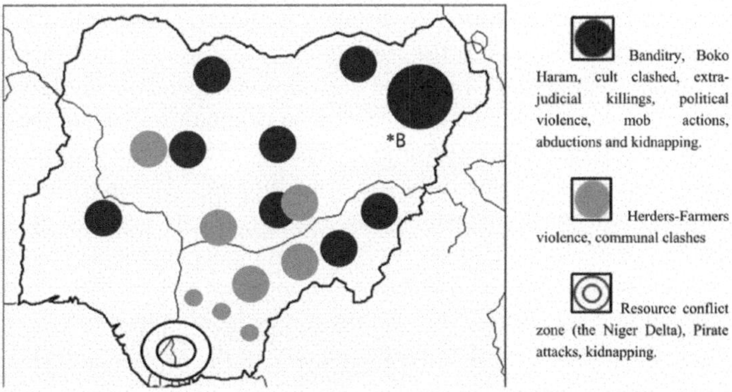

Fig. 4.3 Spatial illustration of the menace of insecurity in Nigeria. *Borno State is the current seat of the politico-religious group, Boko Haram. *Map illustrations mine*

this direction in the Nigerian case, it would become obvious that, perhaps, the greatest bane to this capability remains the tortuous daily experiences of insecurity of lives and properties, which gradually reduces the Nigerian nation to a jungle State. Currently, the freedom of movement enshrined in the constitution as a basic right, as well as the general security of lives and properties, has been hijacked by the unfortunate menace of killer herdsmen's incessant murder and rape, the terrorist Boko Haram insurgency, the excesses and infamy of law enforcement personnel, which has often led to unnecessary losses of lives, especially, those of the youths. Added to these are losses of revenues due to armed conflicts in the resource-conflict zones, where "youth-militants" (if Nigeria ever has the right to qualify Human Rights Activists with such a name) fight for their rights and privileges, etc. Consequently, bodily integrity as a capability issues a red card to the present Nigerian state, indicating a failed policy in this area (Fig. 4.3).

The illustration above shows the current situation of insecurity in the country, with Borno State, in the lead among the other 36 States of the Federation. Other directly affected states include Yobe, Taraba, Adamawa, Kaduna, Katsina, Plateau, Kwara, Bauchi, Nasarawa, Benue, and some South-South/southeastern states like Enugu, Ebonyi, Cross-River and the oil-rich area of the Niger Delta. Statistics show that between January and April 2020, about 1416 victims lost their lives in different violent situations in Nigeria, among them 1141 civilians and about 275 security agents. The whole social atmosphere presents a wounded citizenry under inflicted physical and psychic pains.

4. Category	Means of Evaluation	Social Alert
Senses, Imagination and Thought	Educational institutions, access to information systems.	DEFICIENT

***Senses, Imagination and Thought*—literacy, education and culture, right and access to information**: Access to the right educational system and information is a great asset to any Nation and the social conventions operative in a given society depend heavily on what is being learned, transmitted, enculturated and shared within a given polity. The fourth category among the capacities—the ability to use the senses, imagination and thought fully—is not ultimately lacking in the Nigerian system. The persistent problem, which has a history of its beginnings within the Nigerian state, remains the issue of quality and substance. Nigeria may not be lacking enough school buildings, but the question is about what is left of the general state of the educational infrastructure. Nigeria may not be lacking teachers, lecturers and professors, but the question borders on the quality of the system that produced the academic manpower; even in many cases where the manpower is seasoned and can fit into available world standards, the question hovers around the quality of working conditions being provided for such custodians of the nation's intellectual and character building. Nigeria does not lack pupils or students who have the burning zeal and passion for education and learning. However, the question hampers on the cost of education, as well as on the corrupt and exploitative system of admissions and job opportunities within the educational institutions, where situations of poverty have led many administrations into running the educational institutions as lucrative money-yielding ventures.

A popular theory still current in the Nigerian system is that Nigeria has no *reading public*. Consequently, satirical predators would suggest that the only simple means to hide important information from such polity would be to write it in a book. Such rhetoric, in as much as it may not be one hundred per cent plausible in the case of Nigeria, has at the same time enough concrete elements of truth to the fact that the quality of institutions for the provision of both academic and digital access to information still lacks the required standard and, in varied cases, they are obsolete or corruptly exploited. However, compared to other systems of education around the globe, the Nigerian system qualifies to be considered holistic in approach, as shown below, but there are some socio-political and cultural challenges (Fig. 4.4).

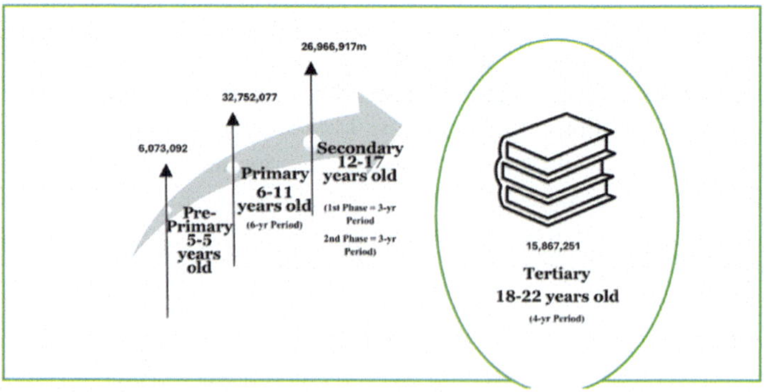

Fig. 4.4 The Nigerian educational system (6–3–3–4 system)

The 6–3–3–4 system comprises six years of primary education, three years of junior secondary education, three years of senior secondary education and a minimum of four years in tertiary education. The number of enrolments presented in *millions* (UNESCO Institute for Statistics) as shown above is an estimated statistic across the whole country. The educational institutions under consideration are the nursery, primary, secondary and tertiary institutions. The compulsory period of education, otherwise known as the basic education period, lasts for 9 years—from primary school (6 years) to the first three years of secondary education. This basic education period is taken to be free and compulsory within the Nigerian constitution. However, there exist sharp differences in the distribution of the literacy rate due to social, economic, cultural and religious factors. The 2018 literacy rate for Nigeria according to zone and gender presented a fact check to support the implicit deficiency within the educational system due to combined factors. With a recognition of the main geopolitical zones, the findings interpret its observations about literacy rate as follows: (a) North-Central—female 49.6%, male 72.9%, (b) North-East—female 31.8%, male 50.5%, (c) North-West—female 29%, male 59%, (d) South-East—female 79.3%, male 86.7%, (e) South-South—female 79%, male 88.5%, (f) South-West—female 80.6%, male 89% (Statistica 2020).

The percentages describe the distributive performances of zones and genders with regard to the constitutional *formal* education system. The sharp disparity between the North and South is vivid due to social, religious and cultural conventions practicable within the individual zones. Thus, while the North is not predominantly open to formal education—the so-called Western education system—due to the dominance of Arabic and Islamic education, the open access given to the missionaries of the nineteenth century's exploration of Africa in the southern part of Nigeria helped to lay the foundations of formal education. The persistent implication of this disparity is the challenges of inconsistencies between the progressive minorities of the North, who promote the idea of formal education as a needful complement to Arabic and Islamic education, and the conservative majority, who predominantly have no formal educational training themselves and are mostly Peasants, Nomads and Bedouins, and who, therefore, see formal education as incongruent to the dictates of Islamic teachings. The activities of the extremists among them have created the sect which goes by the name Boko Haram. In the South, the greatest challenge to the fourth category of capabilities is the dilapidating state of educational infrastructures, especially in most public schools within the rural and semi-urban areas. This has led to the boom of both private and religious commitments in the administration of schools, but the affordability for the common men and women on the streets leaves both the system and sector impact-wise deficient, though not a total failure.

5. Category	Means of Evaluation	Social Alert
Emotions	The level of economic justice, interaction and freedom of self.	DEFICIENT

Emotions—self-transcendence, expanding personal boundaries, reaching out to people and things (nature) one has reason to be close to: Emotions are intrinsic feelings, which are basically the seats of most (if not all) human motivations—motivation towards things and realities outside the individual *conscious self.* This capability, therefore, allows Mmadụ the ability to exercise the conscious feeling of participation, loving and being loved, sharing ideas and having the feeling of making an impact, having a reason for being happy within the community of existence, etc. According to Reed, this capability of emotions is *intrapersonal* since it aims at exploring ways to greater individual person's philosophies of life, values and dreams; it is *interpersonal* because it motivates towards building relationships with others and one's environment (nature); it is *temporal* since one's past and future form a deep conscious process of giving meaning to the present; and it is *transpersonal* because it motivates "... to connect with dimensions beyond the typically discernible world" (Reed 2014).

In the context of our discussion on *Ùgwú-Mmadụ* and the Nigerian case, the capability of emotions is evaluated within the factors receptive to Mmadụ as *a social animal.* Thus, the level of economic justice in a given society would be directly proportional to the quality of interactions among the agents of economics as well as the status of their freedom, emanating from the capability of the individual selves being able to relate, choose and live out the values they have reasons to identify with. The general situation with economic justice in the Nigerian situation, therefore, is a detraction to the capability of emotions if economic justice describes the conscious efforts in policymaking to institutionalise factors and structures of an economy that would be viable enough to ensure equal opportunities for every individual and provide material empowerment for a decent and productive life. The 2019 measure of the variability with regard to income inequality within the Nigerian system using the Gini Coefficient shows *deficient policy response* in the face of persistent economic injustice within the general structure. The findings of the National Bureau of Statistics confirm a distressed Nigerian system with around 40% inequality and poverty levels, where the urban society represents 31.94 Gini Coefficient data and the rural 32.77 Gini Coefficient data.

Interpreting the Gini Index above, the Nigerian National Bureau of Statistics reported:

> In Nigeria, 40.1 per cent of the total population was classified as poor. In other words, on average 4 out of 10 individuals in Nigeria have real per capita expenditures below 137,430 Naira per year. This translates to over 82.9 million Nigerians who are considered poor by national standards. It is important to note, that this number excludes the state of Borno. While the NLSS 2019-19 includes households from Borno, that sample was not representative of the whole state since only households from "accessible" (safe to visit, areas only) were interviewed. Thus, the Borno sample is considered non-random and non-representative.[3]

[3] The Federal Republic of Nigeria. *2019 Poverty and Inequality in Nigeria. An Executive Summary.* National Bureau of Statistics, May 2020. Available via https://nigerianstat.gov.ng/elibrary/read/1092. Accessed November 12, 2020. *Gini Index is a statistical measure of expenditure distribution in the country. It is widely used as a measure of economic inequality. Gini index ranges from near 0, which would indicate perfect equality and 100, which represents perfect inequality.

This situation has really created a Nigerian society where the emotions of self-transcendence in relation to other people and nature are predominantly possessed with varied arts for deceit, greed, swindling, feudal mentality, unhealthy class struggles, superficiality, etc. Emotional freedom is persistently imprisoned in an everyday life of struggling to make ends meet, even at the expense of a neighbour's right to life. In all, therefore, the Ùgwú-Mmadụ remains hunted and debased as a result of both social and economic slaveries implanted in the system.

6. Category	Means of Evaluation	Social Alert
Practical Reason	Freedom of participation, equal access to social and political institutions	DEFICIENT

Practical Reason—**ability to think sensibly, constructively and critically, having a conception of the good, ability to prove qualification for employment and for social and political participation**: One of the essences of Mmadụ is the ability to use the cognitive powers to unlock both himself or herself as well as understand the realities out of him or her. The disposition for and the application of practical reason are, therefore, one of the key attributes of Ùgwú-Mmadụ, human dignity. However, such cognitive powers could be distorted by factors that tend to oppress human freedom, especially in the socio-economic and socio-political realms. Two such distorting factors that are peculiar to the Nigerian system are anchored on the issues of unemployment and exclusive politics.

The youth unemployment crisis in Nigeria is on the increase, predicated as well on the rapid increase in the country's population (cf. Chap. II, Table 3, p. 137). Nigeria is still deficient in the corresponding institutions that should absorb her graduates into the job markets. This is blamed chiefly on the stifled Nigerian economy, which has refused to diversify elaborately so as to decongest the fiscal pressure and over-dependence heaped on the oil sector and market. The most excruciating blow dealt to the Nigerian state due to the youth unemployment saga is the increasing loss of her human resources—the so-called *Brain-Drain-Syndrome*. This means that the quest for the necessary institutions capable of offering young graduates and other manpower entrance to a functional social system and decent life has left so many Nigerian youths frustrated since the general system *ad intra* offers an insufficient solution. Thus, Oforka, with reference to Africa, remarked: "Professionals from different fields are in constant migration to, especially Europe and America, where they believe that their expertise would be appreciated and rewarded. They not only seek greener pastures but also avoid being used for the increase and sustenance of the ever-growing wealth of the cabal that hijacked African politics" (Oforka 2015).

By the third quarter (Q3) of 2018, the unemployment rate of employable Nigerian citizens stood at 23.1% and the rate of underemployed at 20.1% (The Federal Republic of Nigeria 2020). Today, youth unemployment alone has hit 19.67%. While Nigeria is not the worst in Africa, a projected ratio of the employment situation would be incongruent with the projected increase in the size of her population in the coming decade when something concrete is not done to stem the tide. Within the global unemployment statistics, Nigeria ranks 173rd in descending order of *worst affected countries*, with Congo (46.10%) in the lead and Qatar (0.10%) at the bottom of the table as the country with the least number of unemployed persons (cf. Ibid.).

On yet another related level, what is regarded as exclusive politics in the Nigerian system has enthroned a highly tensed, divisive system, which features mixed groups of elite politicians who persist in changing batons among themselves and are ready to eliminate any opposition to their well-designed and imposed plans for protracted enslavement of Nigeria. The situation steals and distorts the natural capability of many Nigerian citizens to exercise their natural powers of practical reason in matters of social and political participation since the stage has been made a jungle laden with predatory detractors with the constant intent to kill and maim. The consequence is an exclusive politics that expels the qualified from mainstream politics, haunts every growing talent considered to be a future challenger, and operates an institution of independent "godfathers", who impose candidates and control affairs when their candidates clinch power. Thus, they infest the whole system with a detractive political and economic capitalism, which translates to oppressive weaponry. Our research, therefore, still sees a deficiency in the proper integration and promotion of freedom to practical reason and its application within the socio-political and socio-economic realms. Also, at this level, the Ùgwú-Mmadụ suffers attack and suppression.

7. Category	Means of Evaluation	Social Alert
Affiliation	Freedom of speech and expression, freedom from intimidation; unimpaired social movement.	DEFICIENT

***Affiliation*—the capability to enjoy self-respect, the feeling of acceptance in the public, functional social interaction**: To realise the self as a self-transcendent being, Mmadụ must do this in an atmosphere of freedom. Thus, the means of evaluating the level of affiliation within the Nigerian context is to examine the workability of her constitutional freedom of speech and expression, as well as the extent to which the citizens themselves feel secure within the general polity. Generally, in social behaviour, when the sense of affiliation stays under attack by societal factors that tend to create both psychic and phenomenal enemies, the Mmadụ himself or herself develops varied inauthentic aptitudes, which serve only as behavioural strategies in order to escape varied reprehensible social situations.

This capability, in the current Nigerian situation, has a history of political and institutional breaches, which play themselves out in various forms of socio-political intimidations and killings. Cases of political killings like that of Dele Giwa—1986; Ken Saro Wiwa and other eight Ogoni Sons—1995; Moshood Abiola—1998 and Kudirat Abiola—1996; Bola Ige—2001; the carnages of political killings and approved organised massacres, both under various military regimes and civilian regimes; the insurgencies of sects with both religious and political affiliations—most horrible in the history of Nigeria remains the experience with Boko Haram; the proscription of non-violent freedom-fighting groups as terrorists, alongside government's culpable silence and indirect endorsement of the killings and maiming of the Killer-Herdsmen,—to mention but these selected few. All of these attest to a distorted form of affiliate life in the Nigerian system. The nature and quality of social interactions and affiliations have wounds in the forms of suspicious relationships, ethnic cleavages that are inbuilt even in matters of national concerns, *Struggling-To-Survive* mentality, confrontational attitudes as escape routes to impose self and subvert another's position, etc. Affiliate social life within the present system, therefore, is deficient.

8. Category	Means of Evaluation	Social Alert
Play	Social and recreational amenities; access to leisure.	DEFICIENT

***Play*—ability to have fun, have access to recreational activities, enjoy peace of mind, vacationing**: Recreation is a necessary part of both physical and psychic developments. Organised societies, therefore, invest in factors, policies and amenities, which in turn help the human person exercise this capability in him or her. Social and recreational amenities as part of public goods and services show the social alert that is *deficient* in the Nigerian case. A hungry human person within the Nigerian structure struggles, first, to beat the potentiality of life-threatening sickness and death itself. Having fun or going on vacation has never made the list of programmes in an average Nigerian family, with many struggling to live within 1 dollar ($ 1) per day. Several other secondary factors militate against the possibility of exercising this capability as well, namely: the general insecurity of lives and properties in all nooks and crannies of the country; the oppressive quality of the general income of millions of the citizenry, which cannot guarantee a minimum decent life; the absence of organised and affordable commercial transportation system and, of course, the underinvestment on leisure and recreational infrastructures as an integral part of social and community development. No doubt that the inability of the Nigerian system to give a proper response to the capability of the citizens to have access to leisure and vocational activities has contributed to the eroding of Nigeria's general Life Expectancy, with the current scientific measure of the human development index pegging it at 54.3 years—representing general life expectancy at birth.

9. Category	Means of Evaluation	Social Alert
Control over one's Environment	Access to the basic needs of daily life; ability of making personal choices and decisions, empowerment programmes.	DEFICIENT

***Control over One's Environment*—the ability to make personal choices, living one's own life and not another person's, right to personal properties**: Freedom to the choices of life is a defining attribute of Mmadụ, the human person. It is the intrinsic power in the exercise of the individual person's autonomy but in relation to one's environment. The concrete means of evaluation concerns predominantly the quality of access, which the general citizenry "equitably" has to the basic needs of

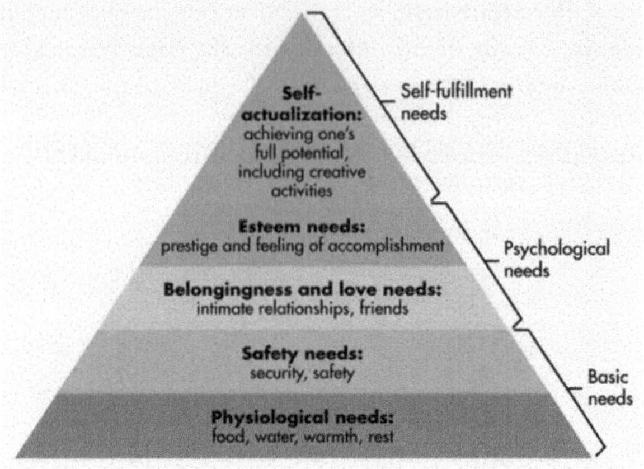

Fig. 4.5 Maslow's pyramid of human basic needs as motivating factors for the functioning of the ninth capability. *Source* SimplyPsychology

daily life, in other words, the quality of access to what our research calls the *basics of a functional life*. Abraham Maslow saw this capability as fundamental 77 years ago with his idea on the hierarchy of human basic needs presented in a pyramid form and discussed in his "A Theory of Human Motivation" (Fig. 4.5).

The above three categories of (a) basic body/physical needs, (b) psychological needs and (c) self-fulfilment needs contribute to the empowerment of Mmadụ in mastering his or her environment through personal choices and decisions, which impacts both the individual person's life and other realities outside of the human person. They are prepotent drives, which, in the opinion of Maslow, must not be thwarted or inhibited since "... Thwarting, actual or imminent, of these basic needs, provides a psychological threat that leads to psychopathy" (Maslow 1943).[4]

The Nigerian situation, which our research met, wallows in a deep social inequality, which places a huge majority of the population under living conditions where the basic needs of daily life are either totally absent or corruptly made unattainable by the actions of a selected few. The most covert consequence of such a living condition is the loss of orientation and perspective since the factors of authentic empowerment are denied. Thus, the *Mmadụ*, who lives in such deficit conditions, degenerates either consciously or unconsciously to a life of psychopathic struggles. Such conditions force the human person himself or herself into the daily life of struggles with endless ambivalences, thus creating both physical and psychic scenes whereby the Nigerian *Mmadụ* seems to be in constant war with both himself or herself and the realities outside of him or her. This leads to watching society degenerate into a daily life of fury and unhealthy (unnecessary) class struggles within the polity, as well as a general life of dishonesty (corruption) being practised almost as a social convention. Such is the picture of the *deficiency* witnessed with regard to this capability in the Nigerian system.

[4] Abraham Maslow (1908–1970) was an American Psychologist. His idea on "The Theory of Human Motivation" was published in the American Journal of Psychological Review (1943).

10. Category	Means of Evaluation	Social Alert
Relation to other species	Ecology, economic policy, climatic policy,	FAILURE

***Relation to Other Species*—having concern and feeling for the realities outside of the self, developing relationships with nature**: The human person—Mmadụ—has the capability to develop a natural relational consciousness about realities (Nature) outside of the human individual selves. The ability to exercise this function led Aristotle to affirm that a person is a social animal. This socialisation is in relation to nature, a part of which Mmadụ is and the resources on which Mmadụ lives, works and, thereby, realises his or her potential. Hence, the means for evaluating the functionality of this capability among a given people would be the examination of attitudes towards ecology, in other words, the assessment of a people's general economic ideology in relation to nature itself that provides the factors of production. Added to this is the level of concern and sense of responsibility that society has towards one of the most current *signs of the time*, namely climate change.

We have had time in this research to discuss the general Nigerian ecological situation and the discussion shows the failure of the government to strike a healthy balance, at least to the barest minimum, with regard to economy and ecology (Nature). The situation of imposed poverty has, in many ways, separated the human conscience from forming healthy and sustainable concepts with the environment within which the person himself or herself lives, moves and has his or her being. The one single factor that has caused this failure remains the socio-economic degeneration of the entire system down to mere *economism.* Thus, when economic policies focus only on absolute *supply and demand*, with only a little or no effort towards regenerative economic sense, human relation to nature itself becomes invariably anthropogenic. Concrete pieces of evidence are identified with Nigeria's over-dependence on crude oil economy, the terrible rate of both private and commercial fossil fuel combustion and the level of carbon emission in the quest for power generation and other factors that have been discussed.

The concluding thesis, therefore, is that Ùgwú-Mmadụ remains in the main under attack, with the presence of all socio-political, socio-economic and socio-ecological factors, which overtly or covertly thwart the individual person's capability to develop a healthy and sustainable relationship with nature's capitals within which he or she realises himself or herself. Bemoaning the gradual dilapidation of access to the exercise of this capacity in the human person, Pope Francis remarked with regard to the global environmental challenges, thus, "[…] We were not meant to be inundated by cement, asphalt, glass and metal (also by toxic crude wastes and spills—*emphasis mine*) and deprived of physical contact with nature" (Francis 2015). In the final analysis, therefore, a social system that plays down on the centrality of the individual person being the central focus for sustainable development loses the battle to a loss of human ecology, which should be the base that unites all other factors of both human and societal developments. Nigeria remains a victim.

4.4 The Unfortunate "Divide": Loss of Human Ecology as Loss of Ùgwú-Mmadụ

In seiner Begründung bezieht sich das Nachhaltigkeitsprinzip zentral auf die besondere Würde und Verantwortung des Menschen als Person, die Voraussetzung und Schutzziel nachhaltiger Entwicklung ist und sich politisch in der Achtung der demokratischen Menschen- und Mitwirkungsrechte konkretisiert.

– Markus Vogt (2009)[5]

The semantics applied in our approach in making the human person, Mmadụ, the central moral agent in both sustainability and developmental discussions, drew from the very concepts that define a people's worldview and socio-cultural conventions. Thus, the approach is not to defend an abstract metaphysical concept but rather to define an ontological basis of a people's concrete existence and experiences. Against this background, therefore, the core consequence of structural sin found within the Nigerian system is a form of "great divide" infringed on the human person within the general structure of Nigerian society. In other words, there is a forced separation of indivisible correlates—"**ndụ**" and "**mma**"—representing a frontal assault and attack on the dignity of the human person. It is an attack on the individual person's right to develop his or her natural capabilities while choosing the life and making decisions he or she has reasons thereof. Hence, the moral agent of development, Mmadụ, within the Nigerian context is *no longer at ease*, using the indigenous expression of the African-Igbo Laureate, Chinua Achebe, since the knife of in-built structural and ideological maladies, the structures of Sin, has remained persistently malignant and detractive in robbing the general polity of people-oriented, integral, transformative and sustainable development. The development concept in Nigeria *falls apart*.[6]

Thus, the above thought-provoking citation from Vogt is explicit in positioning our effort to make *Mmadụ*, the human person, and *Ùgwú-Mmadụ,* human dignity, the central focus within the general approach to sustainable development. The person remains the prerequisite and the protective goal of sustainable development, towards which and upon which all other factors within the development dynamics depend and take their meaning. Hence, the following four theses would be held fundamental in the application of our research findings, namely:

---

[5] Sustainability principle makes central the dignity (Ùgwú-Mmadụ) and the responsibility of the human being as a person the prerequisite and objective of sustainable development. This is made concrete politically in the special attention given to democratic human and participation rights in any given polity (translation mine).

[6] *Things Fall Apart and No Longer at Ease are literary genres of the late Nigerian Laureate, Prof. Chinua Achebe (1930–2013), where he painstakingly reflected on the cultural and institutional conflicts embedded in the Nigerian system, which, to date, remain active and negatively influential, especially, with regard to the human person, *Mmadụ*, his rights and his dignity.

The Centrality of Ùgwú-Mmadụ within the Sustainability Discourse
Thesis 1 The functional union of the attributes of *mma* and *ndụ* is the only meaning of human dignity, Ùgwú-Mmadụ
Thesis 2 Human dignity is the *terminus ad quem* of the sustainability discourse
Thesis 3 Wherever the core attributes of *mma* are either lacking, endangered, abused or forcefully denied, there is no foundation for sustainable development
Thesis 4 Wherever and in whichever manner the attribute *ndụ* suffers man-made secession from its endogenous part *mma,* it would result in a consequential separation that leads to the loss of Ugwu Mmadụ, human dignity

Texts and Illustrations mine

The above-mentioned *theses* are the standpoints of our research in proposing the fact that the proper respect due for the dignity of Mmadụ remains the focal point from which and upon which all socio-political, socio-economic as well as socio-cultural factors work integrally for a society that would not just be based on a utopia, but rather on concrete and concerted efforts to balance interests within intergenerational justice. When the reverse of the above thesis is the case, there would be a degenerative loss of human ecology, which makes any effort for a sustainable human society illusory. The Nigerian case is a signatory to the planned and executed loss of human ecology.

4.4.1 The Consequences of Lost Sense of Human Ecology in the Nigerian Context

A lost sense of human ecology leaves the human person—Mmadụ—in Nigeria balkanised. Helmut Lotti sang about such balkanised situations with regard to general Africa in his album *Classics Out of Africa (1999)* when he remarked that in Africa, and our context, Nigeria, *life may not be certain, but souls are breathing*. The focus of our research, therefore, has been to identify the interrelated causes of the uncertainty of life within the Nigerian system and, perhaps, put forward the feasible proposition that ***even the breathing souls within the man-made balkanised system in Nigeria can still decide to change course and make life certain and sustainable.***

Until a change is legitimised and practised, the current situation of lost human ecology has its consequences, which are socio-political, socio-economic, socio-ecological, as well as socio-cultural. We have made efforts to analyse these varied but intertwined dimensions of the lost human ecological culture in Nigeria through practical, observable and verifiable facts taken from the fieldwork and other statistics presented. But before the work embarks on its final normative inputs towards effective transformation, our central idea of placing the human person as both the object and subject of any sustainable development policy needs a sort of *socio-theological* analysis of what authentic human ecology represents. Attempts would be made as well to emphasise the need to stripe the concept of all false enclosures, which tend to *sentimentalise* the concept itself, with particular reference to the social teachings, thereby attempting to blind the concept from being open to dialogue with the numerous perspectives of modern natural and social sciences.

4.4.2 Setting the Pace Towards Restoration: Human Ecology and the Socio-Theological Contributions of the Social Teachings

Ecology, etymologically, applies to the study, knowledge or science of the environment, where the concept of the environment represents the totality of the existential dynamics involving the physical and external realities surrounding the human person's existence but in relation to him or her. Human ecology, therefore, describes the nature of these dynamics of existence but with particular reference to the human person as a form of *centripetal* factor within the community of existence. In other words, human ecology would examine the nature, quality and centrality of the person's position in relation to the influences, though external to the human person, which constitute a network of both overt and covert factors that exert observable influences on the human person. In the course of our research, this human ecology has been proven to be under attack within the Nigerian polity due to the absurdity of varied social, political, economic and cultural imbalances—Structures of Sin. Relevant excerpts taken from key social documents would help to consolidate our central thesis on the ecology of Mmadụ, as well as pave the way towards a clearer understanding of the *how*, *why* and *consequences* of the battered status of Mmadụ within the Nigerian context.

The social teachings of the church remain aligned with the very priority that should be given to the individual, the *Mmadụ,* in all forms of socio-political, socio-economic and socio-cultural concerns. Thus, over half a century ago, Pope Paul VI, while appraising the challenging situations facing the developing world hued out of the enlightenment, as well as the subsequent decades which saw many of the colonial colonies gaining independence, drew attention to the need to give the right focus to the moral agent of development. In his encyclical, ***Populorum Progressio***, on the development of Peoplesp the Pope teaches:

> Today we see men trying to secure a sure food supply, cures for diseases, and steady employment. We see them trying to eliminate every ill, to remove every obstacle which offends man's dignity. They are continually striving to exercise greater personal responsibility; to do more, learn more, and have more so that they might increase their personal worth. And yet, at the same time, a large number of them live amid conditions which frustrate these legitimate desires. Moreover, those nations which have recently gained independence find that political freedom is not enough. They must also acquire the social and economic structures and processes that accord with man's nature and activity, if their citizens are to achieve personal

> growth and if their country is to take its rightful place in the international community. (Paul 1967)

The reference made by the encyclical with the expression "*conditions which frustrate these legitimate desires*" in Mmadụ alludes to the primacy of the proper focus and positioning of the human person as the focal point in the general development dynamics. This is the human person upon which all societal influences must act in a manner that does not frustrate or detract from the legitimate capabilities in him or her, which constitute the Ùgwú-Mmadụ (human dignity). All manners of approaches should rather integrate efforts and policies towards development fruition, in other words, towards the life an individual has a legitimate reason to choose. Such is, therefore, the practical constitutive element for a workable human ecology. Furthermore, authentic human ecology is nothing but the healthy relational dynamics existing between the individual, with his or her capabilities, and the social and economic structures, which often are procedural to the making or marring of the human person—in other words, to the realisation of authentic or defective human ecology.

During the 20th anniversary of the publication of Populorum Progressio, Pope St. John Paul II, with the publication of *Solicitudo rei Socialis*—On Social Concerns in 1987—reiterated further the need for an undying concern for the respect for who the human person is and the need to channel all aspects of social development towards safeguarding the dignity of the human person. The encyclical speaks about the danger of a form of "naive mechanistic optimism", which has enthroned growing anxiety for the fate of humanity. In other words, there is growing anxiety about authentic human development and authentic human ecology. Thus, the development process, which can only focus on economic gains or mere material accumulation and consumerism, is in crisis since the approach can never be holistic as to recognise the "person" as the moral agent of all social, political and economic development dynamics. The encyclical observes:

> A naive mechanistic optimism has been replaced by a well-founded anxiety for the fate of humanity [….] In fact, there is a better understanding today that the mere accumulation of goods and services, even for the benefit of the majority, is not enough for the realisation of human happiness. Nor, in consequence, does the availability of the many real benefits provided in recent times by science and technology, including the computer science, bring freedom from every form of slavery. On the contrary, the experience of recent years shows that unless all considerable body of resources and potential at people's disposal is guided by a moral understanding and by an orientation towards the true good of human race, it easily turns against people to oppress them. (Paul 1987)

The need for refocusing on the natural potentials at a person's disposal is the clarion invitation to model development dynamics to be both accessible and receptive of Mmadụ, whose potentials (capabilities) to choose, as well as the freedom to transcend the self in a relational affinity, provide the only authentic platform for progress and development strategies that could qualify to be sustainable. John Paul II refers to this as orientation towards the reality and vocation of people, anchored on respect for a person's "interior dimension" (Ibid., no. 29)—the intrinsic freedom to choose the life he or she has a reason for. To integrate well this specific nature of Mmadụ, namely, the interior dimension, with all socio-political, socio-economic and socio-cultural processes, remains the most practical and valid approach to authentic human ecology.

Furthermore, the encyclical, while recalling the teachings of Laborem Exercens, places the human person, or rather, the priority of functional human ecology, as the only *protagonist* of authentic (sustainable) development (Ibid., no. 30). In other words, the encyclical argues that the foundation for the structural base of any society's effort to translate its ideals and policies to reflect not only economic and technological advancements and breakthroughs—*naive mechanistic optimism*—must find a solution in making ideals and policies people-oriented. Such is and ought to be the integral moral character of value-based development dynamics, which possesses the power to make all factors of development durable and development itself sustainable.

The hundredth anniversary of the encyclical, Rerum Novarum, witnessed the publishing of the encyclical *Centesimus Annus* (1991). Deeply influenced by the battered status of the human person and human dignity, as well as an increasingly political and economic dictatorship, which was based on Marxist pure materialism, the encyclical bemoaned an institutionalised disregard for authentic human ecology. The encyclical observed:

> Although people are rightly worried – though much less than they should be – about preserving the natural habitats of the various animal species threatened with extinction, because they realise that each of these species makes its particular contribution to the balance of nature in general, too little effort is made to safeguard the moral conditions of an authentic human ecology. (Paul 1991)

The call to give attention to the moral conditions of the individual is the demand to see Ùgwú-Mmadụ, human dignity, as an indispensable progenitor of real and sustainable development. Hence, all societal and institutional structures must be designed and applied in such a way as to harness and not inhibit the essence of Mmadụ himself or herself and the natural potentials (capabilities) due to him or her. The first proposal, therefore, for practical recovery of the human ecology is to preserve the true meaning of the family as the nucleus of society. The encyclical teaches further:

> The decisions which create a human environment can give rise to specific structures of sin which impede the full realisation of those who are in any way oppressed by them. To destroy such structures and replace them with more authentic forms of living in community is a task which demands courage and patience. The first and fundamental structure for "human ecology" is the family, in which man receives his first formative ideas about truth and goodness and learns what it means to love and to be loved, and thus what it actually means to be a person. (Ibid., nos. 38–39)

The allusion to "what it actually means to be a person" refers to what the authentic life of the human person should mean when the natural capabilities are used unimpeded, thereby translating to functionings, which are functional states and conditions of life at the receiving ends of the individual person's powers and potentials. This single approach, which cares for the authentic life of Mmadụ, would be seen as the only platform for authentic democracy, made possible only by respect for the rule of law and unprejudiced conceptions about the human person (cf. Ibid., no. 48).

Taking further steps forward, John Paul II, in 1995, drew the attention of the world to the need to embrace the culture of life, in other words, the need to see human life in general as ontologically valuable and inviolable. As a result, he issued the encyclical *Evangelium Vitae*, which became an elaboration of the church's stand on the promotion of authentic human ecology as ethics of life. The encyclical places emphasis on the obligatory need to respect the dignity of the human person who experiences the tragedy of what he calls "… the eclipse of the sense of God and of man, typical of a social and cultural climate dominated by secularism, … with its ubiquitous tentacles…." (Paul 1995). Consequently, losing the sense of God, in other words, the sense of the good, the sense of morality, remains a possible catalyst to losing the sense of the human person and the dignity inviolable to his or her being. For authentic human ecology, therefore, there must be the resolve to abandon resolutely the *culture of death* and embrace the *culture of life*. Such culture of life should not just be limited to the spiritual and moral efforts to protect the life of the unborn but to the protection of humanity itself, the human person in all moral, socio-political, socio-economic and socio-cultural dimensions. Thus, the encyclical alludes to what it sees as a social concern for all and to what it calls "… too many signs of death…" (Ibid., no. 77) as a result of human negligence and excesses. This requires the urgent need for all to work for the "… establishment of a new culture of life, the fruit of the culture of truth and of love, (which) is the foundation of a renewed and sustainable society" (Ibid.).

The year 2015 witnessed the issuance of the encyclical, *Laudato Si*, On Care for our Common Home, by Pope Francis. Although the encyclical has as its main focus the urgent need to address the challenging problems of climate change, which the present world must no longer neglect or ignore, it, however, addressed its theme with an integrated approach based upon the clarion call for an integral ecology. It is an appeal that, in praxis, must reflect an ecology of daily life, a practical human ecology—a regenerative and integrative focus on *human and social dimensions* (cf. Ibid., *Laudato Si*, no. 137). For the encyclical, the functioning or malfunctioning of a society's institutions affects the quality of human life (human ecology) and, invariably, the wider environment. Thus, it bemoaned the many situations of our contemporary world, which experiences in many countries "[…] a relatively low level of institutional effectiveness" (Ibid., no. 142).

The encyclical, therefore, calls for a revolutionary approach so as to retrieve the true meaning and value of culture—people's way of life—as an integral part of the general development process since culture itself "… is more than what we have inherited from the past; it is also, and above all, a living, dynamic and participatory present reality, which cannot be excluded as we rethink the relationship between human beings and the environment" (Ibid., no. 144). This will pave the way for an ecology of daily life, in other words, a form of concern for human ecology where authentic (and sustainable) development would "… include efforts to bring about an integral improvement in the quality of human life, and this entails considering the setting in which people live their lives" (Ibid., no. 147). An integral improvement in the quality of human life, therefore, requires the proper institutions and approaches capable of respecting and prioritising human dignity and the dynamic proficiencies intrinsic in Mmadụ in order to live and pursue the type of life he or she has reason to live. To do this in concrete terms, the document demands the immediate resolve of our world to do away with all forms of superficialities in development dynamics and focus on integral policies, politics and socio-economics. For "It is not enough to seek the beauty of design. More precious still is the services we offer to another kind of beauty: people's quality of life, their proper adaptation to the environment, encounter and mutual assistance" (Ibid., no. 150). Such is an authentic human ecology that works for integral and sustainable development.

4.5 "From Human Ecology to Ecological Humanity": Evaluating the 10-Thesis Approach Towards Ethical Regeneration in Human Ecological Discourses

The plausibility of the social teachings' thesis on human ecology can never be denied given the increasing concerns of our modern world with the issue of Anthropocene and the urgent need to integrate the three dimensions of the human ecological living space, namely individual, social and ecological. However, there must be a need as well for this model of thought on human ecology to be open to engaging critically both the dialoguing and opposing voices of all relevant disciplines within the general research spaces with regard to the modern Anthropocene. Thus, the socio-ethical thinker Markus Vogt sees the need for human ecological discourse to assume a form of Critical Modern Anthropology capable of both scientific and pedagogical dialogue with other fields of human, psychological and evolutionary sciences. This would be a sort of coevolution bringing together ecological, socio-cultural and economic systems to bear on the living conditions and spaces of the Mmadụ himself or herself, thereby leading to a new approach and a more holistic concept—*ecological humanity* (Vogt 2016).

This new conceptual approach—ecological humanity—is not contradictory to the mainstream thesis on human ecology. It rather makes the concept more holistic and more domestic in approach to the issue of modern Anthropocene, which, for an authentic interpretation of human conditions, must integrate the *biological*—relating to the natural environment of the human person—and the *socio-ecological*—relating to the cultural spaces of the human person (Ibid., p. 96). Thus, ethical dynamics would be required to cope with the known basic conflicts involving culture and nature, in other words, freedom and causality (Ibid). For the truth remains that the human person has both corporeal and incorporeal as well as both biological and socio-cultural nature. This means that in all ramifications, the person has a dual *interactive* nature (Ibid.). The consciousness of this dual nature must be the focal point in human ecology such that approaches to human conditions would not just seek to place Mmadụ at the centre of life, where the extant capabilities of the human person may as well turn to enslave him or her. The right approach and concern would always be the focus on *how good, how appropriate and how sustainable human*

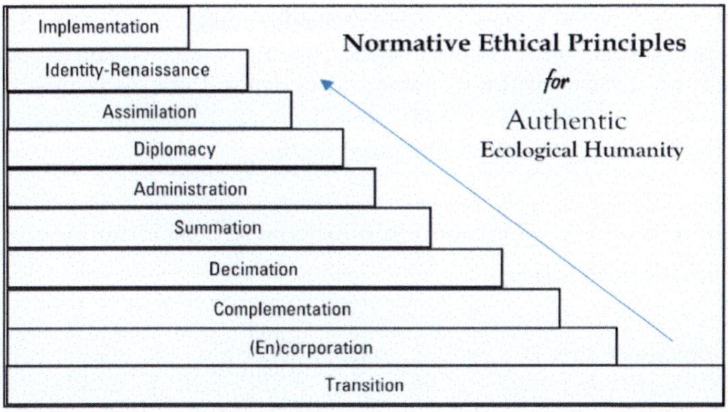

Fig. 4.6 Ethical dynamics for result-oriented human ecological concept. *Illustration mine*

life has turned out to be. Therefore, for a more practical, authentic and not just idealised form of human ecology, a critique of Vogt’s “10-Thesis” approach becomes relevant in the effort of this scientific work to place Mmadụ and Ùgwú-Mmadụ as central in the whole task of sustainable development. This is represented with the diagram below—a form of ethical dynamics for functional human ecology, which, in practice and application, should begin with a transformative-based transition through normative-based implementation (Fig. 4.6).

4.5.1 Transition

Possible ethical transitions would aim to foster practical ethical frameworks that reconcile the complexities of *human-nature interactions.* The core approach is to emphasise humanity as an integral component of the ecosystem—ecological humanity. Thus, efforts towards practical transitions would seek to counter the destructive tendencies of the Anthropocene and the prevailing view of nature as a resource to be exploited, manipulated or technically abused (cf. Ibid., p. 99.).

*4.5.2 (*En*)corporation*

The concept of *en*corporation[7] would be viewed as an ethical process whereby conceptions of human ecology always relate to both the human and non-human realms of nature as a *relational whole.* Vogt calls this *Anthroporelationalität*—a form of relational human-nature anthropology. Thus, a sense of human-nature ethical

[7] *The research work coined the word encorporation or en-corporation, using the prefix “en” to ascribe to the word “corporation” a more relational tone and meaning.

*en*corporation teaches the human person—Mmadụ—that his or her privileged position within the natural entities could only be ethically exercised and properly channelled when the intrinsic value of nature is recognised and respected.[8] To know the intrinsic value of nature is to apply ethical human ability and responsibility in respecting the interdependence of all natural realities.

4.5.3 Complementation

Though complementation is a genetic concept describing the mating capability of two different homozygous recessive mutations, giving rise to the same mutant phenotype, the varied linguistic permutations associated with the word unite themselves with the root meaning of the word found in its verb form, *to complement*—to unite, relate or align with, for a better or more prospect result. Vogt used the German term "*Verknüpfung*", which is best described as "alliance". Applied to the discourse on the needed normative actions for practical human ecology, our approach, following the opinion of Vogt, sees the need for mutual complementation between the natural and social scientific perspectives in a system of *shared interests*. The conflict of interests still existent within the disciplines contributes immensely to the weakness and ill political will that hold the modern cultural and ecological world in bondage. This remains a barrier to making the right decisions in human ecological discourses since the ethical judgement in this regard would depend heavily on how humans themselves view and treat nature within the scientific and research horizons (cf. Ibid.).

4.5.4 Decimation

In an ethical process of decimation and in the context within which we speak, human ecological discourses need to be liberated of all sentiments, which tend to reduce all approaches to mere emphasis on human solidarity and obligation towards "... plants (Pflanzen), animals (Tiere) and landscapes (Landschften)" (Ibid.). The vantage point must be that humanity itself must view its existence as *being in nature, life in nature, life with nature and life from nature.* Thus, human ecological rhetoric must be "decimated" of all superficial attempts to isolate the human existential spaces, in other words, the socio-cultural from the socio-ecological. This would only lead to mere humanistic (sentimental) projections about human ecology. Vogt calls this tendency "sentimental ecology"—*sentimentale Ökologie*. Approaches and rhetoric, therefore, need to be normative enough so as not to be sentimentalising but rather decimating. Focus is expected to be holistic and factual.

[8] Cf. Ibid., *Humanökologie: Neuinterpretation eines Paradigmas mit Seitenblick auf die Umweltenzyklika Laudato Si*, p. 100.

4.5.5 Summation

There would be the need for an ethical summation of what nature within human life's situations really means. Building on such ethical summation would serve as the basis for the required theological and scientific competencies within the human ecological discourses. This means that the foundational competence must be summated to engage nature in its cultural affiliations as a symbolic resource, a natural home, a gifted living space, the home of existence, which humanity, also a creature itself, shares in a mutual manner with other creatures (cf. Ibid.). The right ethical summation, therefore, ensures that with the alternative approach—ecological humanity—creatures of the natural world, humanity itself inclusive, are given the rightful approach as creative and inventive resources of nature. In this way, a more integrative sense of conserving nature would be evolved, which makes humanity itself an active and not passive member.

4.5.6 Administration

A well-structured ethical administration would be the proper response to the economy vs. ecology dichotomy—*the eco-eco dichotomy*. Vogt refers to this as "*verantwortete Haushalterschaft*"—responsible economics. It would be an ethically based and culturally influenced duty within human ecological discourses to critically observe, learn and respect the planetary boundaries, which already, with the experiences of our modern world, are endangered by the growing human population and human impacts. The sub-title of the encyclical, Laudato Si, has already set the tone for such ethical administration with the concept, *On Care for our Common Home* (cf. Ibid., p. 101). A call for a responsible administrative approach would culminate in ecological humanity, which, instead of exploiting, exercises rather a form of stewardship over nature and its capitals.

4.5.7 Diplomacy

Functional ecological humanity needs socio-ecological diplomacy, which sets the modality for grounds of negotiations between law and religion, education and science, as well as economy and culture in human ecological discourses. This would go a long way in highlighting the conflicting areas of interest with regard to the planetary boundaries in the Anthropocene, thereby providing a concrete platform that is plausible for risk management strategies on the one hand. On the other hand, it would apply measures within development strategies to make the sense of resilience inevitable in a learning and development process (cf. Ibid., p. 101–102.). Such a diplomatic process must be as intercultural as it should be multidimensional, yet without being contradictory. It would be unifying, leading to a transformation-based socio-cultural front. The contemporary idea of the "Great Transformation" (integral transformation) can never be a lone process of a divided world policy but rather a diplomatic agenda of a diversified but intrinsically connected world.

4.5.8 *Assimilation*

Human ecological discourses require an urgent form of techno-ethical bionics to help channel human dealings with nature towards becoming more responsible and regenerative. Responsible Anthropocene should abstract from nature and not destroy it. Thus, ethical assimilation based on observation of the complex interplay between social and ecological systems and informed by numerous stored knowledge and information embedded in nature itself should be enlightenment for a new generation of technology that is ready to abstract from and assimilate such in-built natural laws for a sustainable human ecological future. Anthropocene must think ecologically *en route* to Coevolution (cf. Ibid., p. 102).

4.5.9 *Identity-Renaissance*

A practical human ecological discourse only has relevance with an ethical renaissance capable of giving the modern Anthropocene a feasible and sustainable future. Such ethical renaissance means a rediscovery of the identity of the modern man, not just as an intelligent being imbued with natural powers, but rather as a responsible being who is an intrinsic part of the movement and principles of nature—*homo sapiens oecologicus*, an ecologically thinking and acting being (cf. Ibid.). Such a rediscovery of identity would be able to reconstruct approaches within the Anthropocene and make human ecological discourses conscious of the impact-based "interdependence" existing between human life and actions and the natural world itself. This form of ethical renaissance requires to be made a cultural task.

4.5.10 *Implementation*

The ethical implementation needed must be seated on the best strategy to promote the rights and dignity of the ecological human person. Vogt thinks that this stage would be the very agenda to promote the "*Die vierte Dimension der Menschenrechte*", meaning the fourth dimension of human rights. Thus, human ecological discourses need a practically implemented ethics, which is capable of mediating, on the one hand, the fundamental rights of people to the development process and, on the other hand, the socio-political and socio-economic rights of nations to pursue development, even under the known signs and impacts of climate change. In other words, strategies and policies should represent implementable ethics of development capable of safeguarding people's development freedom without compromising at the same time their socio-ecological well-being. This should be understood as the true meaning of *development as freedom* (Amartya Sen)—a form of cosmopolitan duty, which must be made to conform to the dictates of human rights and dignity (cf. Ibid.).

When developed alongside the traditional socio-ethical contributions of social teachings on human ecology, the above-proposed ethical strategy would make the discourses more scientific and practicable. That the embedded key principles in human ecology would be able to offer practical inroads to the liberation of Mmadụ in any given society and, particularly within the Nigerian challenging situations, is not just axiomatic. It is both an option and a mandate in the face of Nigeria's wavering and drifting polity. A resolute strategy is an urgent option for integral socio-ethical transformation. A practical restoration of the status of Mmadụ as the foundational task in rescuing Nigeria's amorphous social structure and system must be able to implement and reflect these normative principles. However, the approach does not have to be merely literal and uncritical. The principles must be indigenised, enculturated and taught. The next sub-topic presents an indigenised blueprint of Nigeria's practical attempt to discuss policy in relation to the moral agent of development—*Mmadụ*, the human person.

4.6 The Place of Human Ecological Ethics Within the Nigerian Proposed Vision 20:2020: A Critique

> By 2020, Nigeria will have a large, strong diversified, sustainable and competitive economy that effectively harnesses the talents and energies of its people and responsibly exploits its natural endowments to guarantee a high standard of living and quality of life to its citizens.[9]
>
> – Vision Statement: Nigeria – Vision 20:2020

The above statement is a pregnant preamble and an excerpt from Nigeria's proposed blueprint for planned economic growth and development strategies for the period between 2009 and 2020. It represents the most current and strategised policy plan modelled to reflect the contemporary development concepts of the United Nations and the African Union. It was and is still envisioned within this platform that such a bold step should pave the way for Nigeria's transitional and transformational story: *a story of a large and strong polity characterised by dynamic, diversified and competitive economic opportunities capable of absorbing and integrating the talents and energies of its people, ndị Mmadụ, thereby improving the quality of general life*

[9] The Federal Republic of Nigeria. *Economic Transformation Blueprint. Nigeria Vision 20:2020.* Ministry of Budget and Economic Planning, December 2009, pp. 19–20. Available via https://nationalplanning.gov.ng/wp-content/uploads/2021/02/nigeria-vision-20-20-20.pdf. Accessed December 17, 2020.

Fig. 4.7 Five guidelines towards achieving results in the Nigeria Vision 20:2020. *Illustration mine*

and living. Ideologically, the blueprint sounds humane, socio-cultural and socio-ecological. Now, within the scope and focus of our work, the planned period of the agenda is almost done, yet the current state of affairs within the Nigerian state, coupled with the greatest threat to humanity and the world's economy in the years 2019 through 2020 and beyond, the Covid-19 pandemic, makes the envisioned milestone an ambition yet to be actualised. While the vision awaited a transition to a service-based economy in 2018, a remarkable blow came in the same year when Nigeria statistically overtook India as the world's poverty capital, with an estimated 87 million of her citizens living almost below the poverty level—less than $1.90 per day—according to the international standard of measuring the movement of development processes. Analytically, the proposed vision has five target guidelines, as shown below (Ibid.) (Fig. 4.7).

The above-drafted key points tend to present an integrated approach to the challenges facing the Nigerian system and structure. Remarkably as well it officially utilised contemporary development concepts and conventions, especially in relation to people's well-being and productivity. Summarising the above key points, the blueprint defines its target thus:

> NV 20:2020 is a rallying call for all Nigerians, regardless of ethnicity, economic status, or religion to unite and stand behind a common cause of placing the country firmly on a path of sustainable growth, and taking it to its rightful place in the comity of nations. The blueprint has, therefore, been designed to reflect accurately the collective interests of the people of Nigeria, using a bottom-up approach that is anchored on a deep understanding of the aspirations of all Nigerian citizens, and knowledge of the future needs of the country. The vision is underpinned by the need to effectively and efficiently mobilise the nation's resources to serve and improve the lives of the its citizens, and to respond appropriately to the growing challenges of an increasingly smaller, mutually dependent, and interconnected world. (Ibid.)

With this articulated optimism, the document proposed four dimensions of aspirations, which can be judged integral enough to apply the ethics of human ecology in "… optimising human and natural resources to achieve rapid economic growth and translating that growth into equitable social development for all citizens" (Ibid.). To this effect, a drafted ***Social Dimension*** would be implemented to work towards a "… peaceful, equitable, harmonious and just society, where every citizen has a strong sense of national identity and citizens are supported by an educational and healthcare system that caters for all, and sustains a life expectancy of not less than 70 years" (Ibid.). The ***Economic Dimension*** is envisioned to transform the economy into a "… globally competitive economy that is resilient and diversified with a globally competitive manufacturing sector that is tightly integrated and contributes no less than 25% to Gross Domestic Product" (Ibid.). An ***Institutional Dimension*** would work for a "… stable and functional democracy where the rights of the citizens to determine their leaders are guaranteed, and adequate infrastructure exists to support a market-friendly and globally competitive business environment" (Ibid.). A socio-ecologically informed realm under the ***Environmental Dimension*** would be implemented to achieve an "… environmental consciousness that enables and supports sustainable management of the nation's God-given natural endowments to ensure their preservation for the benefit of present and future generations" (Ibid.).

The above aspirations, which are models of ideas for objective change, were to be made to reflect on some key national aspirations in what is known as the First National Implementation Plan (1st NIP—2010–13). These key areas are the macroeconomy, agriculture, health, manufacturing, infrastructure, education and general polity. The report taken from the Validation Workshop on the First Four-Year Implementation Phase stated this below (The Federal Republic of Nigeria 2010) (Table 4.2).

An ethical framework for an authentic human ecology is ideologically evident and implicit in the articulated ideas of the blueprint, given its proposed focus on the human person himself or herself, as well as the envisaged strategy of turning all socio-political, economic, institutional and environmental factors to respond to the needs, self-actualisation and the general well-being of Mmadụ. Thus, the plan itself is considered equipped enough to integrate the social, cultural and ecological realms of human existence towards the ideal transformation that can access sustainable development. However, a critical look at the concrete Nigerian situation identifies a source of weakness, which stands as the bane to the effective implementation of not just Vision 20:2020 but also every other known socio-political and socio-economic policy. This bane is identified as deficient governance, in other words, malfunctional administration. The successive experiences of this bane in the case of Nigeria have left the whole structure weakened, corruptly complicated and developmentally exploited.

This leadership failure, which our research holds as the central factor around which all other *structures of sin* revolve, made the Ministry of National Planning acknowledge some key development challenges to the designed policy during the review of the first implementation phase. The challenges are seen in the general failure of governance in the provision of the critical infrastructure needed to drive development; the failure of governance in the fight against youth and graduate unemployment; in empowering and advancing constant contemporary research for development and

Table 4.2 National aspirations within the first national implementation plan

	National aspirations	Visions/prospects
1	Polity	Peaceful, harmonious and a stable democracy
2	Macroeconomy	A sound, stable and globally competitive economy with a GDP of not less than $900 billion and a per capita income of not less that §400 per annum
3	Agriculture	A modern technologically enabled agricultural sector that fully exploits the vast agricultural resources of the country, ensures national food security and contributes to foreign exchange earnings
4	Education	Modern and vibrant education system which provides that opportunity for maximum potential, adequate and competent manpower
5	Health	A health sector that supports and sustains life expectancy of not less than 70 years and reduces to the barest minimum the burden of infections and other debilitating diseases
6	Infrastructure	Adequate infrastructure services that support the full mobilisation of all economic sectors
7	Manufacturing	A vibrant and globally competitive manufacturing sector that contributes significantly to GDP with a manufacturing value added of not less than 40%

innovation; in boosting the agricultural sector for more mechanised and commercial production; the failure of governance in expanding the industrial sectors for more absorption of the working force; in diversifying the economy already monopolised and hijacked by the oil sector; as well as the failure of governance in stemming the tide of the country's incessant borrowing spree plunging the Nigerian nation into deep sub-national debt (Ibid., p. 20.). An increase in sub-national debt without a performing and viable indigenous economy and IGR amounts to a Nation's conscious decision to remain perpetually imperilled.

The above final analysis of the current affairs of the Nigerian system reveals that the nation's problem is not a lack of blueprints in terms of policymaking. It is not as well the lack of human intellectual powers needed in designing her policies. The problem is neither a lack of human and natural resources nor is it the case that Nigeria is landlocked in terms of economic activities. Furthermore, the problem is neither lack of fiscal resources for the provision of the relevant infrastructures as driving forces towards development. Nigeria's problem is in-built structural malfunctions, which are man-made and which serve only the selfish purposes of the corrupt elite class, which feels it has the monopoly of power. They, therefore, keep injecting strategies into the general system that keep it weak and counterproductive. Until those structural sins are approached ethically with the corresponding willpower and in the spirit of *resilience for sustainability*, the persistent negative exploits would continue to rob the nation and its people of sustainable development prospects.

References

Achebe C (2010) The African trilogy: things fall apart, no longer at ease, arrow of god. Everyman. Everyman's Library, USA, p 124

Francis (2015) Encyclical letter *Laudato Si*. On care for our common home, 24 May 2015, no 44

Iroegbu P (2022) The human person in the western and African traditions: a comparative analysis. In: Paper presentation, all saints major seminary and University of Benin, p 67. Available via https://people.stfx.ca/wsweet/00-PCT/Vol.%201%202002%20Iroegbu.pdf. Accessed 28 Aug 2020

Kahiga J, Eberl JT (2013) Personhood: African and western threads of thought. In: Kahiga J, Wanyma MN (eds) Contemporary issues in philosophy, religion and theology. Utafiti Foundation, Eldoret

Maslow AH (1943) The theory of human motivation. J TOC Psychol Rev 50(4):370–396. Available via https://doi.org/10.1037/h0054346. Accessed 23 Nov 2020

Oforka VC (2015) The bleeding continent: how Africa Became impoverished and why it remains poor. Xlibris, USA, p 275

Oforka VC (2016) Afro-Igbo Mmadu and Thomas Aquinas' Imago Dei: an intercultural dialogue on human dignity. Xlibris, USA, p xx

Onwubiko AB (2012) Person and human dignity. a dialogue with the Igbo (African) thought and culture. Peter Lang, Frankfurt am Main, pp 64–65

Paul VI (1967) Encyclical letter *Populorum Progressio*. On the development of peoples, Rome, 26 Mar 1967, no 6

Paul II J (1987) Encyclical letter *Solicitudo rei Socialis*. On the twentieth anniversary of Populorum Progressio, Rome, 30 Dec 1987, nos 27, 28

Paul II J (1991) Encyclical letter *Centesimus Annus*. On the hundredth anniversary of Rerum Novarum, Rome, 1 May 1991, no 38

Paul II J (1995) Encyclical letter *Evangelium Vitae*. On the value and inviolability of human life, Rome, 25 Mar 1995, no 21

Reed PG (2014) The theory of self-transcendence. In: Smith MJ, Liehr P (eds) Middle range theories in nursing, 3rd edn, vol 147. Springer, New York, pp 119–145

Statistica (2020) Literacy rate in Nigeria 2018, by zone and gender. Available via https://www.statista.com/statistics/1124745/literacy-rate-in-nigeria-by-zone-and-gender/. Accessed 14 Nov 2020

Stiglizt JE (2017) Globalisation and its discontents, Revised edn. Penguin, London, p. 6

Tanko M (2013) The political implication of past and present Nigerian financial crises. In: Falola T, Achberger J (eds) The political economy of development and underdevelopment in Africa, vol 186. Routledge, New York, pp 176–191

The Federal Republic of Nigeria (2010) Validation workshop on the first four-year implementation plan for Nigeria vision 20:2020, vol 5. Federal Ministry of Finance, Budget and National Planning, Aug 2010. https://www.nigerianstat.gov.ng/pdfuploads/NV2020%20%20The%20First%20Implementation%20Plan%20(2010%20-%202013)%20-%20HMNP.pdf. Accessed 17 Dec 2020

The Federal Republic of Nigeria (2020) Nigeria, Labour Force Statistics—Volume 1: Unemployment and Underemployment Report. National Bureau of Statistics, Q4 2017-Q3. Available via https://www.nigerianstat.gov.ng/pdfuploads/Q2_2020_Unemployment_Report.pdf. Accessed 15 Dec 2020

UNDP and OPHI (2020) Global multidimensional poverty index 2020—charting pathways out of multidimensional poverty: achieving the SDGs. Report. United Nations Development Programme and Oxford Poverty and Human Development Initiative. Available via https://ophi.org.uk/global-mpi-report-2020/. (Re) Accessed 9 Mar 2021

Vogt M (2009) *Prinzip Nachhaltigkeit: Ein Entwurf aus Theologisch-Ethischer Perspektive*. Oekom Verlag, München, p 471

Vogt M (2016) *Humanökologie: Neuinterpretation eines Paradigmas mit Seitenblick auf die Umweltenzyklika Laudato Si*. In: Haber, Wolfgang/Held, Martin, Vogt M (eds) Die Welt in Anthropozän: Erkundungen in Spannungsfeld zwischen Ökologie und Humanität, vol 97. Oekom Verlag, München, p 67

Chapter 5
Practical Inroads: The Normative Pathways to Integral Socio-Ethical Transformations—Four Propositional Normative Steps

Abstract Ethical norms can create sound normative principles, which must be applied and practised to remain relevant and valuable. In formulating propositions and premonitions, dialectical and interactive ethical methods could be proposed as a practical approach towards recovery and remediation. Consequently, the propositions under reflection take from verifiable ethical possibilities and suggest strategies, new techniques and applications, paradigm shifts, transition methods for both exnovations and innovation, etc. In doing this, emphasis is placed on the fact that possible premonitions in the approach and evaluation of moral situations are to be taken as challenges, chances or opportunities for change and never as ends in themselves with neither options nor choices. In this regard, the leading mantra of approach remains the orientation and mindset towards sustainability, the willpower to ignite concrete transitions and the resilience to translate revolutions into "*re-valuation*" and sustainability into building "*sustained ability*" (capacities) in the human person.

Keywords Ethics · Normativity · Premonitions · Exnovation · Innovation · Practical inroads · Sustainability

When a system is designated as deficient, such an assessment has varied but related meanings: it means that the system performs below the required standard or the system is in need of overhauling or restructuring; it means that the system is not producing results that are commensurate with the objectives of the entire system; it could mean that the system is both ideologically and administratively ill-structured; it could as well mean that the system is infected with detractors, human, ideological or institutional, that compulsively channel the system to forms of selfish and feudal machinations in the hands of the administrators, etc. Again, it is needful to emphasise that deficiency is not the total absence of progress, success or development. It is rather the situation of mismanaged options and factors, which are direct and related obstacles to progress, success and development.

Our research, therefore, stands on the premise that the Nigerian case, evident through the fieldwork and the concrete statistical data presented, is not the case of the total failure of a polity. It is rather the case of inbuilt structural dysfunctions, structural

C. F. Nwosu, *Systemic Remediation*, Environmental Humanities: Transformation, Governance, Ethics, Law 4, https://doi.org/10.1007/978-3-031-61909-0_5

sins, which have long held the general system in bondage and, as a result, deficient. Such structures must be jettisoned if Nigeria ever wants a functional transition to sustainability. Hence, this last chapter introduces some normative pathways that are considered relevant and applicable to remediating the defects within the system. Practical inroads in the Nigerian case must be made an approach and such an approach must be structural. In other words, it must follow defined normative steps so as to be *not just systematic but integrally systemic*. The expected systemic approach would invariably be the invaluable gateway to both transition and transformation to a sustainable polity. The normative powers of the propositional steps would be deeply characterised by action-based systemic dialogue that would always have as its *terminus a quo* the human person, *Mmadụ*. Furthermore, the practical factors that would accrue from this chapter would be made the content, the guiding themes, the resource and focus concepts of the destination of our research, namely an Institute for Ethics, Capacity Building and Sustainable Development.

5.1 Proposition I: The Normative Steps as Approach and Method

In recalling the area of focus of our work, namely the study and analysis of the structure called Nigeria, the factual proof of both structural and system-based dysfunction, its adverse effect being the loss of authentic human ecology and the proposal for practical strategic dynamics of change, four key propositions are considered relevant in setting the pace for integral socio-ethical transformation.

In our approach, Mmadụ, the human person himself or herself remains the focus, the initiator and the actor. The proposed propositional normative steps here begin with the human agents themselves and flow naturally into the system's policymaking and the general learning and administrative processes. The expectation would be a form of organically embedded socio-political, socio-economic and socio-cultural changes. Our research settles for a method that is propositional yet at the same time normative to prove the fact that the field of work, the Nigerian system and structure, presents complex and interrelated challenges that demand policy-based and action-oriented approaches so as to be tackled systemically and integrally. Such steps must be observational and deeply verifiable to better dialogue with all the socio-empirical vectors available and provide the investigative options that would aid decisions, conclusions and actions.

Based on our field of study, Christian Social Ethics, our approach finds these normative steps in the "RJA" of the social teachings of the church, referred to as "... three stages which should normally be followed in the reduction of social principles into practice" (John 1961a), namely: Principle of Reflection, Criteria of Judgement and Directives for Action. It is, in general, a deductive process of abstracting from general principles and applying them to concrete situations. However, in dealing with the individual stages of the RJA, the inductive method needs to be incorporated

Fig. 5.1 RJA method as a socio-ethical approach to action-based social transformation. *Illustrations mine*

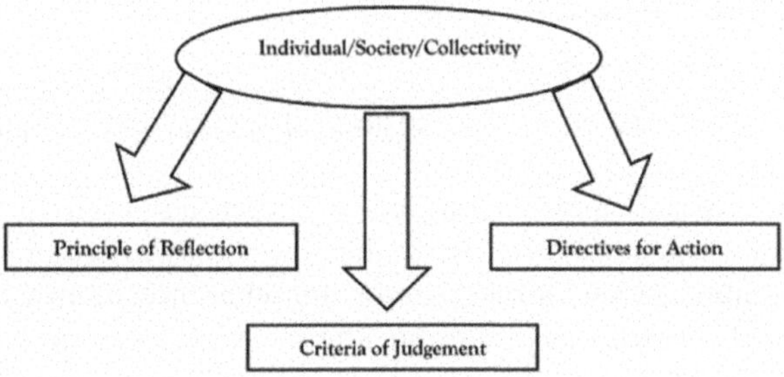

since the stages must first engage individual cases and develop criteria of assessment and judgement before proposing directives for action(s). The encounter with concrete individual cases is a prerequisite for practical inroads. The encyclical Mater et Magistra remarked, "Knowledge acquired in this way does not remain merely abstract but is seen as something that must be translated into action" (Ibid., no. 237). As an adopted method in our research, we represent these principal stages with a schema (Fig. 5.1).

The fundamental motive for adopting this method (step) is to help our research reflect and orient itself towards discerning and suggesting possible and verifiable solutions to the challenges facing Nigeria. Placing emphasis on this need for discernment, Paul VI teaches:

> It is up to (the) Christian communities, with the help of the Holy Spirit, in communion with the bishops who hold responsibility and in dialogue with other Christian brethren and all men of goodwill, to discern the options and commitments which are called for in order to bring about the social, political and economic changes seen in many cases to be urgently needed [….] The Gospel is not out-of-date because it was proclaimed, written and lived in a different sociocultural context. Its inspiration, enriched by the living experience of Christian tradition over the centuries, remains ever new for converting men: end for advancing the life of society. (Paul 1971)

Following the opinion of the encyclical, the common denominator is to reduce all social principles and tools to practical orientations that would lead to the *conversion of men*, seen as the *terminus ad quem* of the life of every human society. The concept of conversion expected as the focal point within a given society, therefore, is interpreted here as the result of initiated concrete transitions and transformations, to be made possible through sets of dynamically related processes and empowered, in our context, by the interplay of the three steps below in a systematic and practical manner:

5.1.1 Principle of Reflection

It may not be out of place to argue that everything begins with the formation of ideas. In defence of the assertion "ideas rule the world", Laurence Summers, former president of Harvard University and economic adviser to former President Barack Obama, argues that what the world has ever known as innovative policies and maverick actions, which, at one time or the other, helped to up-turn status quos and restore workable societal orders, are all rooted in ideas, which began as a small pebble of "a wave that eventually reaches all shores" (Summers 2015). Thus, to begin a critical reflection is to form ideas upon which a planned, concrete and transformative project would be based.

Taken, therefore, as the first stage within the propositional approach, the principle of reflection, in other words, the formulation of the first ideas through *action-based observation processes,* is to be predicated on concrete sensitisation programmes as would be proposed under the key transformative areas. The planned observational and reflective actions would eventually provide the platforms for the formulation of blueprints for transformative inroads predicated on the seven cardinal normative values for social transformation discussed at the end of chapter three as the "*kpim*" (the core) of sustainable social ethical and moral investments. Hence, the systematic and scientific application must obey, incorporate and integrate seven directives for the formulation of transformative ideas for effective social change. Recalling these principles as the reference and requisite points within the first proposition is to preempt Nigeria's problems with regard to the formulation of ideas, policies, laws, rules and regulations, as well as its constitutions. The situation is held to be as sensitive as it is critical (Table 5.1).

5.1.2 Criteria of Judgement

This propositional stage brings concrete experiences and situations with steps to the formulation of ideas in dialogue. It is a sort of *control-experiment* through which all variable factors taken from concrete life situations and experiences within the general polity of a given society are abstracted from. The situations and experiences must be taken in their concrete, past and present statuses for more integrative judgement. In the dialogue that ensues, practical and verifiable middle axioms—intermediate ideas and criteria—for functional, implementable and sustainable approaches would consequently diffuse into policy formulations for the system and institutions. With the criteria of judgement, principles would be applied to concrete social concerns as tools of analysis, verification and evaluation. These criteria identify such concrete situations of concern, analyse them with particular reference to the moral agent, the Mmadụ and make assumptions and propositions for concrete and corresponding actions.

Table 5.1 Seven directives for the formulation of transformative ideas for effective social change—model for the principle of reflection

Normative principles	Transformative ideas
1. Personality	The ideas must be "***Mmadụ-oriented***"—people-oriented, human-value oriented; must be protective of human dignity, of human ecology; they must work to make people agents and not patients
2. Common good	Ideas must be oriented towards civil rights, collaboration, the sense of community, association, charity, democracy, goodwill, equality (equity), rule of law for all and shared responsibility
3. Solidarity	Ideas must reflect both individuality and communality; ideas must relate and integrate individual, society and nature—they have to be integral; interests must be both autonomous and relational; ideas must be symbiotic so as to be objective, spontaneous, re-integrative, flexible and reliable
4. Subsidiarity	Ideas must recognise, respect and promote the capabilities and autonomy of individual citizens—the Mmadụ; ideas must incorporate talents and not destroy them; they must have the character of making the citizenry active and participatory and not a liability; they must promote initiative and creativity; they must avoid over-indulgence in "aid campaign" so as to avoid degeneration to domination, infringement and oppression
5. Justice (intergenerational)	Ideas must speak the language of equal opportunity; they must not be detracting; they must be inclusive so as to be intergenerational; they must be based on values shared for the benefit of all; ideas must recognise human rights as inviolable; ideas must be sound enough to integrate diversity
6. Sustainability as an attitude	Ideas should reflect the nucleus of the general learning process within the human and social institutions; they must be convincing and legitimate enough to reflect the general attitude of the polity; the ideas must not set the human capitals against the natural ecological capitals, an integral part of which *Mmadụ* himself or herself is; ideas must be made to reflect sustainability in stewardship
7. Resilience as a virtue	Ideas must be flexible enough to be resilient; they must be made to be renewable, responsive, adaptive and transformative; ideas must be assimilative of social, political, economic and ecological situations; they should inform the norms of the daily life of the citizenry

Illustrations mine

Furthermore, the dialogue that begins focuses attention on the fact that the concept of judgement here should be more of a process towards discernment. In other words, the judgement criteria *must be open to compromise, especially where conflict of interests arises as a result of cultural, ideological, socio-political or socio-ecological differences.* Thus, in this way, judgement would be immune to setting individual or particular standards as general norms or defining opinions that should be particular as a universal law. The capability to compromise and, therefore, acknowledge uniqueness, differences, freedom of opinions, respect for originality, etc., would all be helpful in arriving at a more sustainable judgement within the ethical and moral approach to given situations. This is the right to cooperation, which should be an integral part of every approach to the development of a sustainable social order. Sustainable goals can never be separated from the integration of options and differences.

5.1.3 Directives (Guidelines) for Action

The directives or guidelines for action are the practical judgements informed by the ensuing ideas, assumptions and propositions emanating from the dynamics between the principles and the evaluating criteria—the concrete social situations and experiences. Actions and applications here are never homogeneous but heterogeneous. In other words, every concrete social concern is taken up and analysed as a unique factor within a series of available pieces of evidence. However, the heterogeneity of the approach does not in any way demean the known normative principles, which apply and guide imperatively regardless of the variance of options. The pre-condition for action here, therefore, is first the resolve to remediate the given concrete social concern by applying the first two propositional normative steps, namely the principle of reflection and criteria of judgement.

Taken as a demand in our search for practical inroads, we maintain that the first critical step in approaching Nigeria as a social, political, economic, cultural and ecological specimen is to unlock and dissect the various facets of her general structure, with the aid of our propositional normative steps. When this is done critically and scientifically—just as this work has set out to do—some system- and structurally based premonitions would emerge as preclusions to the current efforts to rescue Nigeria from the problems of structural and institutional dysfunctions, in other words, the structures of sin. Discussion on such concrete precluding factors would give this work more authenticity.

5.2 Critical Premonitions: Assessment of the Critical Precluding Factors in the General Nigerian Structure

Generally, all models of approach to sustainable development dynamics stand to confront forms and manners of preclusions, which are challenges to plans, ideas and actions. Such preclusions are known to be mental, physical or phenomenal, structural, institutional, as well as ideological. They exhibit predominantly peculiarities that are as variant as social structures and conventions themselves. The preclusions discussed here would serve to deepen the opinion of the research that the Nigerian case does not just deserve a systematic approach but rather a systemic overhaul.

5.2.1 The Restructuring Saga: Nigeria's Geopolitical Conspiracy Theory

The unfortunate but situation-based 1914 colonial policy that saw then Nigerian northern and southern protectorates fused as a union, the so-called Amalgamation, has today become a precluding factor in the geo-structural existence of Nigeria. What was conceived as an administrative convenience has today become one of the most threatening factors to the authentic and sustainable development of the nation itself. Currently, the majority of Nigerians have the feeling that a policy that has turned over 100 years old should be revisited, reformed or completely abandoned. As a consequence, the whole atmosphere is charged with a form of *restructuring saga* and campaign, which sees the country divided into camps of proponents and opponents, progressives and conservatives, hardliners and soft-liners, the campaigners for the maintenance of the status quo and the preachers of radical change, etc. The charged socio-political climate is further submerged into a *conspiracy theory slogan*, namely that "*the unity of Nigeria is non-negotiable*", being popularised by the elite few who are milking from the system.

It is our opinion that geopolitical unions are never to be taken as being sacrosanct. In other words, they are not social and political systems that are so static as never to admit changes and modifications. A twenty-first-century Nigerian society with an estimated over two hundred million population and over eighty different (major) languages is predominantly not the society that was met when the colonial administrations and policies drafted the amalgamation of the northern and southern protectorates, even without adequate representation of the consents and opinions of the affected indigenous people. The more challenging factor is also the fact that getting the political will of any leadership within the present structure to challenge the status quo would be difficult for the following reasons: (a) the leaders from the benefiting zones would never make it an agenda; (b) leaders from the so-called marginalised zones would never be able to garner enough support to move and promote such an agenda. Thus, a national solution working with the present Nigerian system and constitution is a mirage. Historical records and experiences show that Nigeria's 1960 and 1963 constitutional antecedents, even in their very first drafts, contain the more representative and democratic ideologies capable of reforming the general system and structure. However, all those still bent on exploiting the present structural weakness work round-the-clock to stifle their oppressive craft even more. Against this backdrop, our research articulates the first premonition thus:

1st Premonition

The present Nigeria's social structure and the current constitution operative within it are gradually becoming detractors to the general system, which, therefore, calls for a systemic restructuring. Until the whole polity resolves to make it a political project, it remains one of the premonitions to sustainable change.

5.2.2 The Parable of a Fraudulent Federalism

The socio-political unrests, or simply put, the proxy wars that were prevalent in most states of Africa, mostly between pro-Soviet and pro-Western groups of players, subsided with the end of the Cold War. Consequently, more liberal and democratic ideas began a wave of influence in most African states. The remarkable result is that since the beginning of 1990, most Sub-Saharan African countries have become either Federations or decentralised. As federating states, they create a political system where the centre and regions share levels of autonomy, power and responsibility, and such shared responsibilities are enshrined in the constitution. Where decentralisation is practised, the centre devolves powers and responsibilities to the lower or regional administrations, without the lower regions having full-fledged autonomy like the federating system and without this being backed up with a constitutional mandate. Thus, while the two systems sound seemingly distinct, they nevertheless have deeply related objectives, namely: (a) to enhance democracy; (b) to bring good governance and efficiency; (c) to better manage diversity, especially in multilingual, multicultural societies and conflict-prone zones. In this regard, Nigeria is a Federation, but its federal character is problematic and deceptive. It promotes socio-political, ethnic and religious dominance. The map below proves the fraudulence of the federal character (Fig. 5.2).

Approaching the map critically, our research accepts that the transition from 19 to 36 states, which happened periodically in 1987, 1991 and 1996, respectively, was a political attempt to address the inequality of state regions between Northern and Southern Nigeria. However, the geographical difference which makes the three geopolitical zones of the North almost three times bigger than the South-West, South-South and South-East put together presents an irreconcilable geopolitical formula, which has never and can never free Nigeria from consequent political inequality. In addition to this are the endogenous ethic and cultural differences distributed and shared within these zones; the political economy of monthly allocation of revenue to the states from the revenue being solely (over 80%) generated from the same disadvantaged South, as well as the impoverishment of the indigenous people of the resource communities. The wealth of experiences that have accumulated over

Fig. 5.2 Nigeria's six geopolitical zones as sub-regions within the federal character

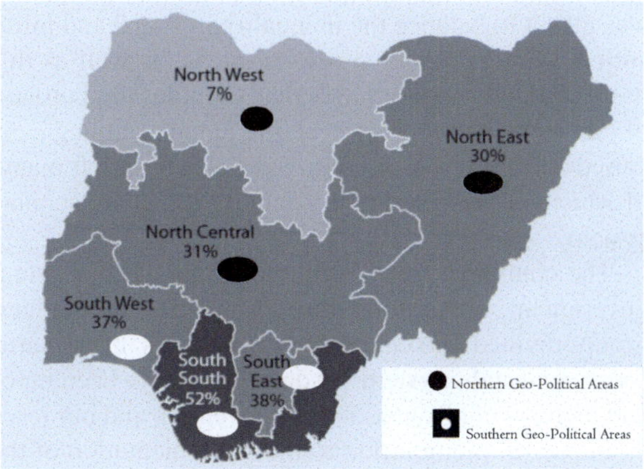

the past decades since after independence shows that the zones do not agree politically, culturally, religiously and ideologically. The reality remains that the geopolitical outfit practised within the present structure called Nigeria would perpetually keep some regions politically disadvantaged and exploited. Therefore, the second premonition is formulated thus:

2nd Premonition

The structure of the Nigerian geopolitical zones purportedly designed to respond to a federal system of governance is grossly deficient, given its harbour of geopolitical imbalances, which are the roots of electoral and political dominance. The present geopolitical structure can never be an "agent" of socio-political development but remains a "patient" still awaiting reformation, restructuring or total dismantling.

5.2.3 *Resource Control*

The issues, conflicts and discussions about resource control in Nigeria have remained socio-politically, socio-economically and socio-culturally endemic. With its dependence almost solely on oil revenue for its existence, Nigeria's manner of administering the crude oil resource areas has continued to be problematic. The derivation principle[1] introduced subsequently by successions of Nigerian Federal Governments

[1] The Federal Republic of Nigeria. The 1999 Constitution of the Federal Republic of Nigeria provided that "… the principle of derivation shall be constantly reflected in any approved formula

was meant to balance the inequality in social and infrastructural development being suffered by the resource reserve areas, as well as to help in the empowerment of the indigenous communities that shoulder the burden of exploration for the benefit of the whole nation. However, the planned strategic manipulations embedded in the varied corrupt resource control policies have left many open wounds, the bandaging of which has assumed both socio-political and economic dilemmas in the Nigerian system.

The complexity within the networks of conflicting interests brings together the government, the multinational companies, the indigenous people and various elite groups divided into camps of beneficiaries, mercenaries and activists who are either for or against the system. To date, there have been fierce tribal, political, institutional and regional confrontations in this regard, and our research identifies the situation as an impactful premonition to the implementation of the sustainability principles for sustainable development dynamics. Thus, the formulation of the third premonition:

3rd Premonition

The Nigerian policy, which vested the exploration and exploitation of natural resources to an unlimited power of the federal government, has been known, from verifiable pieces of evidence and other experiences, to be corrupt and detracting. It stifles the development of indigenous economies, creates agitation and disdain for the strange bedfellows, weakens economic competition and regional innovation, and above all, in the case of Nigeria, it has murdered the sense of diversification. Until such a policy is decentralised, it remains a critical premonition to sustainable inroads to development.

5.2.4 *Imposition of Affinities and Affiliations*

The socio-political network of Nigeria is a conglomerate of ethnic and religious affiliations and affinities. Each of the groups tries to exert influence and, in extreme cases, dominance over the other. There exist both physical and psychological covert and overt tensions in everyday life and activities due to extreme ethnic consciousness, cultural alienation and the major-minor dichotomy in tribal affiliations. Consequently, the Nigerian political scene suffers constant arrays of extreme *affinity consciousness*, which, right from the very foundations of Nigeria's independence,

---

as being not less than thirteen per cent (13%) of the revenue accruing to the Federal Account directly from any natural resources." Proviso to Sec. 162(2) of the 1999 Constitution. Available via http://www.nigeria-law.org/ConstitutionOfTheFederalRepublicOfNigeria.htm. Accessed April 4, 2022.

has remained the root cause of nepotism, election fraud, mediocrity in civil services, as well as internally induced underdevelopment. Our research puts it outright that the rich diversity of the Nigerian socio-cultural space still awaits a socio-political policy that would be able to believe in the diversity so as to have the will and plan to manage it well. The present situation, as it is being witnessed in Nigeria, has a very limited chance of attainable sustainability. Hence, our fourth premonition:

4th Premonition

Until the daily lives and relationships on the streets, institutions, as well as socio-political parastatal structures, are both ideologically and psychically demystified of the complex illicit appeal to affiliations and affinities in the general matter of national interests, every planned approach to re-strategize would always have the hard rock of intertwined internal socio-political and socio-cultural barricade to break. Our research work also would not be exempt from facing this challenge.

5.2.5 *Ethno-Linguistic Conundrum*

The ethnolinguistic diversity of a society is a rich cultural heritage since language itself confers identity and strengthens affinity. When it is managed well, the diversity would be a source of mutual enrichment among the unique identities. Where diversity is a victim of exogenous *imperialisation*, Nigeria as an example, added to internal socio-political impositions, intimidations and disregard, the result is rather an ethno-linguistic conundrum. This factor has led to periodically recorded failures with regard to national issues as well as other numerous attempts to introduce and implement issues of common interests. The present Nigerian structure harbours identities with numerous unique languages, let alone their sub-divisions in dialects. Our research argues that the current socio-political structure present in the Nigerian system remains problematic with regard to the management of ethnolinguistic diversity. As a consequence, the situation constitutes one of the premonitions of sustainable development strategies. The fifth premonition is, therefore, articulated thus:

5th Premonition

The present Nigerian social structure and, therefore, the general system of policy-making is grossly exogenous to the generic socio-cultural and linguistic mentality. Until such generic diversity is made constitutive and homogeneous of the integral structure, ethnolinguistic diversity remains inimical to a sustainable Nigerian society.

5.2.6 Interreligious Conflicts

The concerns about the clashes among religious communities in Nigeria have been raised in chapter one. Nevertheless, it is to be emphasised here again as a critical premonition which has contributed to keeping the system conflicting and problematic. Nigeria's multireligious status, anchored on its three main religions, namely Christianity, Islam and Traditional Religion, has known conflicts and violence. The conflict situations have been predominantly the bipolar confrontations between the dominant Muslim North and the dominant Christian South. Adherents to the pure traditional religion have drastically disappeared following the colonial influence and the work of the Christian missionaries. What remains of the Nigerian traditional religion in the current dispensation is a syncretic attempt to recover and imitate what was lost. Therefore, the reference to interreligious conflicts in the Nigerian context is an umbrella name to describe Muslim-Christian, sectarian, denominational, fanatical and, most recently, terrorist confrontations.

The character of these confrontations in the Nigerian context has become regrettably a corporeal part of its socio-political life. To this end, the strong affiliations to religious faiths continue to rob the system of true democracy, political equity and proper orientation towards competence and excellence. A timeline of interreligious instabilities has given the polity scars of religious and ethnic bigotry, and this has contributed disastrously to the confusion associated with Nigeria's federal character. This means, in effect, that the claim that Nigeria makes of its thirty-six states as federating units among each other has always suffered socio-political incongruencies due to religious (interreligious) and, of course, ethnic prioritisation and the disdain for common and national interests. Furthermore, among individual religious groups, the proliferation of sects and denominations is directly proportional to the incessantness of their internal conflicts over supremacy, money and power. In broader and extreme cases, the conflicts are based on the imposition of religious and cultural values with the demand to promulgate them as laws within a federal and secular state. These forms of conflict situations have given rise to the contemporary forms of violent religious and cultural groups with waring spirits already discussed in chapter one of our work. To this end, we formulate the sixth premonition as follows:

6th Premonition

Until religious identities and affiliations are jettisoned in issues of national interests, such manner of affiliations would always weigh their adverse effects on decision-making processes for the entire polity and, of course, thwart the dictates of the corporate existence of Nigeria.

5.2.7 *Incongruent Population Growth and Development Indices*

Called the most populous nation in Africa and the sixth in the world, Nigeria's population growth, however, is a deep development concern. This is because the growth factor itself exists incongruently with the development vectors available to contain the growth. At the global level (which can also apply nationally and locally), population growth concerns have awakened a special focus on the shortage of natural resources and serious concern for ecological overshoots. However, the climatic and ecological signs of our present world remain predominantly politicised and economically trivialised. The particular case with Nigeria is that of taking pride in mere numbers and statistics with regard to its population growth, yet with deeply insufficient plans and policies to contain the boom. In Vogt's analysis, the United Nations population report of 2011, under the motto, "We are seven billion people with seven billion possibilities", sounded optimistic about this global population growth concerns, with the support for a global approach that should play down the population figures and emphasise the chances and opportunities that are (or maybe) associated with the growth. In paraphrase, the report suggests that instead of the global concern limiting its focus only to the big numbers of the world's population, it should rather focus on policies that would make the world a better place, sustain our growing cities, support the elderly for more active roles in the communities, mobilise better the creativity and potential of the teeming population of the youth, break down barriers of gender inequality and create societies where people can make free choices and attain their full potential (Vogt et al. 2021).

However, the report's opinion should be conversely interpreted in the Nigerian case. The soaring number of its population must be critically monitored given a number of influence factors, namely the increase in the number of poor families and citizens who are in reality hungry and cannot afford a daily three-square meal, the adoption of the *resort to beggary* as a way of life associated with the northern religious cultural group known as "*Almajiris*" and the devotees of which flood the Nigerian streets, towns and suburbs, constituting a societal scene of chaos, indecency and dehumanisation; the increasing number of Nigerian graduates in their thousands who feel dejected and rejected within a state and social structure that has no plan and future for them, the rise in criminality and kidnapping for money which is an aftermath of a failed institution that could not engage and employ its citizens gainfully, the sharp politically, culturally and religiously based gender disparity operational within the Nigerian system, etc. To this effect, growth in population in the Nigerian case must be approached holistically. In other words, both the skyrocketing population figures and the concern for the general deficiency in the transformative policies that would harness and integrate the growth must be treated as mutually related factors. A mere increase in population without a system- and structurally based plans and purposes to absolve the figures and integrate them meaningfully within the general structure remains a danger to the corporate existence of any known ecosystem. Our research would later join the contemporary discussions on the critical factors that exert influence on demographic development, namely the *fertility-mortality-hypothesis* (*Fruchtbarkeits-Mortalitäts-Hypothese*), *pension scheme hypothesis* (*Altersversicherungshypothese*), *education hypothesis* (*Bildungshypothese*), *per capita income hypothesis* (*Pro-Kopf-Einkommenshypothese*), as well as *governance and motivation hypothesis* (*Regierungs- und Motivationshypothese*) (cf. Ibid., pp. 627–628).

The incongruence in the complementarity that should exist between the two factors—rapid increase in population and stark deficiency of development factors and means—causes failure in achieving feasible and viable growth within the general polity. Concrete consequences are seen as predominantly distributed and manifesting in the forms of political and economic instabilities. This is because the growth process has no *driving factors* aimed at achieving three basic objectives,

> […] a) To increase the availability and widen the distribution of basic life-sustaining goods such as food, shelter, health and protection; b) To raise the levels of living, including, in addition to higher incomes, the provision of more jobs, better education, and greater attention to cultural and human values, all of which will serve not only to enhance material well-being but also to generate greater individual and national self-esteem; c) To expand the range of economic and social choices available to individuals and nations by freeing them from servitude and dependence not only in relation to other people and nation-states but also to the forces of ignorance and human misery. (Emmanuel 2016)

With the above-mentioned concerns about the growth and development deficiencies in Nigeria, our seventh premonition is formulated thus:

7th Premonition

The deficient, ill-structured and weak-performing economic, agricultural, and institutional structure that Nigeria has is incongruent with its rapid population growth. Overpopulation is an attack on the food chain, the ecosystem and scarce commodities, predominantly where a reasonable percentage of the available labour force battles unemployment and economic exclusions.

The simplest concrete and clear evidence of the impact of the above-discussed factors remains the constant failures of both national and constitutional conferences in Nigeria, which are usually adopted as ways to respond to critical issues of national interests. Beginning with concerns about the legitimacy of who sets the agenda for the conferences, down to the initial conflicts within the citizenry concerning any agenda that is set, the varied periods of national conferences in Nigeria have been described as *lost decades*. Added to this are the whooping costs of organising such conferences, the recommendations of which, at the end of the day, would not know any implementation. However, our research uses the drawing of attention to impending premonitions to assert that the research carried out is not only meant to be idealised but rather that the work would accordingly meet the Nigerian society expected to be open for dialogue. Therefore, there are clear challenges to be met as articulated and pre-empted with the premonitions. Nevertheless, our work adopts the following as a guiding thesis, namely: ***Premonitions are not ends in themselves. They are verifiable and concrete means and challenges that can lead to and could be translated into concrete changes, transitions and transformations.***

5.3 Proposition II: The Need for the Assimilation of a “Fourth Norm”

The premonitions are already concrete situations awaiting the attention of practical and applicable policies for transitions and transformations. But there is the need to have the will to break away from the Nigerian syndrome of short-sightedness with regard to the general approach to people, institutions and society. The emphasis on a “fourth norm” is the call to integrate the principle of sustainability as an attitude of daily life. Thus, the usage of the word assimilation is to delineate the fact that it is not just enough to know the principles of sustainability, but rather to assimilate it, in other words, allow the knowledge to speak to and influence actions and volitions. Being developed as one of the Church’s social principles, the words of Mater et Magistra become instructive, namely, “It is not enough merely to formulate a social doctrine. It must be translated into reality. And this is particularly true of the Church’s social doctrine, the light of which is *Truth*, *Justice* its objective, and *Love* its driving force” (John 1961b). The control principle is to evaluate every policy within the effects and influence it has on the moral agent of development, the *Mmadụ*, and this is in relation to the community of existence, nature and its other constituents, where Mmadụ remains an endogenous participant. Thus, the fourth norm—sustainability—would always ask three basic questions: (a) what ideas or principles of reflection would better be formed for the benefit and growth of the human person in the community of existence? (b) How verifiable are the proposed ideas with the concrete situations of the human person within a given society, in other words, how verifiable are the criteria for judgement? And (c) how does the conclusion offer solutions to contemporary situations and guarantee justifiable and safe conditions for the generations, life and societies hereafter? In other words, how does the guidance for action guarantee global (integral) justice? The three criteria lead to what we refer to as *sustainability in social reforms.*

Our research proposes that every envisioned policy, strategy or approach, which should qualify to be relevant in finding the practical inroad to authentic and integral development, must be open to assimilating a fourth norm, namely *the sense of, the vision for and the orientation towards sustainability*. Such a sense is an attitude that must be embedded within the system through organised social and educational learning processes. The process of assimilation is itself the combination of dynamics of social, institutional, cultural and educational conventions, which work mutually to influence and shape the mind and character of the moral agent of development—Mmadụ. Thus, emphasis on the fourth norm must be implemented through a series of social, attitudinal and conventional "mantras" such as ***think sustainably, judge sustainably, act sustainably, sustainable enabling, sustaining abilities, capability sustenance, green thinking sustenance***, etc. Our research plans to make these mantras reflect on the learning process through an initial formulation of learning materials, especially at the basic levels—the primary and secondary school levels, since every child is a product of his or her environment. Consequently, the type of institutions and conventions operative in a given environment contribute generically to the shaping of the human person and his or her general social and attitudinal worldview.

With the fourth norm having its focal point as the human person, its key approach must be to interpret the sustainable development goals (which represent its anticipated and achievable objectives) in relation to this focal point, Mmadụ. Nevertheless, its image of the human person must not separate the subject itself—Mmadụ—from the endogeneity and interdependence it has within the community of existence. Thus, the principle of sustainability in this context must be the principle of eco-solidarity and the change of mentality. Sachs explores this as a striking objective of the encyclical Laudato Si in comparison to the 2030 Agenda. According to him, the enthusiasm for the development narrative of the twentieth century has fizzled out. Consequently, it is currently a matter of coming to terms with the decline of expansive modernity. A truly sustainable world can never be realisable if sustainability remains solely identifiable with the West. True sustainability discourse is a focus on the integral aspect of nature and existence, the concern for the biosphere and respect for the planetary boundaries, the revolutionary struggle by all to obliterate inequality both at national and international levels and, thus, delete the widening gaps between the rich and poor, etc. Furthermore, while Sachs believes that both documents agree to the fact that the global economic model is now a thing of the past and, therefore, no longer seems to guarantee the sustainability of our world, some of their specific objectives are discernible. Thus, while the 2030 Agenda wants to make major corrections to the global economic model, the encyclical advocates the de-emphasising of economic hegemony and calls for more ethical responsibility at all levels. Again, while the 2030 Agenda relies on a green economy with a social democratic element (liberal democracy), the encyclical imagines a post-capitalist age based on eco-solidarity and the need to change the mentality (Sachs 2021). In this whole journey, our world's dealing with the present is as critical as its expectations for the future and its generations. To this end, the journey so far, the challenges and reservations with regard to sustainability discourses and implementations need to be highlighted and discussed in our context.

5.3.1 *"The Future is Now": Critical Scientific Reports and Observations About the* Fourth Norm*—The Sustainability Principle—in Connection with the SDGs, 2030 Agenda*

Our research suggests that sustainability as a concept should, of necessity, be an intrinsic part of all efforts to achieve development and record growth and progress. A more direct focus on it and its development as a social principle may have been theoretically latent within extant development histories. Nevertheless, our world's recent epochs have found it inevitable to continue investing in human intellectual and scientific capabilities and breakthroughs, however, alongside the critical and urgent need to sustain what is being achieved for a better progressive and integrative world. The unifying idea is to achieve a world that is capable of living out its full potential within the *present* and still be able to hand over a future that is habitable and sustainable mutually. It is exactly within this ideology that the development of the sustainability concept as a social principle entered pedagogically our world's development discourses, plans, strategies, governance and policymaking. The concrete and most current response with united efforts is identified with the 2030 Agenda described as "[…] both a normative orientation and a guide for action" (United Nation 2019), hence, the emphasis on the principle of sustainability as our fourth norm. In other words, as the uniting and monitoring norm. Again, having discussed the sustainability plans of our world through its umbrella body, the United Nations, in Chap. 3, a look at what has been achieved, the challenges and other reservations would help to articulate the anticipated implementation of our work and make it more focused and better oriented.

The Future is Now, as our sub-topic reads, was the caption of the 2019 Report on the Global Sustainable Development Goals by an independent group of scientists, which contains a scientific analysis of both progress and challenges that the paths of implementation have encountered on the global scene (particularities within individual countries may not have been excluded). The report took a systematic step anchored on three directions, namely the processes that are still on track, the ambivalences and the red zones (the tipping points, 2019 Report: 12) (Ibid., *Christliche Umweltethik, Grundlagen und zentrale Herausforderungen,* pp. 517–518) as shown below (Fig. 5.3).

The "***On Track***" report highlights some of the positive initiatives and steps that countries have taken to incorporate the development goals into national plans and policymaking. Such efforts are projected to enhance and improve socio-political

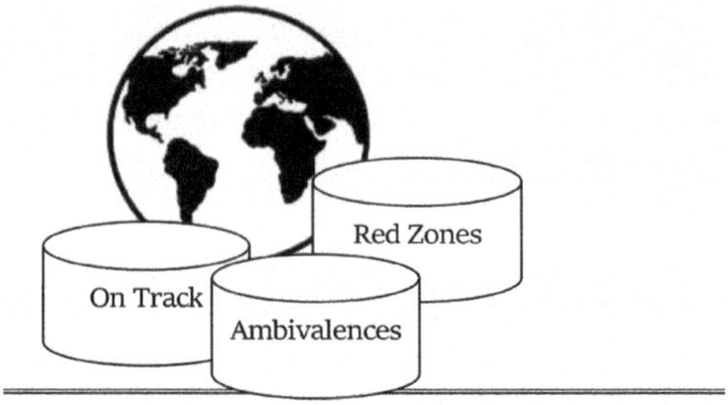

Fig. 5.3 Three focal points on the assessment of the global sustainable development goal, 2019. *Illustration mine*

and socio-economic programming, especially when they are, using the words of the Report, "*coherently implemented*" (cf. Ibid., *The Future is Now. Global Sustainable Development Report 2019*, p. 8). However, this on-track-progress report is, in the main, the combined efforts and responsibility of both governmental and non-governmental actors as well as individuals and groups. Politically striking remains the global consciousness with regard to climate change, regardless of the challenging opposition it receives on the global scene. Nevertheless, the *on-track-report* opines that the negative trends and challenges to implementation should rather motivate our world's spirit to do more and build resilience towards achieving transitions and transformations. The appeal is: "[…] impending policies should urgently be reversed or modified and recent advances that holistically promote the Sustainable Development Goals should be scaled up in an accelerated fashion" (Ibid.). It cannot be denied that there are little and, in countable cases, no efforts within the African circle to get on track with the interpretation and implementation of the development goal. However, combined and related factors continue to exert adverse effects and influences. Given the anticipated field of implementation, the most recent observation of the 2021 Report with reference to the Sub-Saharan African countries and after the global battle with the Covid-19 pandemic is worth mentioning:

> All sub-Saharan African countries continue to face major challenges in meeting most of the 17 SDGs. With continued high levels of extreme poverty in some countries in the region, progress towards socioeconomic goals and access to basic services and infrastructure (SDGs 1 to 9) is poor compared to other world regions. In some countries, insecurity and conflict have reversed gains on various goals, including SDG 16 (Peace, Justice and Strong Institutions). Domestic resource mobilization must be increased across the continent and institutions strengthened to achieve progress on SDG 16 (Peace, Justice and Strong Institutions) and SDG 17 (Partnerships for the Goals). Relatively low levels of consumption have led to somewhat better performances on SDG 12 (Responsible Consumption and Production) and SDG 13 (Climate Action), although vital biodiversity areas lack sufficient protection. The COVID-19 pandemic threatens to reverse the progress that has been made over decades towards reducing poverty and improving socioeconomic outcomes in sub-Saharan Africa. International solidarity and partnerships are needed to respond to the health emergency globally, including through vaccine rollout and strengthened access to financing. (Sachs et al. 2021)

The "***ambivalences***" are the observable reservations and concerns that warn against mere "straightforward projection (that) ignores possible complexities" (Ibid., *The*

Future is Now. Global Sustainable Development Report 2019, p. 11). Such intractable conditions, which tend to slow down rates of application and implementation of policies, especially where concerted efforts are observable and verifiable, elicit ambivalences that threaten optimism and results. The report gave a concrete example as follows:

> [...] the World Bank's 2018 report on poverty found that in the 25-year period from 1990 to 2015, the extreme poverty rate fell, on average, at a rate of about one percentage point per year. However, in the period from 2013 to 2015, the rate fell to less than one percentage point. A similar pattern can be seen with regard to primary school enrolment; an extended period of sharp decline began to level off as the ultimate objective of full enrolment was approached. (Ibid.)

The danger of the ambivalences is that they are syncretic to one another and need tactical discernment to disband their already constituted network of influence. Hence, referring again to the example above, the report noted that the extreme poverty cases and the decline in primary school enrolment opportunities remain predominantly endemic among people, societies and regions that are victims of combined factors like conflicts, weak institutions and very high population growth rates (cf. Ibid., p. 11). This form of ambivalence is experienced where there is at least the effort and will, however, alongside an incommensurate rate of progress. Extreme ambivalences, the report warns, are.

> [...] targets for which recent trends are not even in the right direction, either because implementation of the goals has not yet been able to reverse pre-existing deterioration, or because world recovery from the 2008 economic crisis has brought back negative trends that had for the moment been stalled such as obesity, inequality, greenhouse gas emissions, land degradation, biodiversity loss, wildlife trafficking, absolute material footprints, overfishing and deterioration of coastal waters. Several of those targets, for which even the direction of the trend is wrong, are of particular importance. Not only do they represent trends that are difficult to change (ambivalences), but they also make it harder to reach other goals and targets, sometimes in ways that cascade across the entire 2030 Agenda. (Ibid., p. 12)

The "***red zones***" are referred to as the "[...] negative tipping points, which would lead to dramatic changes in the conditions of the earth system, in ways that are irreversible on time scales meaningful for society" (Ibid.). The identified and most critical red zones of our epoch are seen in CO_2 emissions, the eroding of biodiversity and the persistent inequality within nations. With regard to the case of inequality, Vogt observed, additionally, that given the modern trend of international relations and politics, inequality between nations may be decreasing, but at the cost of a social divide in most countries of the world between globalisation winners and globalisation losers (Ibid., *Christliche Umweltethik, Grundlagen und zentrale Herausforderungen*, p. 518). Furthermore, identifiable with the red zones is the warning that many climate change tipping points seem to be arriving faster than originally thought. As a result, the global sustainability report explicitly states that we are heading down a troubling course, nearing a critically dangerous state that may exhibit unpredictable and chaotic characteristics. However, we still have a window of opportunity in which to plan for change. This window of opportunity will remain open for another 10–20 years (Masserli 2019).

The extended analysis of our *fourth norm*—sustainability—in our context emphasises the central role it has to play as a tool and principle in working out sustainable inroads to confront Nigerian cases and situations. Our research proposes the sustainability principle as serving monitoring, control, uniting and integrative purposes. These characteristics would be the coordinating factors within all our proposed strategies to work with the goals and see to their social interpretation and implementation. Again, following the opinion of Messerli, three inferences need to accompany all further work phases to develop and implement the goals, namely to be aware that (a) the SDGs all have interdependent character; therefore, approaches to implementation need to be synergistic; (b) the nature of the goals themselves and the possibility of realising the envisioned objectives attached to them in our hyper-dependent twenty-first century demands the strong awareness of the challenges of power conflicts and other conflicts of interests and (c) the focus has to be the positive transformation strategies for the human society, not in isolation but rather in relation to nature. Therefore, every transformable process must bring humans and nature together in the journey towards energy decarbonisation, change in patterns of food production and nutrition, transitions and transformation of our concept of well-being as well as a financially workable and equitable network for a more distributive and participative approach to the global commons (cf. Ibid., pp. 518–519).

Conclusively, the six proposed "entry points" as recognised by the Sustainable Development Report 2019 would remain conscious in the further developments of our strategies and approaches as represented below. Forms and strategies of encounter with the entry points are as dynamic as the human societies themselves with their unique experiences, worldviews, cultures, conventions, etc. Strategic uniting factors are proposed as governance, economy and finance, individual and collective action, as well as science and technology (Ibid.) (Table 5.2).

From the above-mentioned entry points, it is supportive and confirmatory to note that the report itself also chose as the very first point of entry the *Mmadụ* and the concern and care for his or her well-being. It is this same human subject that we made our focal point in the journey towards sustainable inroads through ethics and principles of remediation. Our fourth norm would be made more feasible when applied as a guide to ensure that approaches to *Mmadụ* in relation to other vectors of development place him or her always as an agent and not a patient of factors of social enslavement.

Table 5.2 Six entry points for SDGs

Entry points in the implementation of the SDGs
1. Human well-being and capabilities
2. Sustainable and just economies
3. Food systems and nutrition patterns
4. Energy decarbonisation with universal access
5. Urban and peri-urban development
6. Global environmental commons

Illustration mine

5.4 Proposition III: Systemic *Exnovations* as First Transitional *Innovations*

Generally, a social system that is beclouded with obsolete or non-regenerative ideas and policies would always experience difficulties when the need to innovate arises. To be beclouded with non-regenerative ideas defines the situation of being infested with, perhaps, stagnant institutions, which have become incongruent with a newer phase of social conventions. It describes an invention, the services of which may have been supplanted by subsequent discoveries and scientific pieces of evidence and ideas, be it political, social, economic or cultural, the functioning and services of which are gradually beginning to prove less effective and obsolete. Such situations within a social structure build up some forms of structural obstructs which, as experiences have shown, militate against efforts to innovate, reform and transform, as well as against general transitions to sustainability. Of recent, the need to turn attention to such militating factors and eventually get rid of them has pulled together scholars from varied fields to reflect on the concept "*exnovation*" as a prerequisite step towards plausible and practicable innovation. To get access to successful inroads to the Nigerian situation, our research proposes the need for an organised project of an initial exnovation process, the character of which must not just be systematic but rather predominantly *systemic*.

5.4.1 A Systemic Model of Approach

The socio-structural problem that Nigeria has is better described as systemic in the sense that each aspect of the dysfunctional parts always connects other chains of intertwined and related factors, which attack the general structure aggressively and simultaneously. What this means in practical terms is that for the corruption in the executive arm of government to be relatively persistent, it has to build itself a network of collaborators within the other two arms, namely, the legislature and the judiciary. In other words, there is always a network of allies who must speak the same language, take the same oath of dishonesty, agree on protecting and defending one another's impropriety even at gunpoint, as well as unite in the resolve that the only way to keep the general system and structure hijacked would be to weaken all the parts that make up the whole not just systematically, but rather systemically—attacking all at the same time.

The above-mentioned exposition is the very situation in the Nigerian case. Thus, we propose a systemic approach that does not need to be ***unitary*** but ***unilateral***—in other words, a *Unilateral Systemic Approach.* We clarify: with the present structure known as Nigeria and with reference to the premonitions we have raised, there can be no successful Unitary Systemic Approach since there exist strong complexities with language, cultures, affinities and affiliations, as well as religious groups and sects. Hence, all known unitary efforts to address the ailing structure have failed to date to remedy the situation. A Unilateral Systemic Approach would be anchored on the distinct identities of the different nations and cultures that make up the British coinage, Nigeria. Such a Unilateral Systemic Approach would be developed within the known unique and varied languages, worldviews, mentalities, beliefs and conventions in Nigeria.

The first approach or process of identity recognition is to make the first touch on the central agent of development, Mmadụ, the human person. The systemic approach, therefore, does not just begin with the person himself or herself. It rather believes in the human person, who has qualities and capabilities upon which and within which all the three dimensions of sustainable development dynamics revolve, namely economic, social and environmental. Taking this as a reference point, a systemic approach to exnovation and transition considers a given structure as a whole which cannot exist endogenously without the functional parts. Hence, in the process of exnovation, especially with regard to social systems and policies of a given society, our research proposes a kind of *exnovative relational quantum mechanics*[2] at social and policy levels, in the sense of interwoven networks of processes and factors—socio-political, economic and environmental. This would get the *exnovator* more informed and guide the exnovation policy towards understanding the interrelatedness among factors of social institutions and governance.

Concerning the key innovation and transformation platforms for Nigeria, a systemic exnovation must be approached in such a way as to address an energy-power revolution that must, at the same time, be ecologically regenerative. It must work to address socio-political and economic emancipation that must never neglect and isolate agro-cultural revaluation; it must take up an ecological regeneration that should never treat the human person (Mmadụ) as a being outside the very concept of ecology; it must be made to understand economic emancipation as an indispensable correlate to empowerment evolution; its agro-cultural revaluation must approach soil with resource justice, since the moral agent of development himself or herself—*adam, earthling; adamah, soil* (Vogt 2019)—is an endogenic part of it and depends heavily on it, etc.

One real advantage of the systemic approach model, as proposed by our research, is the ability to approach socio-structural dysfunctions—social structural sins—holistically. In other words, it attempts to destroy not just a particular visible social problem but rather has the ability to follow up a network of chain reactions, which build up interrelated factors that keep the problem cancerous. In this way, the admonition of St. John Paul II remains evergreen, namely, "We should not limit ourselves to deploring the negative effects of the present situation of crisis and injustice. What we are required to do is destroy the roots (the interrelated factors) that cause these effects" (Paul 1995). With this in mind, our research commits to offering knowledge through the already studied development principles for an exnovation strategy taken

---

[2] Carlo Rovelli initiated the development of a system of views now called relational quantum mechanics. This concept has at its foundation the view that all systems are quantum systems and that each quantum system is defined by its relationship with other quantum systems with which it interacts (cf. Wikipedia, accessed February 12, 2021).

in an interrelated form and cutting across other relevant and related factors. An indigenously designed blueprint of such a systemic approach known as ANIDS, Anambra Integrated Development Strategy, based on the then Millennium Development Goals (MDGs), was strictly implemented amidst fears, opposition and optimism by the then Governor of Anambra State, the Southeastern part of Nigeria, Mr. Peter Obi. The very details of the programme are not predominantly within the scope of this work. However, the principles that guided the programme to fruition, cutting across both public and private sectors and which led to concrete transitions and transformations in the state in question, would be studied and taken up further by this research work as tools for practical inroads.

5.4.2 The Normative Power (Factors) of Exnovation

To understand better the normative aspect of exnovation, it is pertinent to gain a working understanding of what the concept represents and defines. The coinage of the word "exnovation" is not meant to serve as an opposite of "innovation" but rather as a *correlate* of the word. This is because the very act of exnovating is itself an innovative approach. Therefore, it cannot contradict itself. Thus, amidst growing concerns about the so-called innovation bias, the emergence of the word exnovation came as a policy response. Innovation bias is the opinion that though innovations are important drivers of sustainability transformations, they, however, have not proved sufficient enough to replace unsustainable structures. Hence,

> [...] The promotion of renewable energy, for example, has been insufficient for pushing coal out of the energy market. The prevalent "innovation bias" should be overcome by complementing innovation politics and research with a stronger occupation with the purposive termination of unsustainable technologies, products and practices. (Heyen et al. 2017)

With the emphasis on "purposive termination", the concept of exnovation began to take root in organisational management pieces of literature and within the German debate on sustainability dynamics. Kimberly, therefore, representing the original reflection on the concept, defined exnovation as "[...] the removal of an innovation from an organisation. Exnovation occurs when an organisation divests itself of an innovation in which it had previously invested" (Ibid.). In divesting, a system tries to make modifications, corrections, or, in extreme cases, the total abandonment of what has been in vogue, especially when all verifiable pieces of evidence show that time, experiences and other better approaches have taken it over.

Understood from the background of the Latin verb root, "novare"—making new, exnovation would not just mean a means towards renewing (or renewal). It rather connotes more the direct concepts like "change", "transform", "break", "overthrow" (Ibid.), etc. Thus, the very beginnings of sustainable development discourses applied these direct concepts to discuss exnovation by focusing on the need to break away from "[...] technologies and products which are economically feasible (some supported by subsidies) but should be phased out on the grounds of ecological sustainability" (Heyen 2017). Consequently, the initial development of the concept concentrated more on socio-technical structures rather than on societal (policy) discourses. However, further development of the concept discusses the need for *Exnovation Governance*, which approaches the concept more integral. In this context, governance "… denotes the coordination of actions and steering of behaviour towards an explicitly public goal" (Ibid., *Out of the Comfort Zone! Governing the Exnovation of Unstainable Technologies and Practices*, p. 326). Furthermore, society's approach to coordinating actions and steering behaviours within the system is heavily dependent on policies and planned strategies. In this regard, our research proposes a form of exnovation approach, which calls for change, transformation, transition, discontinuation, radical breakaway and radical revisiting of the varied policies currently operative within the present structure, Nigeria—*Policy and Institutional Exnovation*. This would be based on the yet-to-be-discussed five cardinal institutional platforms for systemic and transformative approaches.

A more recent application of the concept, in a broader sense, talks about *policy dismantling*. It represents the "[…] cutting, diminution, or removal of existing policy" (Bauer et al. 2013), especially when such a policy begins to turn against or detract from the system. The emphasis here is on the need for proper *strategising* so as to contain risks and blame. It follows, therefore, that strategic planning is predicated on the backdrop of driver-policies that serve as catalysts for strategic implementations. The proponents of policy dismantling identify it as the hallmark that should be fundamental to the politics of "hard times" occurring mostly during periods of economic austerity. However, we import the idea and propose it as befitting in addressing the economic dimension of Nigeria's quagmire and as an inclusive concept in approach to the general system, policies and structure. Thus, when policy dismantling is made a strategy, and with a focus on exnovation, three (3) normative strategies must be seen as guidelines, namely: *Preferences*, *Strategies* and *Effects*[3] (Fig. 5.4).

As shown above in policy dismantling, the three normative orientations in exnovation policy serve as guiding steps to initiating executable exnovation governance. Our research proposes the orientations as practicably adaptive within the socio-ethical dialectics for social transformation, namely, the principle of reflection, criteria for judgement and directives for action.

The ***Preferences*** are to be formulated with the principle of reflecting such that they reflect the best policy options that have the human person himself or herself—the Mmadụ—as both the subject and object of focus. They must contain ideas that are practicable in the community of existence—an integral justice. The ***Strategies*** must be informed by the criteria for judgement based on evaluative experiences. No authentic policy dismantling is possible or conceivable without being preceded by periods of application, experimentation, evaluation, perhaps possible amendments and then decision-making. The strategy must be based on experiences and projections. The ***Effects*** are the projected directives for action. They must not necessarily

[3] Preferences, Strategies and Effects are concepts taken from the subtitle of the book, "Dismantling Public Policy: Preferences, Strategies and Effects" as cited above.

Fig. 5.4 Normative strategies of policy dismantling—policy exnovation. *Illustration mine*

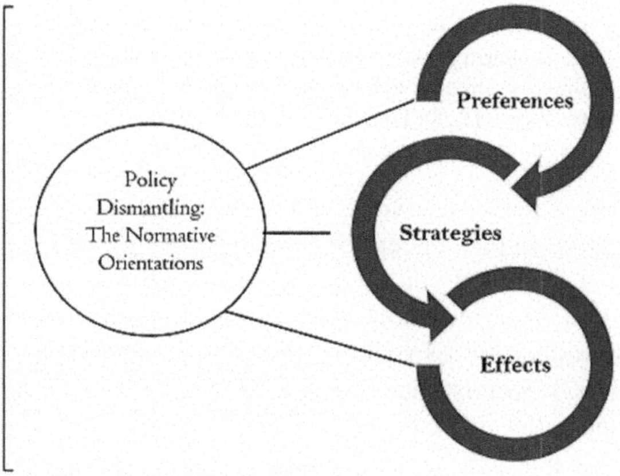

be concrete. They rather can be projected, referential or comparative. They must be either concretely evidenced or referentially projected to be integral in responding to socio-political, socio-economic, socio-cultural, as well as socio-ecological policy demands of a given societal system and structure. Concrete ways to do this would be proposed while presenting the five key areas of urgent policy dismantling in the Nigerian context.

5.4.3 Resistance and Coalitions in Exnovation Governance

Policies and institutions are factors of governance predominantly within the social, political and economic sectors of a given society. Within these sectors, there are chains of players and representatives of varied models of opinions, strategies and interests. Thus, exnovation governance as a strategy does not come without challenges. Consequently, two concrete terms are associated with the challenging experiences within exnovation governance, namely, *resistance* and *coalitions.*

Resistance and coalitions, or more succinctly, anti-termination coalitions (cf. Ibid., *Out of the Comfort Zone! Governing the Exnovation of Unstainable Technologies and Practices*, p. 330), occur since policymaking and governance involve chains of interrelated actors, interests and institutions. Within these factors, there are always the phenomena of "gainers" and "losers", the "mainstream beneficiaries" and the "participative beneficiaries", or, in deep conflict situations, the "non-beneficiaries". The different groups represent diverse interests that often challenge the approach to revisit, change, dismantle, disband, transform, abandon or exnovate what has been in vogue, and which at the same time is proving obsolete, adverse or detractive to the system and structure. The chains of actions and reactions amount to systemic change resistance, where,

> [...] in addition to actors who are directly affected, suppliers of products or expertise as well as politicians and bureaucrats – (the policymakers) – feeling responsible for the sector in

> question can be part of such a coalition, sharing the same problem perception and values [….] Fearing losses, vested interests are especially motivated, and also especially strong as they are already well organised and have usually spent years building up relationships and resources [….] (They unite to build up) instrumental, discursive, material and institutional forms of power and resistance. (Ibid.)

The well-chosen words of the encyclical Solicitudo rei Socialis of John Paul II with regard to the concept of Structures of Sin, which remains one of the main concepts that informed our research, articulated this same persistent presence of resistance and anti-termination coalition when he referred to individuals or groups "[…] who introduce these structures, consolidate them and make them difficult to remove" (Paul 1987). Amidst the expected challenges, our work still sees exnovation governance and policy dismantling as achievable possibilities, which must be approached through both political and economic processes. The approach requires the *will* to action and dialogue, added to an integral sense of justice to purify agreements and compromises of vested interests. This would be made a focus in our institute.

5.4.4 The Socio-Ethical Interpretation of Exnovation

Interpreting exnovation socio-ethically is to propose the need for a conscious practical departure from socio-political, socio-economic, socio-cultural, as well as socio-ecological structures and policies—structures of sin, which, as the encyclical puts it, once they are inbuilt in a structure or system "[…] they grow stronger, spread, and become the source of other sins, and so influence people's behaviour" (Ibid.). Socio-ethically, the emphasis is placed first on the will to strategise and, secondly, on the criteria, methods and instruments to engage in exnovation process, whether as socio-technical or policy-based dismantling, change, transformation, termination, reparation, renewing, abandonment, etc. In all, the common denominator is the need to change base and transit to something better, something more beneficial and more integral for a productive system and structure. Thus, applying the concept *specifically in policymaking* and our study of the Nigerian context so far, we would socio-ethically introduce the concept, ***Policy Remediation***. This would be a key focus within the objectives of the proposed institute. A working definition is given thus:

Policy-remediation is the socio-ethical will and resolve to impact a transformative change on already existing policies operative in a given society, with the practical normative intent to remove, remake, remodel, reinvigorate or reinvest in policy patterns that are more responsive to an existing

The socio-ethical orientations, as would be taken up by the research work as practical inroads, focus on, first, analysing existing policies and evaluating them with the "**Five-Rs" of Policy Remediation,** namely: Remove, Remake, Remodel, Reinvigorate and Reinvest. The research work would be convinced that a normative application of this model of approach would eventually make inroads into a transformative change, which, under three evaluative criteria, would produce a practical result through an *ETT-Approach*. This would be taken up in the institute as a model of approach.

ETT-Approach = Exnovation – Transition – Transformation

(***Exnovation towards Transition for radical Transformation***)

5.4.5 *The Three Evaluative Criteria for Systemic Policy Remediation*

Policy remediation using the Five Rs would set its goals and objectives right with three evaluation criteria, which should stand at the base of all approaches to sustainable development dynamics. Our research formulates these as the focus areas for people-oriented and integrally based social policies, namely, *Mmadụcology*, *Eco-eco Dichotomy* and *Integral Justice* (Fig. 5.5).

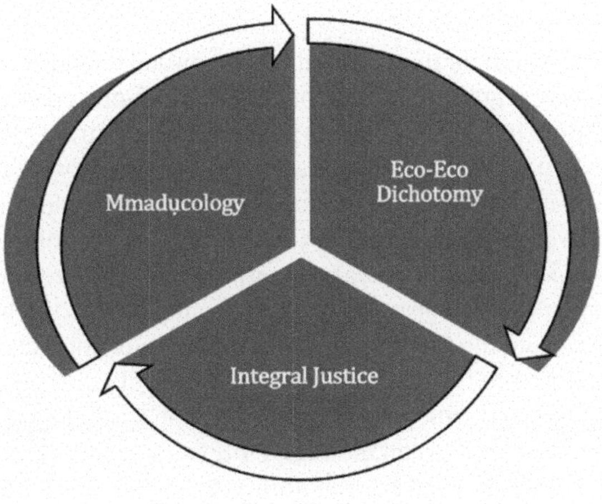

Fig. 5.5 Three criteria for effective systemic policy remediation (Policy remediation must be pro-Human Ecology—*Mmadụcology*; it must maintain a healthy relationship between economy and ecology; it must be trans-generational—integral justice—so as to be sustainable). *Illustration mine*

Mmadụcology is not a new concept in terms of meaning but rather an Africanised semantic for human ecology, which has been discussed extensively as a prerequisite for the authentic development of any given society considered to be sustainable. As a criterion, therefore, policy remediation must examine the related effects of policies on the moral agent of development, who remains the determinant factor in the general decision-making in matters of conventions and institutions. The Eco-eco Dichotomy must be employed to juxtapose the Anthropocene and the need for balanced interests between pursuing economic goods and the relevant precautions against their extremely adverse effects on nature and ecology. This must be reflected in decision-making and policy application. Furthermore, appeal to ***Integral (Inclusive, Intergenerational) Justice*** would be taken as a norm meant to inform and influence the decisions of today for a regenerative tomorrow (future)—for the sustainable life of the society, for sustainable development.

Within each criterion, a method of formulating ideas and questions *for or against* the object of evaluation would be adopted. The objective would be to exhaust every known ideological and verifiable option available in order to arrive at decisions that would not, in principle, turn policies into detracting tools. However, experiences show that even people-oriented policies could be twisted in the hands of corrupt leaders and policymakers. The *strategised twist* would, at the same time, be clothed with a form of promulgated or imposed legality designed by the corrupt leaders themselves. Such experiences are only premonitions that our approach holds in the know and would work, through the institute, to confront and expose within a Five-Key-Area of social concerns in the Nigerian context.

5.5 Proposition IV: Presenting a Five-Key-Area Scheme of Policy Remediation for Transformative Socio-Ethical Recovery

Our search for the practical ethical inroad would be without content and purpose when concrete areas of social concerns are not presented as concrete integrative spheres of action. In other words, the principle-based strategy identified with the *ETT-Approach* would never happen in a vacuum but rather with a given. Similarly, the normative orientations towards policy remediation become inert when their principles have no verifiable pieces of evidence and applications. In this regard, we present below five key areas of social concerns within the present Nigerian structure, which have urgent needs for policy remediation. We use the term "**socio-ethical recovery**" to delineate the fact that the Nigerian case is not lack of manpower, ideas, policies and opportunities. It is rather the need to apply critical policy remediation to recover the system and structure and put it back on track. Hence, the research work uses the term "subsistence" as an *indicative prefix* attached to the nomenclature of the schemes to buttress the need for the ability to transit from what has been in vogue to a better level that would be more productive and result-oriented. Consequently, in approach to the five key areas, we use ideological and development terms that give a sense of the will towards recovery, namely: *evolution, revolution, revaluation, regeneration and emancipation.*

5.5.1 Subsistent Socio-Political/Economic Emancipation

As we have argued under the premonitions and with regard to the general Nigerian structure and system, the present federal system operative in Nigeria gives no future for that country. The concept of its federal character, therefore, must be revisited, redefined and, if necessary, disbanded and abandoned for political, economic, cultural, religious and developmental reasons. More than sixty years of experimental independence is enough to convince Nigeria that the present structure is defective. The geopolitical imbalances with the resultant ethnic tensions, the varied religious unrests and fanaticisms, the increase in the activities of *affiliation groups* yelling for emancipation within a sovereign nation, the daily losses of lives and properties; the increasing situation of destitution within the biggest economy of Africa; the growing challenges of terrorist groups; the most current and unbearable tensions in the schools due to kidnapping activities of varied banditry groups, etc.—all of these combine to send signals on the urgent need for policy remediation towards subsistent socio-political and economic emancipation.

Both national and international economic and policy observers, including our research, unite with the opinion that there exists a deep administrative, socio-political and socio-economic "disconnect" in the Nigerian system. The 2015 National Human Development Report summarised this thus:

> This disconnect between development priorities and resource allocation, among others explains the challenges and deficit. The parlous situation is aggravated by the fact that the Nigerian economy is largely not diversified, is inefficient and lacks equity in allocation. These features have continued to impact negatively on both human security and human development of her citizenry. Nigeria's economic performance since attaining independence in 1960 has been profoundly disappointing [….] Nigeria faces an ongoing challenge of making its decade-long sustained growth more inclusive. (But) poverty and unemployment remain prominent among the major challenges facing the economy. (UNESCO 2015)

Known existing efforts in response to the "disconnect" at the Federal level gave rise to three economic blueprints designed to be implemented at the national, state and local government levels. The blueprint integrates the three concepts below (Fig. 5.6).

At least conceptually, the above-represented Nigerian model was designed to reflect the modern sustainable development dynamics associated with the current Sustainable Development Goals (SDGs). The two directional arrows symbolise the "up-down" and "down-up" policy remediation expected to energise the blueprint in an act of mutuality within and among arms of government, institutions, policymakers, as well as other non-governmental parastatal structures. Yet, an analysis of the concrete Nigerian situation reveals that the blueprint still has a very long way to go in the journey towards achieving its objectives, namely, the emancipation of

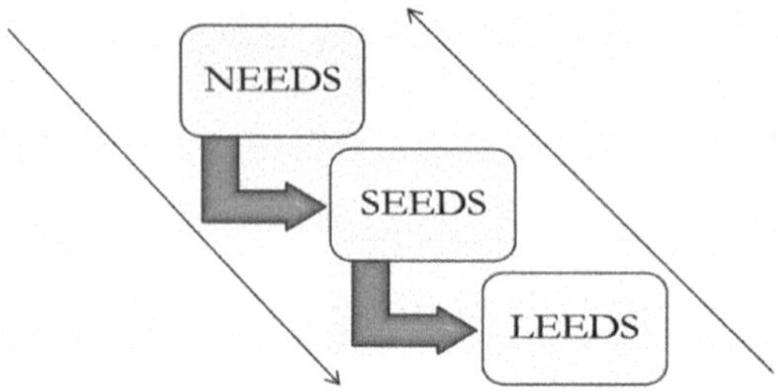

Fig. 5.6 Nigeria's national, state and local levels' economic blueprint. **NEEDS**: *National Economic Empowerment and Development Strategy*; **SEEDS**: State *Economic Empowerment and Development Strategy*; **LEEDS**: *Local Economic Empowerment and Development Strategy. Illustration mine*

the moral agent of development, Mmadụ, the human person. Thus, according to the development report, "As laudable and well-intended as these economic blueprints are, the associated challenges have been poor implementation as evidenced by the weak plan-budget link since 1999" (Ibid., p. 21).

Our research considers it pertinent to observe as a problem the lack of *regionalisation* or *localisation* of the already conceptualised blueprints by experts, intellectuals and social, political and economic scientists. In other words, the real targeted audience for the objective of the blueprint—given the existing status of the Nigerian society regarding literacy rate—remains unattended to and unincorporated in the whole policy dynamics. The strategy of our approach would be different. First, there would be no outright discard of the existing blueprint but rather a radical reading of and remediation of the same. Secondly, our strategy would take the basics to the right audience—the grassroots, the regions and localities—and tries to speak the language of the concerned indigenous people.

Key Policy Remediation Steps

- **Radical restructuring, dissolution or abandonment of the present structure and system.**
- **Strict Decentralisation Campaign.**
- **Pure Regional Politics and Optional Coalitions.**
- **Teaching of Identity and Cultural Freedom.**
- **Initiatives for Competitive Economy.**
- **Clamour for Regional Economic Independence.**
- **Forming coalitions of legal and human rights activists for the building of radical voices against corrupt leadership systems.**

5.5.2 *Subsistent Agro-Cultural Revaluation*

The present Nigerian system faces the horror of being reminded that nations that are heavily dependent, with regard to excessive importation of food and basic consumer products, are remote from being sustainable states. The catastrophic fall from the 1960s and '70s agro-cultural boost of the economy to a near-exclusive reliance on the oil and gas economy, in other words, a free and catastrophic degeneration to a mono-economy, is a national fraud imposed on the general system by successive political miscreants. There can be no sustainable development success within the present Nigerian system if no remedial policy uses ***radical diversification*** as its language. That the Nigerian artificial complacency with a mono-economy is a time bomb needs just two simple facts to be proved, namely (a) the sporadic volatility of the oil market, which has recorded at various times several plunges by the economy into recession within the Nigerian experience; (b) the verifiable fact that economic crude oil itself is an *exhaustible natural capital and supplant-based technology*. With a great wealth of potential, the lack of economic diversification stifles regional potentials and capabilities, and this has become a bane in the Nigerian system. Policy remediation must be strategised to work towards *revaluation* of this sector of the economy. Our research, through the institute, would prefer, as a criterion, the launching of revaluation seminars and workshops at the local and grassroots levels, whereby the workshops would speak the indigenous language of the local people.

Key Policy Remediation Steps

- **Radical Grassroots Agronomy through seminars and workshops.**
- **Founding, educating and empowering regional and local union groups for agro-cultural revaluation.**
- **Push for Techno-Partnership for minimal essential agro-mechanisation.**
- **Creation of functional and accessible Micro-Credit Institutions at the local and grassroots levels.**
- **Teaching modern irrigation techniques.**
- **Driving the Will for a radical transition to modern and scientifically improved animal husbandry.**
- **Teaching primary soil knowledge and methods of improvement.**
- **Campaign for both governmental and non-governmental partnerships for better storage, processing and cost-effective transportation.**
- **Encouraging cooperative and standardised local market price control.**

5.5.3 Subsistent Energy-Power Revolution

Nigeria needs urgent energy and power revolution. Sustainable economies, industries and productions are generically dependent on the availability and capability of the energy and power supply system of a given society. The horror of underproduction, insufficiency or total unavailability of the energy system in Nigeria has robbed the nation of investors, progress and development. Yet a rich resource of potential energy-power possibilities remains untapped. This has pushed the whole nation to the abuse and consequent adverse effects of private fossil fuel energy and power generation (as presented in the fieldwork). We would embark on grassroots' sensitisation on the need for cooperation, coalitions and partnerships towards available modern energy transitions. The map showing the distribution of identified renewable energy potentials in Africa reveals four readily available potentials in Nigeria, namely, bioenergy, wind energy, solar energy and hydro-energy (International Renewable Energy Agency 2022) (Fig. 5.7).

Granted that the political economy of energy transition belongs to the macroeconomic sector of a polity, our strategy would be to dedicate the energy, first, to sensitising the grassroots, localities, individual persons and unions on the need to change patterns with the approach to combustible fossil fuel energy generation. To this effect, and since the Nigerian situation has been unnecessarily persistent in its

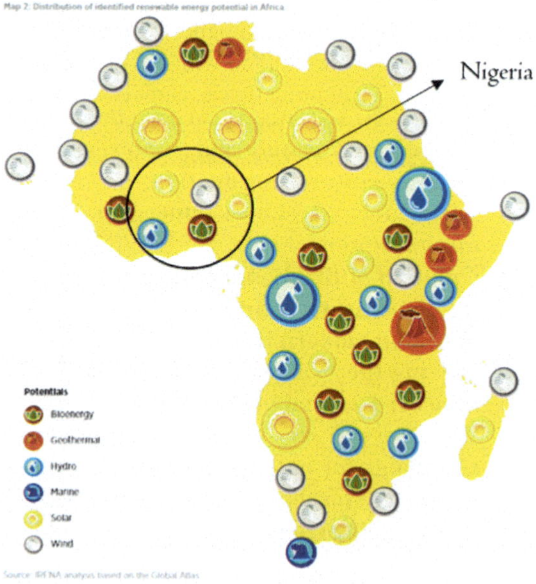

Fig. 5.7 Main energy potentials in Nigeria

Source: IRENA International Renewable Energy Agency

indolence in improving the situation, we propose the terms coalitions and partnerships as redemptive concepts in approach to this area. This means, in real terms, the orientation towards forms of cooperative power generation systems instead of the private individual power generation system, which would eventually help to reduce the length of emission, noise pollution, health risks and loss of revenue. This would be the first step towards the transition to other potential renewable energy systems. As noted in chapter two of our work, Nigeria is the biggest market and business partner for Chinese combustible power generating sets in the whole of the African continent, if not in the whole world.

Key Policy Remediation Steps

- **Grassroots sensitisation programmes for partnership and coalitions in both private, commercial and quasi commercial power generation.**
- **Simplified awareness programmes on health risks with regard to combustible power generating sets.**
- **Solar Energy Campaign.**
- **Building coalitions for partnering governments and institutions for stringent emission taxes.**

5.5.4 Subsistent Ecological Regeneration

Nigeria's ecological struggles took a head-long plunge into becoming an emergency situation with the impropriety happening in the oil and gas sector and the moribund state of the energy sector, which has led to a proliferation of individual/private energy/power generation through combustible generating sets. Our research plans to initiate plans for socio-culturally based ecological regeneration. However, the question is: what is it concretely that needs to be ecologically regenerated? To answer this question, we first emphasise that regeneration is both revitalisation and liberation. An ecological regeneration is to respect the dynamics of nature and, perhaps, assimilate or abstract from them and never destroy them. The ecological politics in the Nigerian system, aided by the activities of both national and multinational partners, is an affront to nature and its capitals. The current practices and the consequences thereof have been laid bare in the preceding chapters, sub-chapters and fieldwork. We intend, in this regard, to initiate ideas and methods that would aid the inroad towards changing the current ecological narratives associated with the present structure and system in Nigeria. Such a change of narrative would be socio-culturally articulated to form part of the learning process, with the basic truth that the destruction of nature and ecology amounts to the destruction of the human person himself or herself. As one of the strategies, our approach would teach and develop further the spirit of Laudato Si articulated in its Seven Cs of approach to integral ecology (Fig. 5.8).

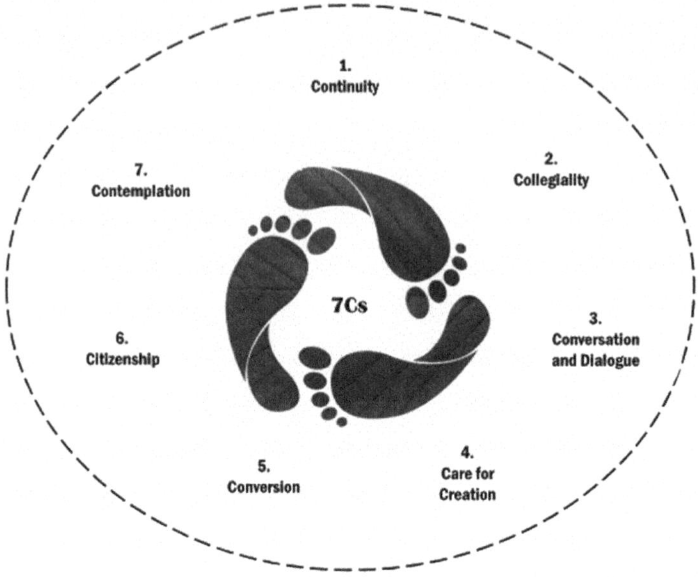

Fig. 5.8 Ecological seven Cs of Laudato Si. *Illustration mine*

The ecological symbol represented by *human footsteps* emphasises the fact that the expected approach towards subsistent ecological regeneration demands a step-by-step attitude on the part of our human world. Such a step-by-step approach would be further developed in our institute through a process of indigenisation of the above "7Cs", in other words, reading the seven concepts within the African context in general and Nigeria in particular. Thus, in ***Continuity,*** Nigerian society would be keen to connect its environmental history from the past to the present to understand the need to restructure or completely abandon unsustainable attitudes towards ecology. Connection to a chain of diverse environmental histories would open a new phase to the contemporary and expanded understanding of ecology to include the human person himself or herself, society, peace and well-being—Integral Ecology. The sense of ***Collegiality*** must be explored so as to make Nigeria's approach to ecological discourses inter- and multidisciplinary, holistic and more scientific. ***Conversation and Dialogue*** would be interpreted to mean the concrete legitimate effort to get the society involved, convinced and thereby ready and equipped for dialogue—a form of dialogue based on the united interest of all to do good for humanity, which, on the other hand, is to be seen as a public good. The concept, ***Care for Creation,*** would be further interpreted to reflect the Afro-cultural understanding with regard to the sense of stewardship, which is embedded in the African sense of responsibility for family, community and *communality* (commonality). ***Conversion*** as one of the 7Cs would be targeted as a goal whereby the institute would work to reshape and redirect attitudes and minds towards ecology and environment—the modalities would be socio-ethically structured. The concept, ***Citizenship***, would be seen as the anticipated objective whereby the attitude towards ecology must have taken place in the individual lives of the citizens themselves as a cultural, attitudinal and educational revolution. In this sense, Nigerian citizens would understand the protection of the environmental and ecological capitals as an integrative responsibility of protecting life itself in its entirety—human life inclusive. Lastly, the whole steps would lead to ***Contemplation***—a spiritual union with nature and creation that can only elicit appreciation for nature, responsibility and stewardship towards natural capitals, abstraction from and assimilation of the innate functions and principles found in nature, not the detraction and destruction of the same.

Key Policy Remediation Steps

- **Sustainability awareness creation at the local and grassroots levels.**
- **Sponsoring and sensitising locally based coalitions of "*Ecological Stewardship*".**
- **Introducing a process of "Taxonomy" for economic trees and community-shared biodiversity.**
- **Radical campaign on "Greening", "Plant a Tree, plant Life", "Attack on nature, attack on Life", "Abused Ecology of today is the revengeful Mortality of tomorrow", "Ecology is Economy", "No Ecology, no Economy", "Abused Ecology, Lost Economy".**

5.5.5 *Subsistent Empowerment Evolution*

Nigeria's present structure suffers the constant loss of human resources since many of its citizens feel excluded, constrained, marginalised, depredated, defrauded and unwanted. The rate of emigration and other forms of human capital flights to other continents, nations and economies leaves the general structure and system with so many open dilemmas to be resolved. In contemporary terms, Nigeria suffers a deluge of brain drain at present, and the number of people preparing to leave is in geometrical progression and succession. The situation is directly proportional to socio-political and socio-economic insensitivity with regard to capacity building and empowerment. As our research argued while proving the need for the recovery of human ecology as obligatory for authentic sustainable development, there can be no successful inroad to integral development when the agent of development himself or herself—the Mmadụ, the human person—is debased to being a "patient" of both institutional and policy-based societal ineptitude. Thus, we see empowerment and its correlate, capacity building, as non-negotiable for the liberation of Nigeria. The approach to this area—empowerment and capacity building—would be the ***centripetal base*** of our proposed institute.

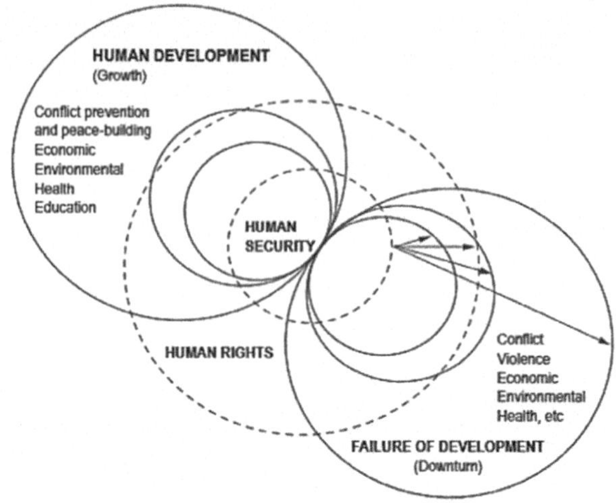

Fig. 5.9 Interactive growth and downturn diagram of human securitisation as approach to empowerment. *Human Security as the securitisation of human development (growth) to protect from threats and failures of development (downturn). Source* @IIEP-UNESCO

Further development of the approach to subsistent empowerment evolution would be to define and handle empowerment always as an inextricable part of the concept of human security and insecurity. Thus, "… achieving Human Security includes not just protecting people but also empowering people to fend for themselves" (Sen 2015), remains a powerful contribution to the fact that the known *functional human security is directly proportional to human empowerment*. It is, therefore, a public good that should be both protected and promoted. Hence, our research work would develop the approach to empowerment by focusing on tackling the varied failures and threats to development as exemplified by the "growth and downturn" diagram, according to the study carried out by Tadjbakhsh and Chenoy (2007) (Ibid.) (Fig. 5.9).

The above-represented securitisation diagram reveals, predominantly, dynamics of development empowerment, which the present Nigerian structure and system needs. Thus, the model of empowerment envisaged must be integrated to suit the subjective experiences of people at the micro-levels since the conditions and situations within which people or communities find themselves have deep roles to play with regard to their *sense of security*. Of course, other objective factors apply more generally. The Human Development Report puts it beautifully thus:

> The concepts of 'security' and 'insecurity' have relative connotations in different contexts. For some, insecurity comes from sudden loss of guarantee of access to jobs, health care, social welfare, education, etc. For others insecurity stems from violation of human rights, extremism, domestic violence, spread of conflicts, displacement, etc. To be meaningful, therefore, security needs to be defined as a subjective experience at the micro level in terms of people's experience. For example, 'security' for a farmer in Benue State, a food basket in Nigeria, is the livelihood he gains from selling his crops, but this form of security is very different from the 'security' interests of a Lagos indigene who is keen to become a financial giant. For a school teacher in Ekiti State, security is the possibility to educate his children and invest in the construction of his house, confident that the little he has today would not be taken away from him tomorrow – a different matter from that of the coalition troops in Borno State, fearful of a suicide attack or a renewal of insurgency by the Boko Haram. (Ibid., *Human Security and Human Development in Nigeria*, p. 12)

Empowerment generally is fundamentally not just an objective policy to be achieved from without. A concrete example remains the unending struggle to successfully implement the many externally based empowerment programmes being imported under the nomenclature known as *Help for Africa.* Of course, a whooping majority of such projects have failed since the manner and policy of implementation have always met deeply rooted indigenised or collaborated corruption and misappropriation machinations. Therefore, our focus would be integral and holistic in order to change the narrative and positively channel resources, whether foreign or national, towards educational, skill and vocation empowerment programmes, talent discernment, identity and character building, learning-based emphasis on the power of excellence, etc. These areas would be made evolutionary in the sense of, first, embarking on a process of rebranding what the status quo has been and trying to reconceptualise what minds and conventions have followed and obeyed, but which has rather detracted the whole system in general and the human person in particular. In this way, we would be responding to human insecurity as not just the presence of terrorism, violence and crime but rather as the struggle against the "[…] deprivation of the good things of life (particularly economic, political, personal, food and health security issues), mostly as a result of poor governance and the nation's failure to harness its rich natural resources and considerable growth potential for broad-based development" (Ibid., p. 13). To this end, therefore, the evolutionary sense would follow the ETT-Approach to *exnovate* (policy remediation), transit and transform.

Key Policy Remediation Steps

- **Pull resources and coalitions together to totally obliterate child-labour as a result of family poverty.**
- **Develop extra-curricular empowerment learning programmes on local, community and church levels.**
- **Establish Talent-Hunt agencies at the grassroots levels.**
- **Develop Union-Based or Institution-Based artisan employment and public service providers to checkmate Mediocrity.**
- **Develop Complaint-Bureau and Ethics Commission at the local and community levels to deter sub-standard public services.**

5.6 Critical Motivation: The *Will* to Transition

The varied concepts like "will to power", "will to pleasure" or "will to meaning" represent the various attempts by Friedrich Nietzsche, Sigmund Freud and Viktor Frankl, respectively, at affirming that all socially based phenomena and activities, with all the psychologically based behaviours, are all dependent on distinct motivational force(s)—driving force(s)—inherent in the human person himself or herself.

They are described as innately based forces applied through the single act of "willing". To *will*, therefore, is to decide to go into action for something, to bring interest or desire to fruition, or to make critical decisions for concrete transitions.

Our research sees "critical motivation" in the approach to the practical inroads, the innate and constant will of the agent of development, Mmadụ, towards transition. This "will", though innate and natural to the human person, tends to work at its peak, most especially when the *moral agent of development*, the human person, feels debased, oppressed, exploited, marginalised, socially, politically and economically excluded, impoverished, etc. The motivation instinct to find meaning and make life and situations functional for the well-being of the moral agent himself or herself triggers the "*will to transit*"; in other words, the motivation towards change. At the societal level, on the one hand, this *will to transition* can meet acceptance and inclusion and, thus, become innovative. On the other hand, it can meet oppression, opposition, tyranny and exclusion and become offensive and radical. This explains exactly the motivations and ideologies behind protests, revolutions, strikes, civil disobedience, agitations, different forms of activism, etc.

However, the exercise of the *will to transit* can be done for positive and progressive purposes, like in the cases of concrete plans and policies for effective change for the common good. It can also be exercised in a negative and detracting manner, like in the cases of planned political hijacks, institutional annexations, legal and policy-based repeals to aid corrupt and detracting practices, etc. The active *will to transit* in its varied forms has become an integral part of Nigerian daily life and they are proofs, in our context, of the fact that the most relevant (if not the only) approach to practical ethical inroad to sustainable development, is to pave way for the purposive exercise of this power (will) in the moral agent—capacity building. However, this *will to transit* has been under the heavy pangs of suppression and oppression, beginning with the imposed impacts of colonialism, which continue to rob many Africans of the courage to self-determination, as well as through other forms of non-indigenised political independence concepts, which have left so many African states socially, politically and economically fragile. Furthermore, many African states have, in the last few decades, experienced the devastating onslaught of military rule and forms of dictatorial regimes, which have kept democracy drifting in many African nations.

In the effort to contain the pressure coming from the sub-version and subjugation of this will, the Nigerian case has given rise to a network of coalitions that fight either for or against forces considered enemies of the *will* or enemies of the ideologies of the group initiating the fight. Such groups are either governmental, non-governmental, institutional, cultural, tribal or religious. They build up the socio-political, socio-economic and socio-cultural dissonance seen within and among the groups like the Odua People of the Southwestern Nigeria, the Arewa People's Congress in the North, the Independent People of Biafra (IPOB) of the Southeastern Nigeria, the Movement for the Sovereign State of Biafra (MASSOB) also in South-East, the Movement for the Emancipation of Niger Delta (MEND), the Niger Delta Forum (NDF), the Niger Delta Avengers (NDA) with their allies: Red Egbesu Water Lions, Joint Niger Delta, Liberation Force, Niger Delta Red Square, Niger Delta Greenland Justice Mandate (NDGJM), all in South-South; the insurgency of the terrorist and fundamentalist group, Boko Haram, agitating for Islamic state and Sharia law; the current so-called "fulanisation" agenda setting most regional states in conflict with the Fulani herdsmen, added to the consequent killings, destruction of farmlands, kidnaping and rape; the unending Nigerian story of ASUU strikes which is dragging quality, excellence and substance in the Nigerian Institutions of Higher Learning to the mud, as well as the most recent revolution and violent clashes in protest against corruption, brutality and inhumanity of some security forces, particularly the Special Anti-Robbery Squad (SARS). The whole Nigerian atmosphere is tensed up, with the consequence of being socially, politically, culturally and tribally belligerent. Thus, our research, through the proposed institute, would focus on investing in the existent *will to transition* already manifest in varied forms and group ideologies and would try to use the relevant ethical and sustainability principles to make them effective, remedial and result-oriented.

5.6.1 The Will to Transition *as Concept Cum Cultural Revolution: A Socio-Ethical Approach*

Our work sees the great need to change the Nigerian *narratives*. The practical approach would be to emphasise and teach the concept of "cultural revolution". To interpret this socio-ethically, we refer to the powerful words of Pope Francis in the Apostolic Constitution, Veritatis Gaudium. Thus, talking about an expected model of change that would stand the test of time and that would bring transformation through the ecclesiastical universities, the Pope opined:

> [...] today we are not only living in a time of changes but are experiencing a true epochal shift, marked by a wide-ranging 'anthropological' and 'environmental crisis'. Indeed, we daily see signs that things are now reaching a breaking point, due to the rapid pace of change and degradation; these are evident in large-scale natural disasters as well as social and even financial crises. In a word, this calls for 'changing the models of global development' and 'redefining our notion of progress'. Yet 'the problem is that we still lack the culture necessary to confront this crisis. We lack leadership capable of striking out on new paths.' This vast and pressing task requires, on the cultural level of academic training and scientific study, a broad and generous effort at a radical paradigm shift, or rather – dare I say – at 'a bold cultural revolution'. (Francis and Gaudium 2017)

Reading meaning into the opinion of the encyclical, this research work interprets the **will to transition** as not just the will to change policies but rather the innate will to change attitudes, characters and conventions—a cultural revolution of attitudes, an epochal shift. It means the revolution of the way of life, the manner of approach, a culturally-based conversion and re-orientation—"a bold cultural revolution" (Ibid.), in other words, a process of changing the narratives towards ethics of and in daily life, etc. In all, it is a learning process leading to a life lived out of conviction and not just out of varied social and psychological (mostly negative and involuntary) *conditionings*. The *will to transition* must be considered the aptitude for social change (*Kunst gesellschaftlichen Wandel*) (Schneidewind 2018). Applying the "criteria for judgement", experiences would show that the *will to transition* can be creatively exercised or abusively over-exercised, manipulated or stalled. Thus, a socio-ethical approach emphasises the need for legitimised and structured "leadership" capable of using ethical and sustainability principles to govern the agent of development himself or herself—the Mmadụ, towards the legitimate exercise of the *will to transition*. The needed governance must engage the learning process towards capacity building, which is envisioned as the application platform, where ethics would meet and shape capacities and capabilities, while sustainability principles would work to maintain, invigorate and keep them *green* (sustainable).

Only within this socio-ethical sense could one speak about the *will to transition* that would eventually bring about Transformation—a radical integral form of change. It would be the *will to transition* that is not just exercised with an untamed spirit of a mere revolution but rather with the sense of a revolution that is substantively *cultural*; in other words, a sort of *behavioural re-valuation*. Thus, in this sense—"re-valuation"—the revolution begins in the individual selves, characters and attitudes and devolves towards the external. Again, the will to transition approached socio-ethically would lead to integrally based sustainability to be known as "**sustained-ability**", first, in the moral agent himself or herself. Secondly, this "sustained ability" would be further experienced within the existence outside of the moral agent, namely nature, but through and in relation to the moral agent, in other words, a *sustained form of ability (capability) in the community of existence*. The two concepts—revaluation and sustained-ability—would be introduced and developed in our proposed institute to delineate the functional relationship that should exist in the dynamics within Ethics *versus* Capacity Building and Ethics *versus* Sustainable Development, respectively.

5.7 The Proposed Practical Inroads as a Model of the "Great Transformation" Governance

> *Analytisch liegt dem Singular "Große Transformation" die These zugrunde, dass es sich bei den Wandlungsprozessen der Gegenwart nicht um isolierte Einzelphänomene handelt, sondern um Elemente einer umf*ssenden Veränderung. Der Zusammenhang dieser Elemente bedarf jedoch einer exakten Beschreibung und Analyse: Wie können Wechselwirkung zwischen Klimawandel, Ressourcenknappheit, Finanzkrise, Arbeitsunsicherheit oder demografischem Wandel – um nur einige Aspekte der gegenwärtigen Umbruchprozesse zu nennen – angemessen beschrieben werden? Lassen sich daraus Erkenntnisse über die Unterscheidung von Ursachen und Symptomen ableiten? Wie lässt sich in Bezug auf zentrale Wandlungsprozesse das Zusammenspiel globaler, kontinentaler, nationaler und regionaler Veränderungen beschrieben?.[4] (Markus Vogt, Wandel als Chance oder Katastrophe, 2018)

[4] Vogt, M. *Wandel als Chance oder Katastrophe*. 2018, München: Komplett Media, pp. 14–15. * English translation—Analytically, the singular "Great Transformation" is based on the thesis that the change processes of the present are not isolated individual phenomena but elements of a comprehensive change. However, the connection between these elements requires an exact description and analysis: How can the interaction between climate change, resource scarcity, financial crisis, job insecurity or demographic change—to name just a few aspects of the current processes of

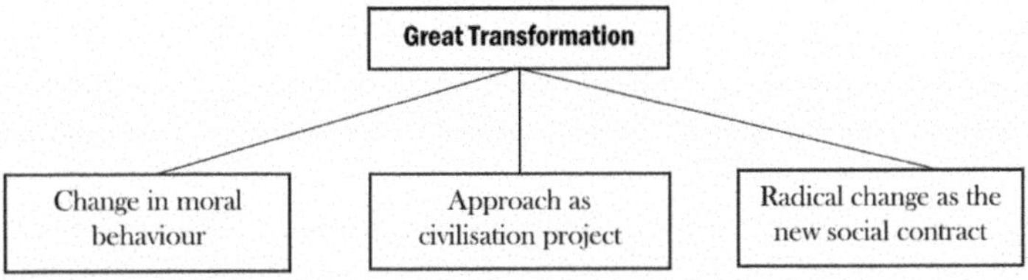

Fig. 5.10 Main features of the "Great Transformation Discourse". *Illustration mine*

Vogt's assessment of the "Great Transformation" concept reaffirms our proposed method of approach to the Nigerian situation, namely, the *systemic form of approach,* which views the society as a united whole, with its structural elements, social, political, economic, cultural and ecological, taken to be integral and interrelated, hence, the expression, "*not isolated individual phenomenon*". In this way, the contemporary approach, which the term "great transformation" represents, becomes a scientific form of *radical change* (Umbruch) with the objective of a thorough, deeper and relational analysis of social systems, structures and functions for more sustainable policy remediation and general social transformation. Anticipated is a form of "breakaway" from different forms of political, economic and lifestyle-based status quo, which holds the society in bondage. It needs to follow an organised process of learning, training and re-training, in other words, a form of cultural rebranding capable of strengthening identity for sustainability.

In the light of the above, the world is currently experiencing concerted efforts like never before, and this is in the forms of the coalition of governances, leadership conferences, exchange of scientific findings and knowledge, discourses on the future of our world with regard to transitional and transformable technological exnovations and innovations, practical ways to increase efficiency, as well as sustainable routes to making human lifestyle more adaptable to the changing conditions of our contemporary world. These efforts have been summarised in the concept credited to Uwe Schneidewind,[5] namely, "Zukunftsart" (German)—art for the Future—the art to transit to and transform for a better, more sustainable future. As an art, our research identifies three fundamental features of the anticipated "great transformation" approach, namely: (a) change in moral behaviour; (b) approach as a civilisation project; (c) radical change as a new social contract (cf. Ibid., *Die Große Transformation: Eine Einführung in die Kunst gesellschaftlichen Wandels*, pp. 25–26) (Fig. 5.10).

The *change in behaviour* depicts the need for radical upgrade and anchoring of the various policy-based societal institutions towards value orientations that are transformative, regenerative and sustainable. It is a conscious process of liberating

change—be adequately described? Can we derive knowledge about the distinction between causes and symptoms? How can the interplay of global, continental, national and regional changes be described in relation to central processes of change?

[5] Uwe Schneidewind is since 2010 the President of the Institute for Climate, Environment and Energy, and Professor of Innovation Management and Sustainability at the University of Wuppertal, Germany.

the mind from ignorance and enslavement to inimical "comfort zones", which stifle progress and sustainable development. Such value orientations depend heavily on a conscious process and project of civilising the mind, identity and character of the moral agent himself or herself, Mmadụ, a *civilisation project*. Taken as a civilisation project, therefore, the sense behind great transformation works to connect the past, present and future since the history of humanity reveals that moral revolution has always been part of the general experience: democratic struggles, abolition of slavery, the struggle for women's rights (cf. Ibid.), etc. As a rule, the envisaged transformation must be seen as a new form of *social contract* involving all and integrating all levels of the one ecology—human, social, political, economic, cultural and environmental. Following the opinion of Bruno Latour, the transformation should lead to the social contract principle, where all—human and social—should be integrated into finding a habitable common ground for a new universality (Ibid.).

Our work would, through the planned institute, apply these three related concepts in the various approaches to translate sustainability principles into policy remediation so as to rescue the plaguing Nigerian situation. To do this effectively, the mega-trends in great transformation discourses encompassing ***wealth and consumption, energy transition, resource efficiency, mobility, nutrition, urbanisation and industrialisation*** must be developed and adapted to speak the language of the local people and reflect their indigenous experiences and worldviews. These individual areas would be taken integrally within our Five-Key-Area scheme of work and would be seen as both inter- and interrelated.

5.7.1 Avoiding the Danger of "Utopia": The Mega-Trends and the Current Social Epoch in Africa

While the anticipated objectives of the great transformation remain a process in progress, models within this same process should be seen as endogenous to the whole social, political, economic and institutional strategies in approach to our modern world and experiences. In this regard, we consider it a major area of attention when the various discourses and approaches to "great transformation" movements recognise that the status of the mega-trends, with regard to people, culture and society, is predominantly complex and diverse. Consequently, the mode of discussion, the social, political, economic and ecological objects of focus, as well as the social and developmental epoch of the society in question etc., all play important roles in shaping the narrative about great transformation politics. Thus, it remains the true submission of this research work that the concept of great transformation would be nothing but a *utopia* when the current mega-trends are not strictly reduced, interpreted, developed and adapted to the realities within the African continent and its (still struggling) states.

The above-mentioned view informs the reason why our approach has chosen to abstract and emphasise more the principle behind the great transformation discourse—a systemic, integral approach, not *isolated individual phenomena*. Hence, while the mega-trends and standard objectives attached to the individual trends reflect more the designed *change-process* of the developed economies, and partly, for a number of developing economies, especially in the West and Asian countries, the current formula of approach to the mega-trends must confront some hard socio-political, socio-economic, socio-cultural and socio-ecological complexities, if not incompatibilities in the African case in general and Nigeria in particular. Hence, while the average person in developed economies invests more in maintaining and stabilising their current well-being and, perhaps, avoiding waste, the millions of destitute in some developing and underdeveloped economies have yet to experience what socio-economic well-being means. Again, while the developed economies currently feel they have had enough of the model of energy supplies that have been in trend till date, with the consequent climatic hazards and changes they have brought, not only for the inventors and mega consumers themselves, but for the whole human world, millions in the poor nations are still struggling to even have a steady experience of the current trend of energy supply, and in some areas, political, economic and colonial reasons have made this a social "stillborn". The same disparities affect other areas like resource politics, mobility economics, affluence and waste alongside malnutrition, urbanisation and destruction of indigenous and local economies, as well as industrialisation without indigenisation.

The above-expressed concerns are meant to pre-empt one of the major objectives of our research, namely a focus on and taking hold of the human person, *Mmadụ*, within the Nigerian situation and putting him or her back into his or her natural worldview. This would be done by adapting ethical, sustainability and development principles that could help to transit and transform the general structure and system for authentic and more sustainable development of the human person. A look at the proposed concept for the work and application tool for this research work—*The Institute for Ethics, Capacity Building and Sustainable Development*—would, perhaps, give this research work a focal and practical conclusion.

5.7.2 Introducing the Integrated[6] Structure for Action and Application: The Institute for Ethics, Capacity Building and Sustainable Development (INECSus)

As already seen in the course of our research, practical inroads cannot just happen. They are concepts that must be reflected upon, written, processed and applied. Therefore, there is a need for a micro-structure that would design ethical ideas capable of building characters for a sustainable society. This idea for an institute is new, propositional, indigenous and process-based. The central focus is the human person and the need to redeem the true identity of the Afro-Nigerian Mmadụ as a prerequisite

[6] The word "integrated" is used here in anticipation that the proposed institute would be the platform to meet the society at its macro, meso and micro levels.

for the indigenisation, assimilation and adaptation of the sustainability principles for an authentic life and sustainable development in the Nigerian polity.

The researcher and the research itself see this as an effective way of enlightening the target audience—the Nigerian populace at the grassroots—to wake up to the epoch within which our world finds itself. As a teenager then attending a post-primary educational institution, this researcher battled passionately a psycho-social linguistic expression (we avoid giving it the name, convention), which continued even to the time of this research work to influence the minds and characters of the Nigerian populace, namely that "*Nigeria has no reading public*". Since then, many have always tried to ask the pertinent questions: how can there be hope for a reading public when millions of children—the so-called future generation—have no access to education? How can there be a reading public when there are no quality libraries and the creation of awareness programmes for their uses? How can Nigeria constitute a reading public when the leadership itself at various levels has no interest and the leaders themselves do not read? How can there be a reading public when, for over 60 years of independence, Nigeria's National Budget has always reflected disinterestedness in educating the minds and forming characters, etc.? The Nigerian case, we repeat, is not and remains not the total absence of institutions, policies and manpower. Our problem revolves around the quality of leadership and governance.

Proposal for an institute, therefore, is a way of contributing to creating awareness of the fact that the pseudo-ascription, "Nigeria has no reading public", has remained derogatory of the nation's capabilities and needs to be approached both ideologically and practically. A conceptual presentation of what the institute should be and focus on is given below (Fig. 5.11).

It is striking at first sight and, with regard to the schema, that our research has three interrelated and inseparable focuses, namely, Ethics, Character or Capacity Building and Sustainable Development. The three themes are further ascribed, namely Ethics

Fig. 5.11 Schematic conception of the proposed institute for ethics, capacity building and sustainable development—INECSus

for BLUE, Character for RED and Sustainable Development for GREEN. This is done as pre-emption to the basic driving principles that should guide the themes: blue—representing inspiration, wisdom and reliability that should go into ethical formulations for policy remediation; red—representing the anticipated well-being of the human person himself or herself and the restoration of the passion and love for the human persons to having the freedom to use the capabilities to live the type of life they have reason to live; and green—representing fertility, fruitfulness, progress, sustainability. The Institute would develop sustainability principles as *principles of greening*. These three themes would give rise to the three anticipated departments of the institute as represented below (Fig. 5.12).

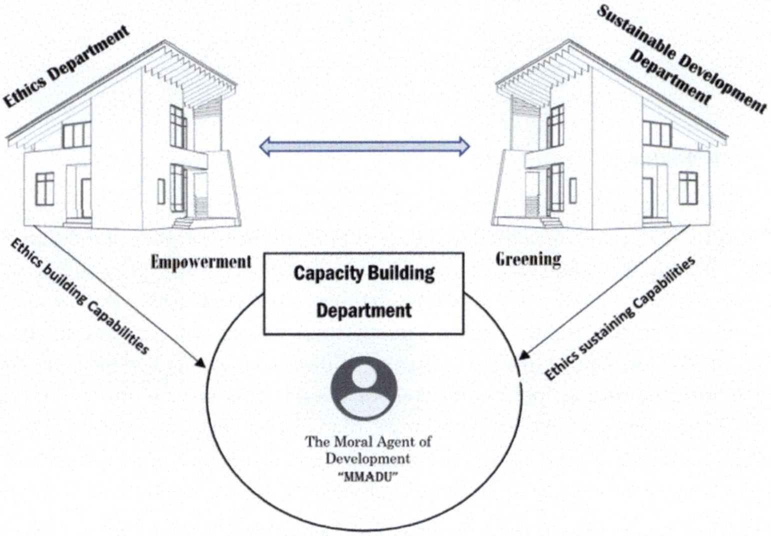

Fig. 5.12 Schematic prototype of architectural structure and design of the institute. *Illustration* What we envisage here is the corporative work of three schemes of ideas in using ethics to plan and empower the moral agent of development—Mmadụ, the human person. Again, what is empowered would be further monitored through the principle of "greening"—the sustainability actions, the application of the "4th Dialectic"—to ensure authenticity, adaptability, regeneration, etc. *Sources* (a) Symbol for the Mmadụ, human person: ClipDealer Royalty Free Media (b) Architectural Prototype: depositphotos

5.7.3 The Idea Behind the Logo

Illustration mine

The initial development of the concept observed and reflected on the process behind seed planting and germination. Hence, a representation of the first letter of the proposed institute's initial—*INECSus*—with a pictorial young growing tree, blossoming with green leaves. This young, healthy tree represents the fundamental strategy of the institute, beginning with its micro-structure, the Institute itself, where the will to transition would inform a change of narrative to, first, view and take hold of the Mmadu, the human person himself or herself as a seed plant in the making. Thus, with the concept that in every seed plant lie a series of potentials considered innate to the seed plant and awaiting further development amidst other overt conditions that can make or mar the development, the institute would depend on this approach to recover the lost human ecology, which has remained the bane of authentic sustainable development in the Nigerian case. This is the "planning stage". Thus, the planting stage depicts the stage of applying the planned socio-ethical policies through awareness and empowerment programmes—Capacity Building, all in an atmosphere of freedom and self-determination that would, first, motivate the agent himself or herself towards self-reliance, belief in his or her natural capabilities and avoidance of playing "the victim". The empowerment stage would be integrated with the study of both the conditions that better aid the exercising of capabilities and conditions that better manage political, social and economic risks when confronted on the fields of application.

The next phase is the stage of evaluating the empowerment approach with the sustainability principles—*the greening concept.* Since both departments, Ethics and Sustainable Development, represent the two sides of a coin in designing policies and ideas, the sustainability principles cannot easily be separated from the initial formulation of ideas for capacity building and social transitions and transformations. In other words, the greening concept is a fundament for the initial development of ideas and policies and remains the *monitoring and evaluative livewire* within the whole process. Summarily, therefore, the whole process represents the planting of a seed, which would eventually germinate and develop into a tree with different branches—the branches representing "capabilities", the sustenance (greening) of which goes on as a process—watering, pruning, manoeuvring, fungal protection, propagation, etc.—a general process of sustaining what has been planted.

5.7.4 The Integrated Objective of the Institute

The objective of the proposed institute as the practical response to what our research sets out to achieve is integrated into the specialities of the three departments as represented below:

Ethics Department

Reflecting on ethical and sustainable ideas that would lead to forms of policy remediation that would help to save the drifting human ecology in the Nigerian case. Plans and ideas must have the human person, Mmadụ, as both the object and subject of focus. Thus, the ideas must be made ethical enough to keep the human person always as an agent and not a patient within the social structure and system.

Sustainable Development Department

Ethical and empowerment ideas must be evaluative enough to reflect integral justice. The moral agents of today must live and think responsibly to avoid developing a society of moral patients for tomorrow. Therefore, the sustainable development department focuses on constant evaluative function with regard to ideas, policies, empowerment and awareness programmes – all in an effort to make sure they respond purposefully to sustainability principles.

Capacity Building Department	*Here is the meeting point, the sphere of application and action. In the Capacity Building Department, the objective is to see the guideline for action as having a single focus, namely, concern for the lost dignity and rights of the human person within the socio-structural defects and dysfunction seen in the Nigerian structure and system. The focus would be on restructuring and recovering the lost human ecology since the human capacity built today remains tomorrow's wealth and sustainability.*

The above three articulated and integrated objectives would give rise to the proposed *action-plan* to be known as "**Action-Plan-PPG**": Planning (Ethics), Planting (Capacity Building, Empowerment) and Greening (Sustainability, Sustainable Development)—PPG.

Evaluation and Conclusion

The political, economic, socio-cultural and ecological data provided are the most current during the research and they represent the current debate, concerns, feelings, agitations, doubts, disappointments and expectations within the entire social structure and system. Within the period of writing this conclusion, the entire social, political and economic security of Nigeria has deteriorated massively. The kidnapping of school children of both ordinary and political persons for ransom is becoming an unfortunately lucrative business. The situation is further complicated by the failures of leadership of various affected political persons, who think that the best policy is to negotiate with terrorists, murderers and bandits. There are losses of lives and properties, added to psychological trauma being suffered by ordinary citizens, who live their daily seconds, minutes and hours under the tension of insecurity and predation due to the menace of the Fulani Herdsmen. What the very next moment in the life of every Nigerian citizen means and brings about remains appallingly uncertain. The power and energy supply system and quality remain massively underproductive. As a consequence, investors have absconded from the situation, many companies have closed down, others are downsizing and a good number have relocated to neighbouring countries.

Lifestyle in the capital city, Abuja, the economic city, Lagos and other mega and semi-mega towns remain predominantly ostentatious, corrupt and aristocratic, while millions who reside in the villages and suburbs struggle, on a daily basis and under varied dehumanising conditions of life, to keep alive. In this regard, and in order to checkmate over-crowding and over-population in these so-called *viable* areas, the cost of living is inflated for the common men and women while the corrupt politicians and their allies hijack the cities as their *reserved kingdoms*. Various private, non-governmental and religious groups are struggling to rescue the educational sector. However, the cost-demand still defeats the aim since millions of families among the willing populace remain trapped, incapacitated and impoverished and thus unable to afford the services due to the Nigerian oppressive socio-economic structures.

The current social atmosphere is charged with a teeming number of youths in their millions, who move about without perspective, with destroyed visions and confused identities; street crimes are on the increase; graduates face varied confusing options—either he or she finds a job, or he or she plans to run away from a home that has no plans and opportunities for him or her, or, when this fails, he or she discovers overnight the call to be a pastor, man of God, prophet or prophetess, etc. In this way, a victim of social inequality and injustice begins, under religious pretence, to exploit other victims more vulnerable than him or her. To this effect, Nigeria has become one of the countries where, alongside the incessant proliferation of Churches, religions and faith, that same religion serves as a viable weapon of deception, manipulation and exploitation.

It is exactly against this background and the current events within the Nigerian situation and structure that our work begins a Christian socio-ethical approach, which introduces not just a *theory-based* approach to the Nigerian circumstances but is resolved and determined to embark on practical and *action-based* ethical recovery and policy remediation for inclusive socio-political, socio-economic, socio-cultural and socio-ecological transitions and transformation—The Institute for Ethics, Capacity Building and Sustainable Development. This proves and offers to be the best formidable ethical front towards a new indigenised global (integral) solidarity, following the opinion of Laudato Si, able to dispel the attitude of denying the real problem or being indifferent to it (Ibid., *Laudato Si*, no. 14). Hence, the coalition of minds, people and institutions must hit the grassroots, speak the indigenous languages of the people and take root in the general process of learning and character formation.

Our research would be decisive in settling down to work towards translating principles into actions so as to help redefine the approach to Mmadụ himself or herself, namely, to see the general approach as *no more an approach to a thing but rather an approach to a person*—a moral agent (Ibid., no. 65). This means that the recovery of the general ecology within the Nigerian structure and system—political, social, economic, ecological and cultural—needs a new anthropology of the human person himself or herself. It needs a leadership formula with an inclusive policy, not a detractive one. To this effect, the Catholic Church herself, within which the researcher lives and works and whose social doctrines and teachings provided resource materials for the research, must grasp the opportunity now to show leadership and competence. Such leadership and competence must go beyond mere "moralising", and Vogt's strong words on the bulk of responsibilities facing the Church herself in an epoch of radical social change and with regard to nature and ecology deserve attention:

> In the face of such high demands for social change, it seems to be important to neither overestimate nor underestimate the role of the churches: they are no "moral agencies" undergirding socio-ecological imperatives with theological pleas. Instead, their specific competence lies in a "moral beyond moralizing". In the given context, the ecclesiastical starting point of reflection and advice has to be the message of liberation together with the praise of creation and the awareness of its/her beauty. This way, it brings forward the longing for lost integrity,

> justice and fullness of life. What we need is not a rearmament with moral outrage but a re-centring on the grand narratives of guilt and responsibility, anthropology and the perception of nature, as well as culture and technology, in order to allow us to cope with the challenges of today's life. It is foremost this narrative and anthropological integration that constitutes the surplus value of Christian theology of creation compared with secular ethics of nature. (Ibid., *Perceptions of Soil in Catholic Theology*, p. 358)

The reference to "re-centring the grand narrative" amounts to the call to restructure what the status quo has been—a demand for a conscious, determined and practical transition away from the *comfort zones*. A determination to re-centre and change the narrative would enthrone a new epoch of development, which would become the new name for peace.[7] It would be peace embedded in the freedom of the human person to live the life he or she has justified reasons to live. Peace, in other words, functional social security developed and defended within the dictates of human rights and dignity. This requires a capability approach capable of evaluating all socio-political, socio-economic, socio-cultural and socio-economic ecologies, only with direct reference to their impacts and contribution to capability expansion in Mmadụ, the human person and the maximisation of chances for good quality of life. This is the practical focus of the proposed institute.

References

Bauer MW, Jordan A, Green-Pedersen C, Hertier A (2013) Dismantling public policy: preferences, strategies, and effects. Oxford University Press, United Kingdom, p 203

Emmanuel M (2016) Federalism in Nigeria, between divisions in conflict and stability in diversity. Globethics, Geneva, p 110

Francis A (2017) Apostolic constitution *Veritatis Gaudium*. On ecclesiastical universities and faculties. Rome, December 27, no. 3

Heyen DA (2017) Governance of exnovation: phasing out non-sustainable structures. Working Paper 2/2017. Öko-Institut e.V., p 10. Available via www.oeko.de/fileadmin/oekodoc/WP-Exnovation-EN.pdf. Accessed February 10, 2022

Heyen DA, Hermwille L, Wehnert T (2022) Out of the comfort zone! governing the exnovation of unsustainable technologies and practices. GAIA-Ecol Perspect for Sci and Soc 26(4):327, 326–331. Available via https://doi.org/10.14512/gaia.26.4.9. Accessed February 10, 2022

International Renewable Energy Agency (2013) Africa's renewable future. The path to sustainable growth. Available via https://www.irena.org/-/media/Files/IRENA/Agency/Publication/2013/Africa_renewable_future.pdf?rev=0ed4dcd2daa24c389c3ec63ddc63cf9a. Accessed April 14, 2022

John XXIII (1961a) Encyclical letter *Mater et Magistra*. On Christianity and social progress. Rome, May 15, no. 236

John XXIII (1961b) Encyclical letter *Mater et Magistra*. On Christianity and social progress. Rome, May 15, no. 226

Masserli P (2019) Der global sustainable development report 2019. Vortrag bei der Wissenschaftsplattform 2030, Berlin. In: Vogt M (ed) Christliche Umweltethik, Grundlagen und zentrale Herausforderungen. Freiburg im Breisgau, p 518

---

[7] Paul VI (1967) Encyclical letter Populorum Progressio. On the development of peoples, Rome, 26 March 1967, 76–77.

Paul J II (1987) Encyclical letter *Solicitudo rei Socialis*. On the twentieth anniversary of Populorum Progressio. Rome, December 30, no. 49

Paul J II (1995) World day of peace message. In: Women: teachers of peace. Rome, January 1, p 2. Available via https://www.vatican.va/content/john-paul-ii/en/messages/peace/documents/hf_jp-ii_mes_19841208_xviii-world-day-for-peace.pdf. Accessed March 16, 2021

Paul VI (1971) Apostolic letter *Octagesima Adveniens*. On the occasion of the eightieth anniversary of the Encyclical Rerum Novarum. Rome, May 14, no. 4

Sachs W (2021) Papst vs. UNO, Sustainable Development Goals und Laudato Si: Abgesang auf das Entwicklungszeitalter? In: Vogt M (ed) Christliche Umweltethik, Grundlagen und zentrale Herausforderungen. Verlag Herder, Freiburg im Breisgau, p 517

Sachs J, Kroll C, Lafortune G, Fuller G, Woelm F (2021) Sustainable development report 2021. Cambridge University Press, London, p 30. Available (Online) via https://s3.amazonaws.com/sustainabledevelopment.report/2021/2021-sustainable-development-report.pdf

Schneidewind U (2018) Die Große Transformation: Eine Einführung in die Kunst gesellschaftlichen Wandels. Fischer, Frankfurt am Main (Expression adapted from the caption of the book)

Sen A (2015) Human security now. In: Soka Gakkai International Quarterly, United Nations Educational Scientific and Cultural Organisation. UNESCO, National Human Development Report, p 12. https://planipolis.iiep.unesco.org/en/2015/national-human-development-report-2015-human-security-and-human-development-nigeria-6340. Accessed April 14, 2022

Summers HL (2015) Ideas rule the world, not kings and corporations. In: Lecture series with big think edge. Available via https://bigthink.com/videos/innovation-revolution-and-moral-advance-ideas-rule-the-world-not-authority-figures/. Accessed August 20, 2021

United Nations (2019) The future is now. Global Sustainable Development Report 2019, p 8. Available via https://sdgs.un.org›gsdr›gsdr. Accessed February 12, 2022

United Nations Educational Scientific and Cultural Organisation, UNESCO (2015) Human security and human development in Nigeria. national human development report, pp 20–21. Available via https://planipolis.iiep.unesco.org/en/2015/national-human-development-report-2015-human-security-and-human-development-nigeria-6340. Accessed April 14, 2022

Vogt M (2019) Perceptions of soil in catholic theology. In: International yearbook of soil law and policy, Vol 2019. Springer, Cham, pp 357–367. First Online, March 25, 2021. Available via https://doi.org/10.1007/978-3-030-52317-6_17. Accessed December 5, 2022

Vogt M (2021) Christliche Umweltethik, Grundlagen und zentrale Herausforderungen. Verlag Herder, Freiburg im Breisgau, pp 617–618

Appendix A

Questionnaire

Summary of a Research and Case Study:

Inefficiency in Power Supply and the Proliferation of Private/Independent Power Generation – The China Connection

(Proving the structural dysfunction in the Nigerian system)

Method: Data Collection; Quantitative and Qualitative Analyses

Location: Eastern Nigeria, Onitsha, Nnewi, Anambra State

Targeted Audience: Traders, Families, Organisations, Companies, Factories, Onitsha Main-Market, Nnewi Main-Market

Purpose:

To prove inefficiency as a structural dysfunction within the Nigerian polity, using the power/energy sector as specimen and to propose a possible, alternate and workable solution to constant power/energy inefficiency. Call to develop and invest in Alternative and Clean-Energy Technologies

By:
Chijioke Francis Nwosu, M.A

Doctoral Research Student, Faculty of Catholic Theology, Christian Social Ethics, Ludwig Maximillian University, Munich, Germany. Affiliate Doctoral Research Student, Rachel Carson Institute for Environment and Society, Munich, Germany.

C. F. Nwosu, *Systemic Remediation*, Environmental Humanities: Transformation, Governance, Ethics, Law 4, https://doi.org/10.1007/978-3-031-61909-0

The General Situation of Power/Energy in the Country

1. Reasons for Power/Energy Inefficiency

a. The general problem of mangled megawatts in the energy sector

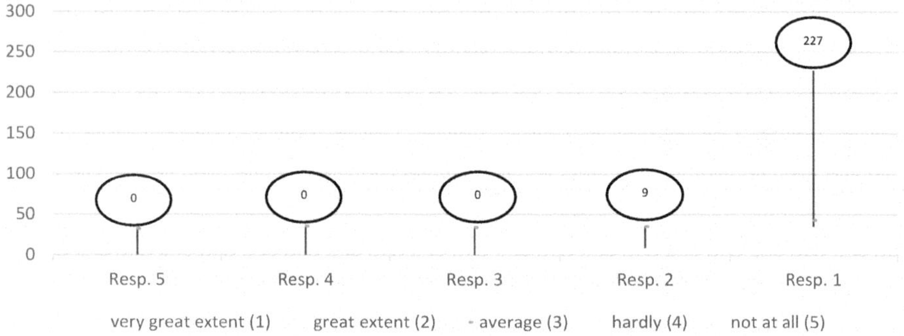

b. Poor and sub-standard power infrastructure

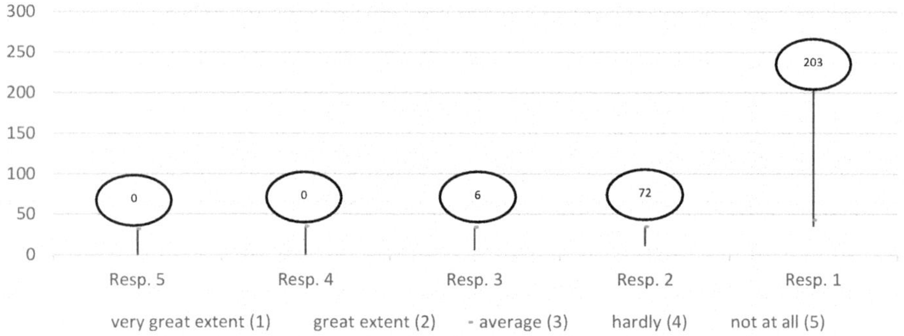

c. Corruption

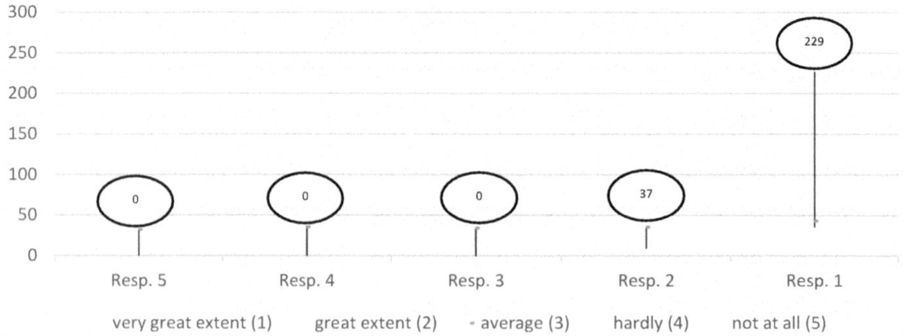

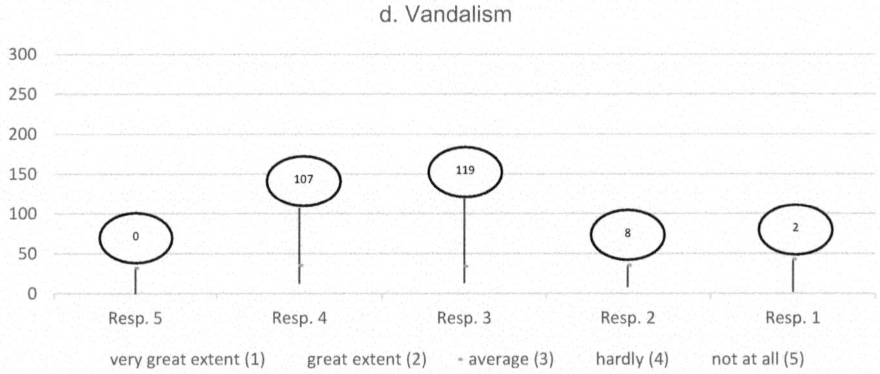

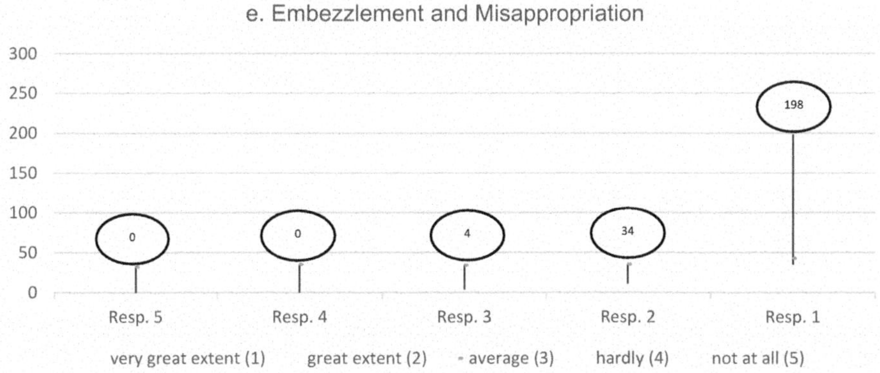

2. **Effects of Power/Energy Inefficiency**

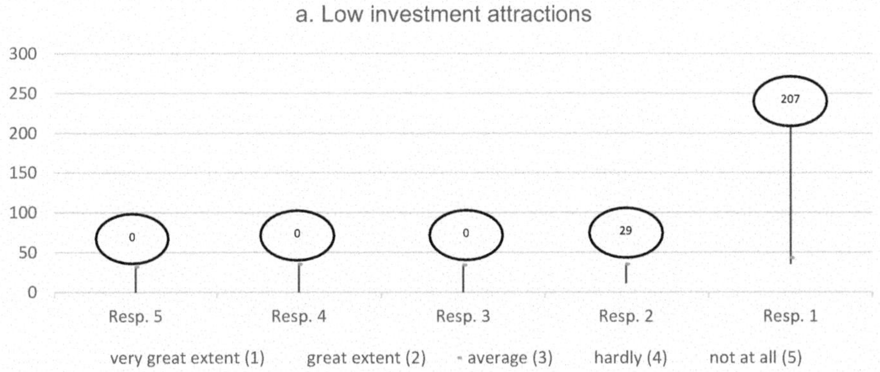

b. Low industrial output

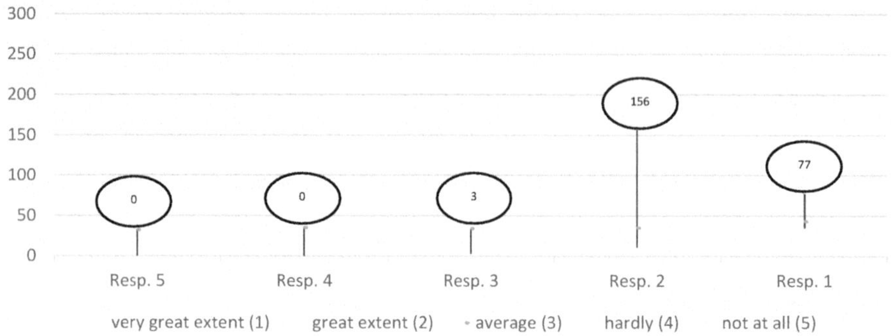

c. Increase in cost of production

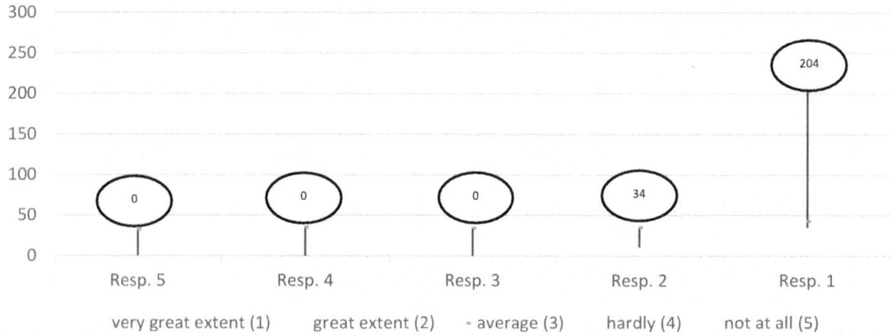

d. Techno-social and infrastructural malfunction

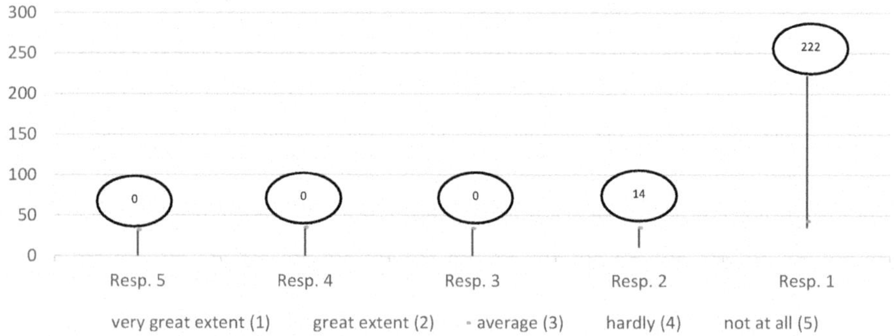

3. Reasons for Private Power/Energy Generation

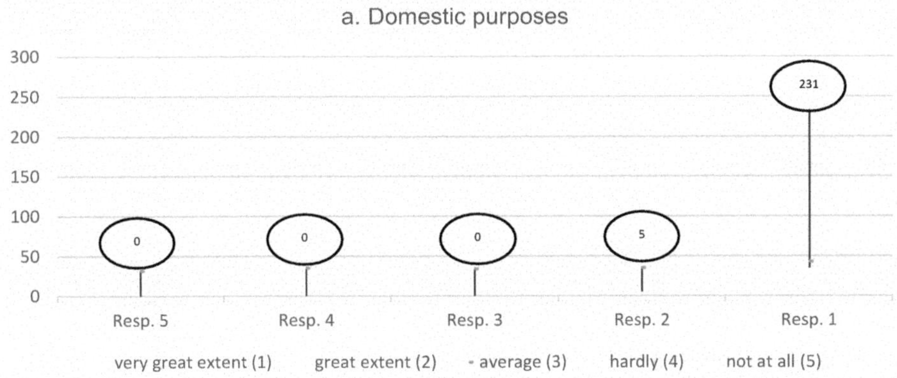

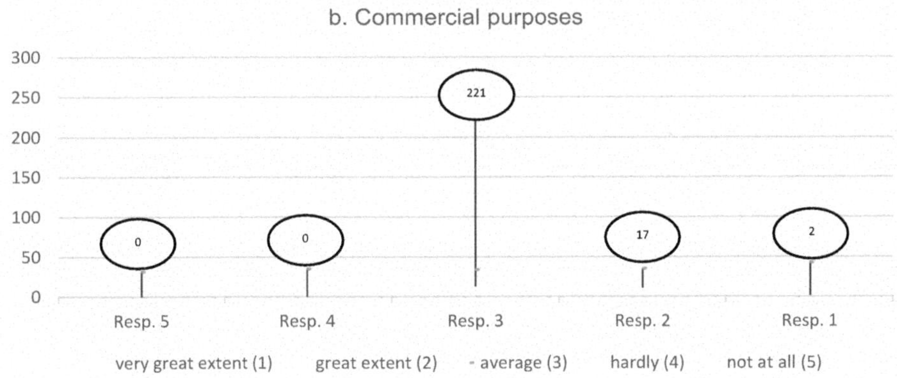

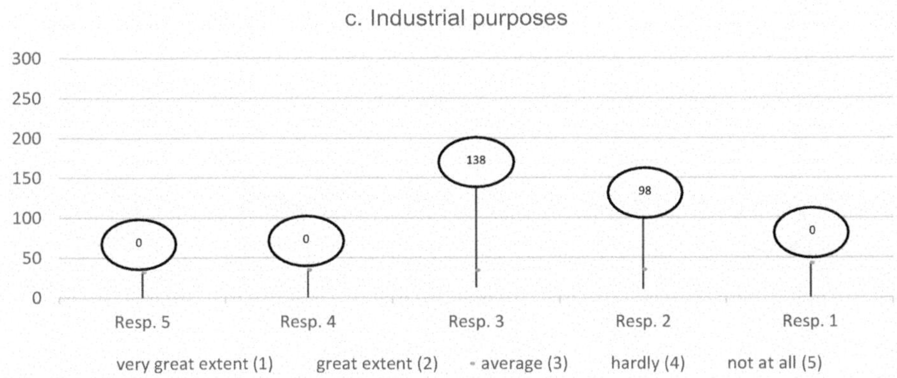

d. Institutional purposes

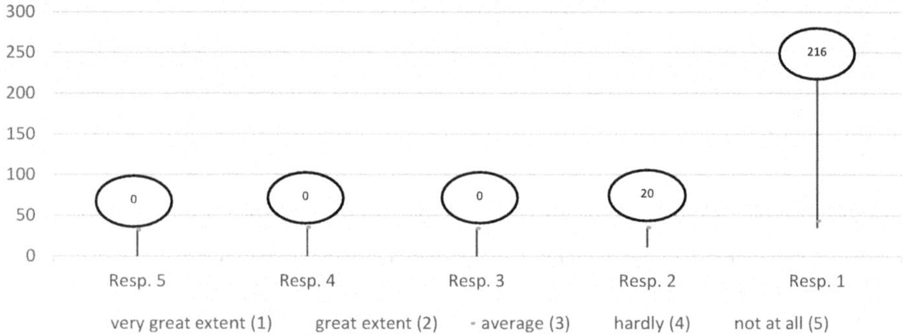

4. Sources of Power/Energy Generation

a. Generator sets

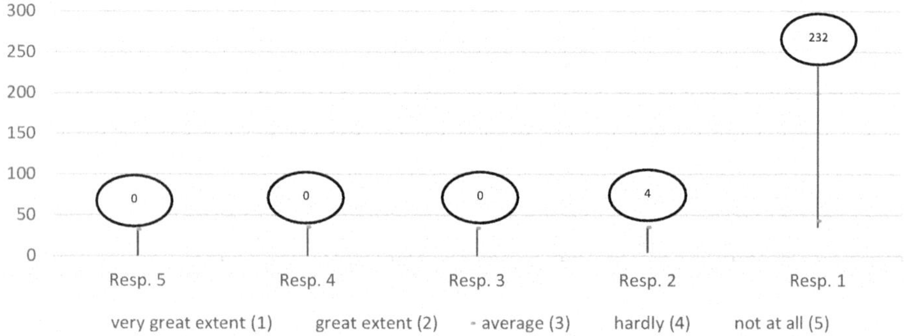

b. Solar Panels

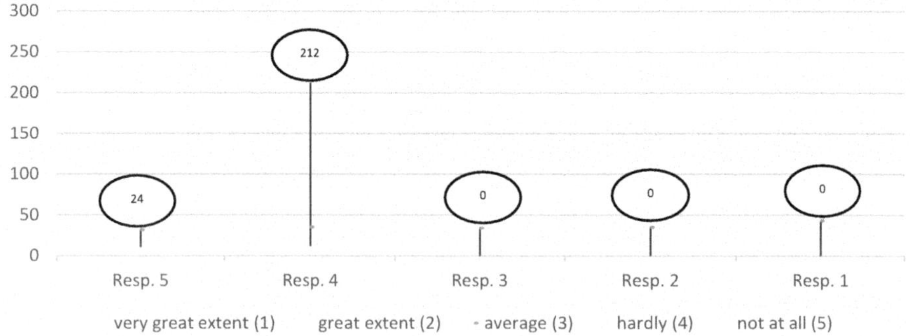

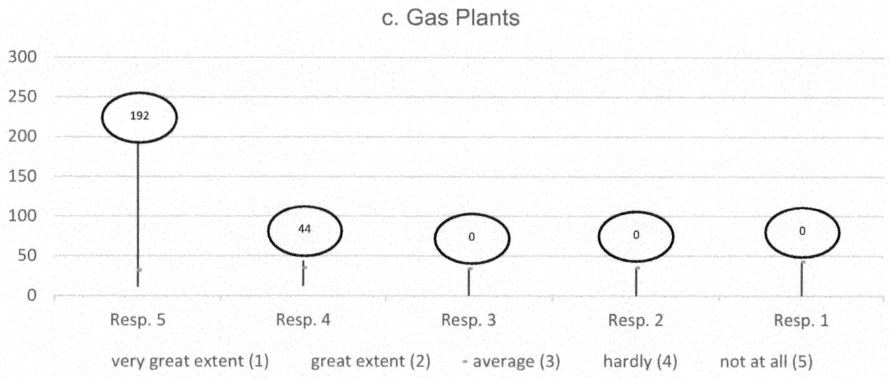
c. Gas Plants
300
250
200
150
100
50
0
192
44
0
0
0
Resp. 5
Resp. 4
Resp. 3
Resp. 2
Resp. 1
very great extent (1)
great extent (2)
average (3)
hardly (4)
not at all (5)

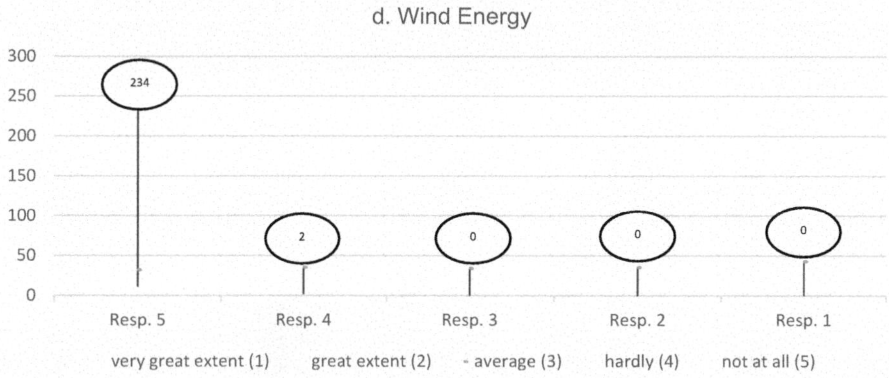
d. Wind Energy
300
250
200
150
100
50
0
234
2
0
0
0
Resp. 5
Resp. 4
Resp. 3
Resp. 2
Resp. 1
very great extent (1)
great extent (2)
average (3)
hardly (4)
not at all (5)

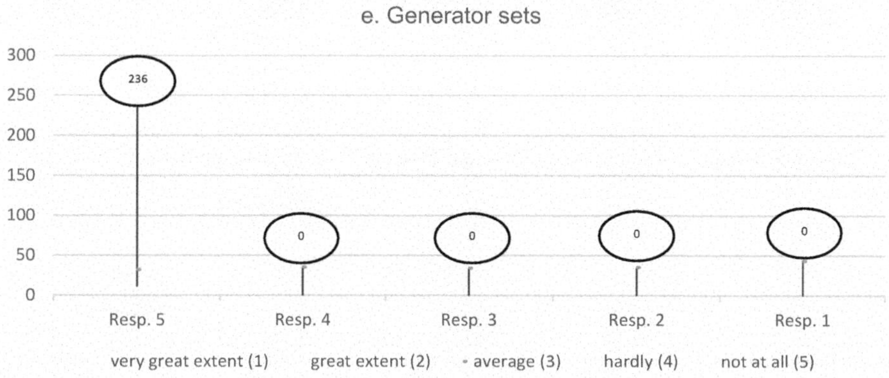
e. Generator sets
300
250
200
150
100
50
0
236
0
0
0
0
Resp. 5
Resp. 4
Resp. 3
Resp. 2
Resp. 1
very great extent (1)
great extent (2)
average (3)
hardly (4)
not at all (5)

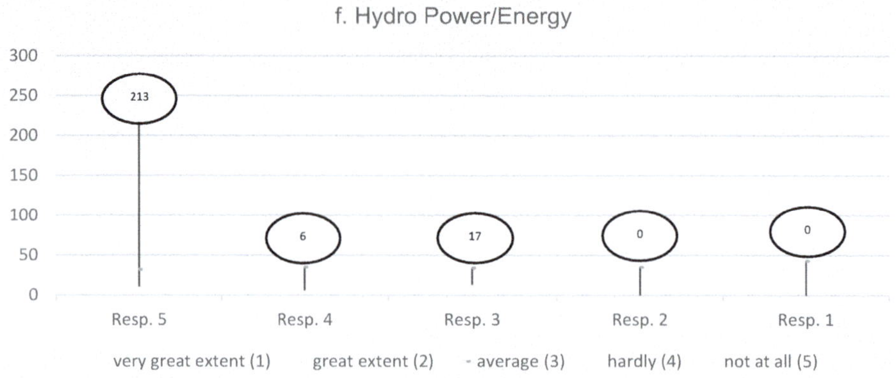

5. Cost Statistics of Commercial and Domestic Uses of Generator Sets

0-1 = not at all; 1-2 = hardly; 2-4 = average; 4-5 = great extent; 5-6 = very great extent
(0 = no respondent; 6 = 300 repondents)

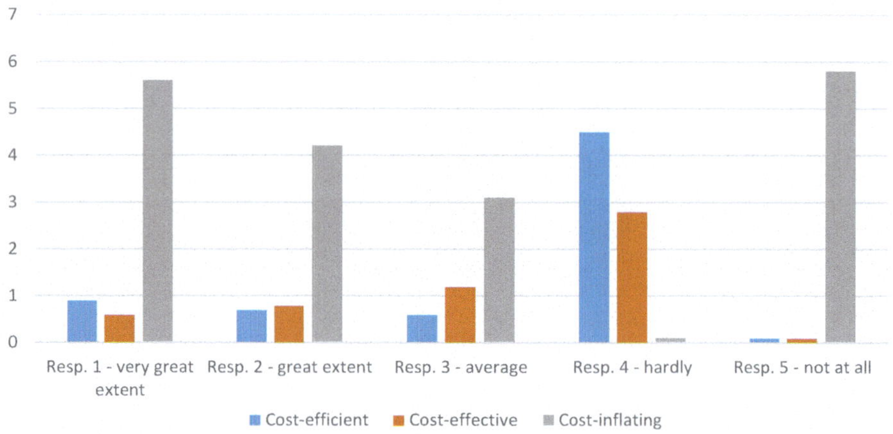

6. Adverse Effects of Excessive Private Power/Energy Generation

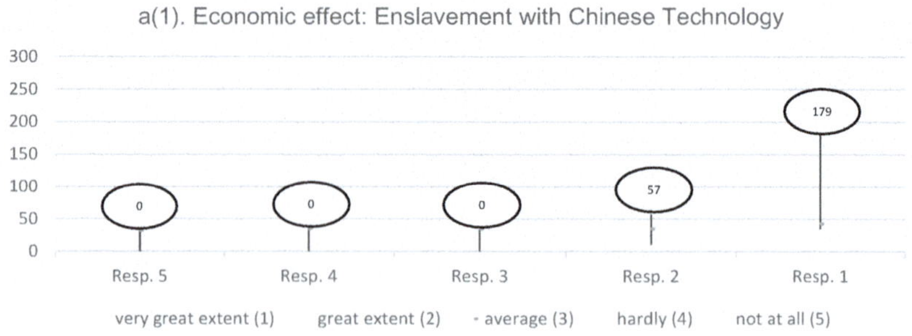

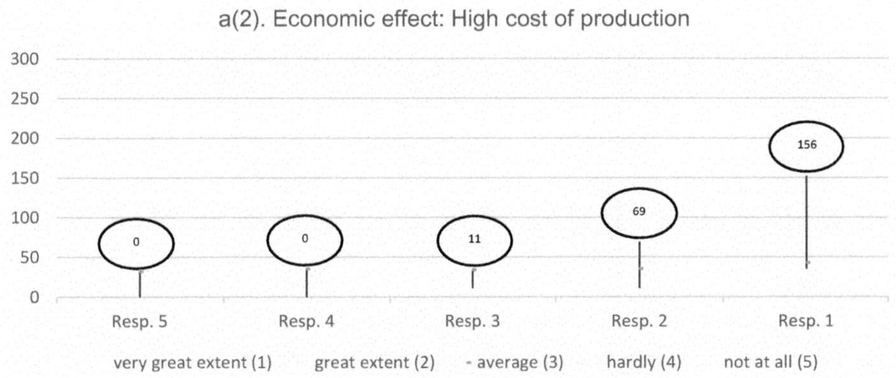
a(2). Economic effect: High cost of production
300
250
200
150
100
50
0
0
0
11
69
156
Resp. 5
Resp. 4
Resp. 3
Resp. 2
Resp. 1
very great extent (1)
great extent (2)
average (3)
hardly (4)
not at all (5)

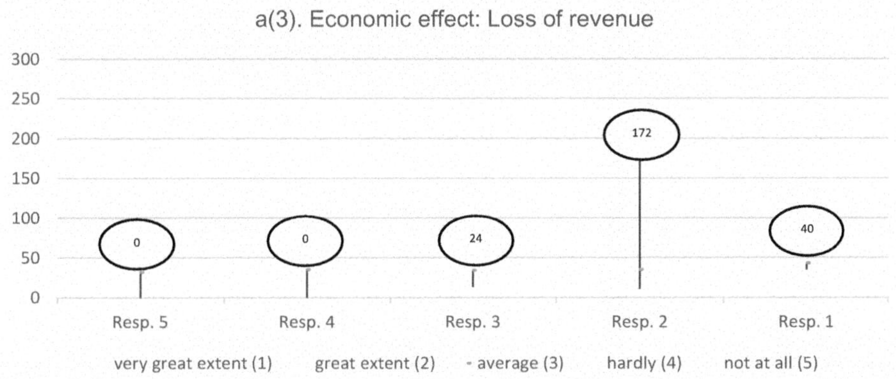
a(3). Economic effect: Loss of revenue
300
250
200
150
100
50
0
0
0
24
172
40
Resp. 5
Resp. 4
Resp. 3
Resp. 2
Resp. 1
very great extent (1)
great extent (2)
average (3)
hardly (4)
not at all (5)

b(1). Socio-environmental effect: Noise Polution

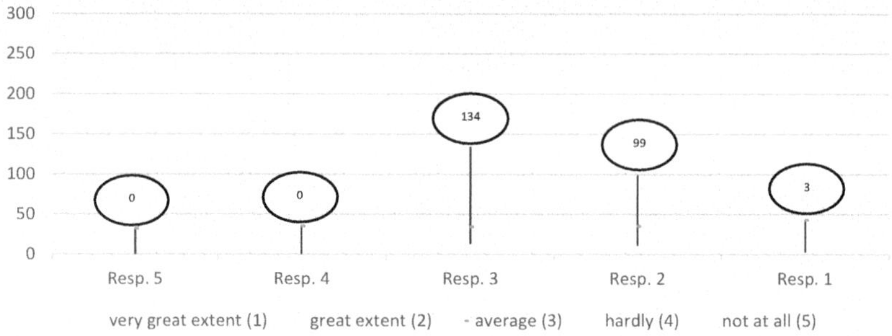

b(2). Socio-environmental effect: Air Pollution

b(3). Socio-environmental effect: Environmental Pollution

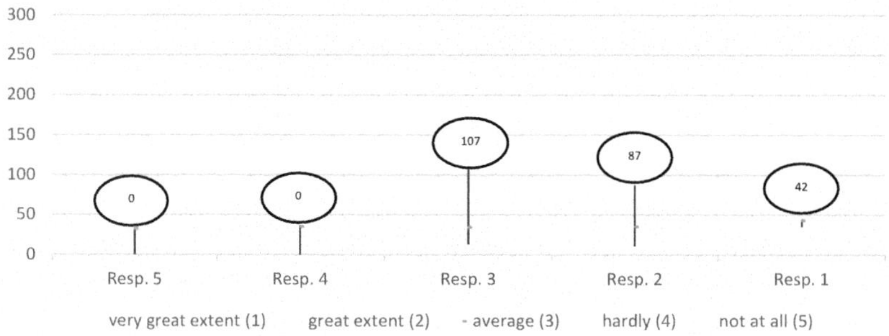

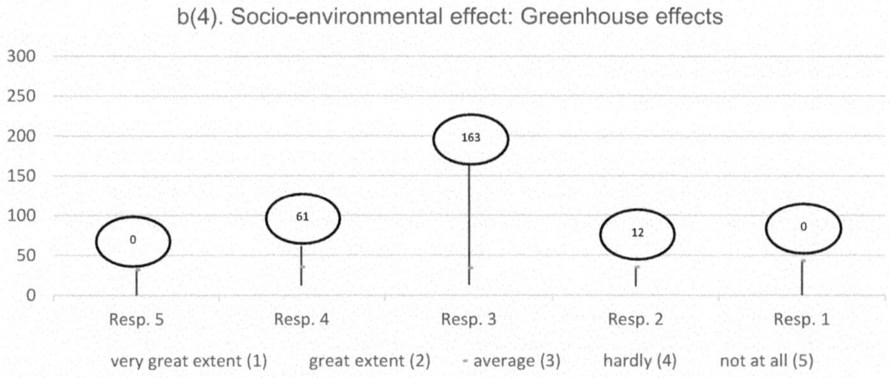
b(4). Socio-environmental effect: Greenhouse effects
300
250
200
150
100
50
0
0
61
163
12
0
Resp. 5
Resp. 4
Resp. 3
Resp. 2
Resp. 1
very great extent (1)
great extent (2)
average (3)
hardly (4)
not at all (5)

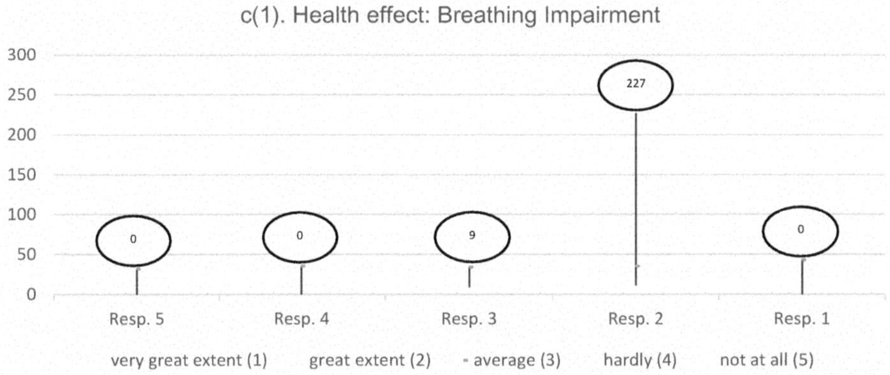
c(1). Health effect: Breathing Impairment
300
250
200
150
100
50
0
0
0
9
227
0
Resp. 5
Resp. 4
Resp. 3
Resp. 2
Resp. 1
very great extent (1)
great extent (2)
average (3)
hardly (4)
not at all (5)

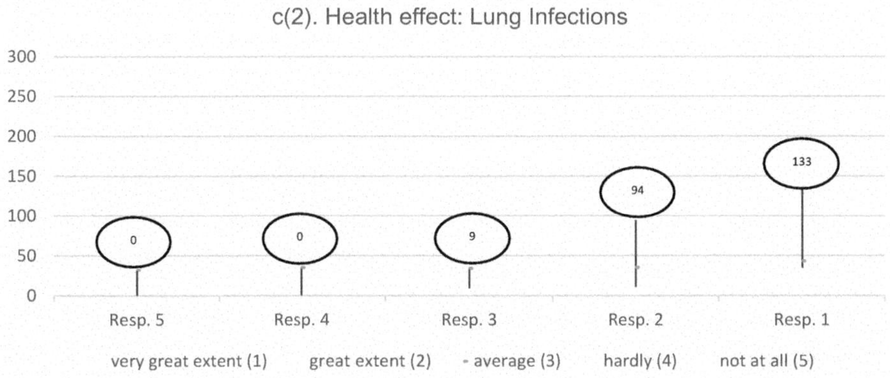
c(2). Health effect: Lung Infections
300
250
200
150
100
50
0
0
0
9
94
133
Resp. 5
Resp. 4
Resp. 3
Resp. 2
Resp. 1
very great extent (1)
great extent (2)
average (3)
hardly (4)
not at all (5)

Appendix B

Statistics of Civilian and Military Government Regimes in Nigeria

Regime/ republic	Period in office	Type of government	Head of government	Remarks
1	01/10/1960–15/01/1966	Civilian	Sir Abubakar Tafawa Belewa	Ousted by a bloody coup
2	15/01/1966–29/07/1966	Military	Gen. Aguiyi Ironsi	Ousted by a bloody coup
3	29/07/1966–29/07/1975	Military	Gen. Yakubu Gowon	Ousted by non-bloody coup
4	29/07/1975–13/02/1976	Military	Gen. Murtala Mohammed	Ousted by a bloody coup
5	13/02/1976–01/10/1979	Military	Gen. Olusegun Obasanjo	Successful transition
6	01/10/1979–31/12/1983	Civilian	Alh. Shehu Shagari	Ousted by non-bloody coup
7	31/12/1983–27/08/1985	Military	Gen. Mohammadu Buhari	Ousted by non-bloody coup
8	27/08/1985–27/08/1993	Military	Gen. Ibrahim Babangida	Ousted by civic pressure and protest
9	27/08/1993–17/11/1993	Civilian	Chief Ernest Shonekan	Ousted by non-bloody coup
10	17/11/1993–08/06/1998	Military	Gen. Sani Abacha	Ended with the sudden death of the dictator
11	08/06/1998–29/05/1999	Military	Gen. Abdusalam Abubakar	Successful transition
12	29/05/1999–29/05/2007	Civilian	Gen. Olusegun Obasanjo (Rtd.)	Successful transition

(continued)

C. F. Nwosu, *Systemic Remediation*, Environmental Humanities: Transformation, Governance, Ethics, Law 4, https://doi.org/10.1007/978-3-031-61909-0

(continued)

Regime/ republic	Period in office	Type of government	Head of government	Remarks
13	29/05/2007–29/05/2011	Civilian	Alh. Umaru Musa Yar'adua/Goodluck Ebele Jonathan	Witnessed the death of the incumbent president and the continuation of the regime through the vice president
14	29/05/2011–29/05/2015	Civilian	Goodluck Ebele Jonathan	Defeated in the polls after first tenure in office
15	29/05/2015–29/05/2023	Civilian	General Mohammadu Buhari (Rtd.)	Successful transition
16	29/05/2023–till date	Civilian	Bola Ahmed Tinubu	

Appendix C

The Unfortunate Cases of Suffocation and Death Due to the Inhalation of Poisonous Fumes from Generating-Sets

No	Date	Case	Source
1	July 30, 2018	Decomposing bodies of eight (8) family members found in their apartment in Cross River State, Nigeria: The investigation reports that the tragedy is a result of generator fumes from the power generating set left to run throughout the night	Ike Uchechukwu, Vanguard Nigeria, Monday, July 30, 2018
2	June 8, 2018	Pandemonium in Ogun as a family of five, friend die in the new house: Investigation suspects intoxication by poisonous chemicals or generator fumes	Adekunle, The Tide, Nigeria—Friday, June 8, 2018
3	August 2, 2017	Generator fume kills a family of four in Kogi State	Yinka Oladoyinbo Lokoja, The Punch, Nigeria—Wednesday, August 2, 2017
4	October 30, 2017	Delta: Couple, 3 children suffocate to death from generator fumes	Ochuko Akuopha, Vanguard Nigeria—Monday, October 30, 2017
5	June 20, 2018	Edo: Mother, 5 children dead; father in coma from generator fumes	Emmanuel Okogba, Vanguard Nigeria—Wednesday, June 20, 2018
6	May 22, 2017	How generator fumes killed three in Edo	Alexander Okere, The Punch, Nigeria, Monday, May 22, 2017
7	June 20, 2018	More Heart-breaking Photos of Family of 7 Killed by Generator Fumes in Benin	Odinaka, Tori New Nigeria—Wednesday, June 20, 2018

(continued)

C. F. Nwosu, *Systemic Remediation*, Environmental Humanities: Transformation, Governance, Ethics, Law 4, https://doi.org/10.1007/978-3-031-61909-0

(continued)

No	Date	Case	Source
8	August 14, 2018	Generator fumes kills family of 5 in Rivers State	Ebube Emelogu, The Punch, Nigeria—Tuesday, August 14, 2018
9	July 16, 2008	Generator fumes kill 17 at Nigeria prayer meeting	Anamesere Igboeroteonwu, The Reuters, Nigerian—Wednesday, July 16, 2008

Appendix D

National Public Institutions and Stakeholders Active in the Energy Market in Nigeria

No	Institution	Established	Role
1	Federal Ministry of Power (FMP)	–	Responsible for ensuring the establishment of a robust power sector that fully supports the socio-economic needs of the nation
2	Nigeria Electricity Regulatory Commission (NERC)	2005	To monitor and regulate the electricity industry of Nigeria and ensure compliance with market rules and operating guidelines
3	Rural Electrification Agency of Nigeria (REA)	1981	Aims at connecting all local government headquarters and selected neighbouring towns and villages to the national grid…and manage the Rural Electrification fund
4	Electricity Management Services Limited of Nigeria (EMSL)	2015	To guarantee efficient and reliable production and delivery of power as well as the safety of lives and property in the electricity sector
5	National Power Training Institute of Nigeria (NAPTIN)	2009	To respond to the training needs in the power sector
6	Federal Ministry of Environment (FMENV)	1999	To protect the environment against pollution and degradation and to ensure the conservation of natural resources for sustainable development in Nigeria…. Coordinating all climate change matters under its Department of Climate Change, which should follow the objective of fostering renewable energy and energy efficiency

(continued)

C. F. Nwosu, *Systemic Remediation*, Environmental Humanities: Transformation, Governance, Ethics, Law 4, https://doi.org/10.1007/978-3-031-61909-0

(continued)

No	Institution	Established	Role
7	Federal Ministry of Science and Technology (FMST)	1980	Through its department on Renewable and Conventional Energy Technology, the company's focus lies on nuclear, renewable, and alternative energy sources, energy efficiency, and R&D activities addressing energy-related problems associated with environmental degradation, pollution, and climate change
8	Energy Commission of Nigeria (ECN)	1988	Charged with the responsibility for the strategic planning and coordination of national policies in the field of energy in all its ramifications
9	National Agency for Science and Engineering Infrastructure (NASENI)	1992	To promote local manufacturing of renewable energy technologies such as solar modules, small hydro turbines, pole-mounted transformers and wind turbine blades
10	Federal Ministry of Lands, Housing and Urban Development (FMLHUD)	2010 (as an independent Ministry)	Aims at providing adequate housing for all Nigerian—Plays a strategic role regarding energy efficiency in buildings which it lives up to by e.g. including energy aspects in the ongoing review of building code
11	Federal Ministry of Water Resources (FMWR)	2010 (Reformed)	Through its Department of Dams and Reservoir Operations, the FMWR is involved in numerous hydropower projects
12	Federal Ministry of Industry, Trade and Investment (FMITI)	–	Supervises products, processes and companies in the energy industry and supports and enacts renewable energy and energy efficiency measures. It oversees the production of component parts of solar panels and is responsible for policies regarding the blending of biomass and provides industry incentives for renewable energy applications
13	Standard Organisation of Nigeria (SON)	1971	Its functions include: (a) Preparing Nigerian Industrial Standards and ensuring the compliance of products and methods with such standards; (b) establishing a quality assurance system including certification of factories, products and laboratories; (c) fostering interests in the formulation and adherence to standards by industry and the general public; (d) assessing the conformity of imported products in the port of origin (pre-shipment verification)

(continued)

(continued)

No	Institution	Established	Role
14	Nigeria Bulk Electricity Trading Plc (NBET)	2010	Has the mandate of engaging in the purchase and resale of electricity and ancillary services from independent power producers and the successor generation companies
15	Nigerian National Petroleum Corporation (NNPC)	1977	With regard to renewable energy and energy efficiency, NNPC explores the use of biofuels (mainly ethanol and biodiesel) for mixing with conventional fuel and seeks to reduce its internal consumption
16	Presidential Task Force on Power (PTFP)	2010	Constituted to drive the implementation, monitoring, and performance evaluation of the power reform agenda, it acts as an inter-agency interface to ensure that every milestone in the power sector reform roadmap is accomplished
17	Nigerian Governor's Forum (NGF)	1999	Aims at promoting a cross-state platform to discuss public policy issues and share experience on good governance…Also meant to be a non-partisan forum promoting democratic values, good governance and sustainable development in Nigeria

Source www.giz.de

Index

C. F. Nwosu, *Systemic Remediation*, Environmental Humanities: Transformation, Governance, Ethics, Law 4, https://doi.org/10.1007/978-3-031-61909-0

Q

R

SPRINGER NATURE

GPSR Compliance

The European Union's (EU) General Product Safety Regulation (GPSR) is a set of rules that requires consumer products to be safe and our obligations to ensure this.

If you have any concerns about our products, you can contact us on ProductSafety@springernature.com

In case Publisher is established outside the EU, the EU authorized representative is:

Springer Nature Customer Service Center GmbH
Europaplatz 3
69115 Heidelberg, Germany

The manufacturer's authorised representative in the EU is Springer Nature Customer Service Centre GmbH, Europaplatz 3, 69115 Heidelberg, Germany. If you have any concerns regarding our products, please contact ProductSafety@springernature.com

Printed and bound by CPI Group (UK) Ltd, Croydon, CR0 4YY
01/10/2026
02225969-0015